Handbook of SURFACES and INTERFACES

EDITOR

Léonard Dobrzynski
Laboratoire des Surfaces et Interfaces
Centre National de la Recherche Scientifique
Institut Superieur d'Electronique du Nord
Lille, France

ADVISORY BOARD

J. Beeby
Department of Physics
University of Leicester, Leicester, England

J. Benard
Ecole de Chimie de Paris, Paris, France

G. Ertl
Institut für Physikalische Chemie
der Univesität München, München, West Germany

R. Kubo
Department of Physics
University of Tokyo, Tokyo, Japan

W. H. Weinberg
Chemical Engineering, California Institute of Technology,
Pasadena, California, U.S.A.

Volume 1
CONTENTS

Volume 2
CONTENTS

Handbook of SURFACES and INTERFACES

Volume 3: Surface Phonons and Polaritons

Edited by

Alexei A. Maradudin and Richard F. Wallis

Department of Physics
University of California
Irvine, California

and

Léonard Dobrzynski

Laboratoire des Surfaces et Interfaces,
Centre National de la
Recherche Scientifique
Institut Superieur
d'Electronique du Nord
Lille, France

GARLAND STPM PRESS
New York & London

15 14 13 12 11 10 9 8 7 6 5 4 3 2 1

Library of Congress Cataloging in Publication Data

Maradudin, A.A.
Surface phonons and polaritons.

(Handbook of surfaces and interfaces ; v. 3)
Includes bibliographical references and indexes.
1. Phonons—Handbooks, manuals, etc. 2. Polaritons—Handbooks, manuals, etc. I. Wallis, Richard Fisher, 1924– joint author II. Dobrzynski, Leonard, joint author. III. Title. IV. Series.
QC173.4.S94H36 vol. 3 [QC176.8.P6] 541′.3453s
0-8240-9855-2 [539.7′217] 78-27392

Published by Garland STPM Press
136 Madison Avenue, New York, New York 10016

Printed in the United States of America

CONTENTS

TABLE OF CONTENTS

PREFACE

This book provides a summary of our understanding of surface phonons and polaritons which is as up-to-date and complete as possible. The history of the subject dates back to the work of Lord Rayleigh in 1887 who demonstrated the existence of an elastic wave localized at the surface of a solid. For many years the investigation of surface elastic waves was carried out mostly by seismologists. The study of surface vibrations of a discrete lattice of atoms was started only around 1950 and has had an explosive development in the last decade. The development of experimental techniques, such as elastic and inelastic low-energy electron diffraction, has made available experimental information concerning the dynamics of surface atoms. The use of surface elastic waves in many applied devices has led to much activity concerning the continuum theory of lattice dynamical surface waves. In this volume, the theory of surface modes of vibration in crystal lattices is developed primarily from the lattice point of view, although a new continuum approach is also presented.

The effects associated with the interaction of particles and radiation with crystal surfaces, especially the interaction of phonons, are discussed. The interaction of electromagnetic waves with surface excitations gives rise to surface polaritons whose experimental study, which began approximately ten years ago, has stimulated extensive development of the subject. The theory of surface polaritons in dielectric and magnetic media is developed in this volume.

PREFACE

Surface Phonons and Polaritons was written with special attention to its pedagogical value. We believe that it fills a need both as a text for advanced graduate students and as a reference for investigators having either little or extensive knowledge of the subject.

A. A. Maradudin
R. F. Wallis
L. Dobrzynski

Chapter 1

INTRODUCTION

Some twenty-five years before the pioneering work of Born and von Karman (1) on the dynamical properties of infinitely extended crystals, Lord Rayleigh (2) had shown on the basis of the theory of elasticity that a semi-infinite, elastically isotropic medium bounded by a single, stress-free plane surface can sustain surface vibration modes. These are modes which are wavelike in directions parallel to the surface of the medium, but whose amplitudes decay exponentially with increasing distance into the medium from the surface, with a decay length which is comparable to the wavelength of the wave parallel to the surface. Thus, these modes are localized in the vicinity of the surface. They are acoustic in nature, in the sense that their frequencies depend linearly on the magnitude of the two-dimensional wave vector which characterizes their propagation parallel to the surface of the medium, and consequently vanish with vanishing wave vector. Their frequencies also lie below the continuum of frequencies allowed an infinitely extended elastic medium for the same value of the two-dimensional wave vector. Such surface vibration modes have subsequently come to be known as Rayleigh waves, after their discoverer.

Subsequently, the vibrations of semi-infinite or finite crystals began to be studied from a lattice dynamical, as opposed to a continuum, point of view, beginning with the pioneering work of Rosenzweig (3) and Lifshitz and Rosenzweig (4). It was discovered that in addition to surface vibration modes of acoustic character, surface modes of optical character can exist in crystals with more than one atom in a primitive unit cell. The earliest modes of this type to be studied have frequencies lying in the gap in the frequency spectrum of the corresponding infinitely extended crystal between the acoustic and optical branches, if one exists, which tend to nonzero limits as the two-dimensional wave vector characterizing them tends to zero. They also differ from acoustic surface modes, in which all the sublattices comprising the crystal move with essentially the same amplitudes, by the fact that the sublattices vibrate against each other, with amplitudes which decay with increasing distance into the crystal from the surface. Subsequent investigations have revealed the existence of more complicated types of surface vibrational modes, even in Bravais crystals (5). Such surface modes clearly have no counterpart in a continuum theory of surface waves, and will be called optical surface modes. From considerations of the translational periodicity of a crystal in directions parallel to a free surface, Lifshitz and Rosenzweig (4) concluded that only optical surface modes corresponding to infinitely long wavelengths parallel to the surface can be optically active, provided that additional restrictions on these modes imposed by crystal symmetry are also satisfied. They suggested that such optical surface modes may lead to additional lines in the infrared absorption and Raman spectra of crystals.

INTRODUCTION

The latter effects are examples of another way in which the presence of surfaces can alter the physical properties of crystals: selection rules governing physical processes in crystals, which have their origins in the symmetry properties, translational and/or rotational, of an infinitely extended crystal, can be relaxed for finite crystals or for atoms in the surface layers of crystals, for which these symmetry properties no longer hold.

The interaction of optical vibrational modes with the electromagnetic field leads to a coupled excitation known as a polariton. In the presence of a surface it is possible to have a localized or surface polariton in which the magnitudes of the electric and magnetic vectors decrease exponentially away from the surface. Definitive information concerning surface polaritons can be obtained by means of optical experiments.

We have noted above that a crystal surface can give rise to exceptional vibration modes which are spatially localized in its vicinity, namely, the acoustical and optical surface modes. However, the introduction of a crystal surface also modifies the frequencies and displacement fields of the wavelike modes of an infinitely extended crystal, even if for some reason it does not give rise to surface modes. The difference between the frequency distribution functions of a finite crystal and of a cyclic crystal possessing the same number of degrees of freedom arising from these two sources is of the order of the ratio of the number of atoms in the surface to the number in the interior (if the distribution function is normalized to unity), and has the character of the distribution function of a two-dimensional crystal. It follows that the presence of surfaces will alter the temperature dependences of the vibrational thermodynamic functions of a crystal from their bulk forms, and will give rise to distinct size effects, which can become significant when particle sizes become small enough that the ratio of surface area to volume is no longer negligible.

The preceding effects of a surface on the dynamical properties of a crystal are examples of effects associated with the surface as a whole. There is, however, a class of effects associated with the modification of the dynamical properties of individual atoms by the presence of a crystal surface. The crystal structure in the surface layers may be different from the structure in the bulk of the crystal. The forces acting on atoms in the surface layers will be different from the forces acting on atoms in the bulk since an atom in the surface layers has fewer nearest neighbors, next nearest neighbors,..., than an atom in the interior of a crystal. One would therefore expect that dynamical properties of individual atoms, such as their mean square displacement and mean square velocity, are different for atoms in the surface layers of a crystal from what they are for atoms in the bulk of the crystal, both in magnitude and in their symmetry properties. It is important to take these differences into account in the interpretation of experimental data obtained by the techniques of low-energy electron diffraction from crystal surfaces (6), or the Mössbauer effect for resonant nuclei plated on or diffused into the surface layers of a crystal (5), for example.

Although the preceding examples of the effects of a crystal on the dynamical properties of individual atoms were presented in the context of an otherwise ideal crystal, the presence of a crystal surface can affect dynamical properties of impurity atoms when the latter are in the vicinity of the surface. If an impurity atom gives rise to localized or resonance modes when it is in the interior of a crystal, any degeneracies these modes possess can be lifted when the impurity atom is brought up to the surface of the crystal, due to the reduction in the symmetry of the crystal to which a surface gives rise. The mean square displacement and mean square velocity of an impurity atom will also be different, both in magnitude and in their symmetry properties, when the impurity is in the vicinity of a crystal surface, from what they are when the impurity is in the interior of the crystal.

It is conceptually convenient, and often computationally useful, to regard a crystal surface as a defect in an infinitely extended, cyclic crystal, created by equating to zero all interatomic forces intersecting a fictitious plane which bisects the crystal but contains no atoms itself. (In fact, this procedure creates the two parallel surfaces of a semi-infinite crystal slab, when the cut cyclic crystal is "unrolled." The resulting crystal slab of course remains cyclic in directions parallel to these surfaces.) It is well known, by now (8), that two defects in a crystal can interact with each other through the modification of the vibration field of one defect by the presence of the other. From the preceding considerations it follows that there is an energy of interaction of an impurity atom or a lattice defect with a crystal surface, which is a function of the distance of the crystal imperfection from the surface, so that the surface exerts a force on the imperfection. The sign of this force depends on the nature of the imperfection. In a similar fashion, two atoms adsorbed on a crystal surface interact with each other as a consequence of their mutual modification of the normal vibration modes of the semi-infinite substrate. Finally, just as point defects can scatter acoustic waves propagating through a crystal, so can an extended defect, such as a crystal surface. This scattering of phonons by the boundaries of a crystal becomes the dominant mechanism responsible for lattice thermal resistivity in insulators at very low temperatures, when all other resistive mechanisms become inoperative, and the phonon mean free path becomes comparable with the linear dimensions of a crystal.

All of the physical effects described so far, in which crystal surfaces play the determining role, exist in the harmonic approximation for the lattice vibrations. When anharmonic terms are retained in the expansion of the crystal's potential energy in powers of the displacements of the atoms from their equilibrium positions, new surface-induced effects arise, which are absent in the absence of anharmonicity.

Since the mean square displacements of atoms in the surface layers of a crystal are larger, in general, than those of atoms in the interior, due to the different sets of forces acting on these two kinds of atoms, it is reasonable to expect that anharmonic effects associated with atoms in the surface layers of a crystal will be larger than they are in the bulk of the crystal.

It is well known that the phenomenon of thermal expansion of solids has its origin in the anharmonic terms in the crystal's potential energy. For the reason just given, it can be expected that crystals will display a differential (larger) thermal expansion in the vicinity of their surfaces compared with its magnitude for the bulk of the crystal.

Another consequence of the anharmonic terms in a crystal's potential energy is the coupling it introduces among the independent normal vibration modes of the harmonic approximation. This coupling permits energy to be exchanged among these modes, which leads to their attenuation, or damping, in time. In particular, an important mechanism for the attenuation of Rayleigh surface waves in insulators is their anharmonic interaction with the other normal vibration modes of the crystal along which they propagate.

In the chapter on Dynamical Theory of Crystal Surfaces we present the theory underlying the surface-induced vibrational properties of crystals which we have just described briefly, as well as experimental results bearing on them. In writing this section we have tried to emphasize results which have a general validity, that is, which are independent of particular choices for the crystal symmetry and structure and of assumptions about the range of interatomic forces. However, in the chapter on Examples we will illustrate many of the general methods and results of Chapter 2 by applying them to the determination of various dynamical properties of crystal surfaces on the basis of a very simple crystal model, which permits such calculations to be carried out simply and analytically.

In the chapter on Surface Polaritons we present the general theory of surface polaritons as well as many illustrations of specific cases. The theoretical investigation of what we now call surface polaritons dates back to the work of Zenneck (9) and Sommerfeld (10), who studied the propagation of an electromagnetic wave along the plane interface between two media. The recent development of new experimental techniques such as attenuated total reflection (ATR), inelastic scattering of low-energy electrons, and Raman scattering, has stimulated a sharp renewal of interest in both the experimental and theoretical aspects of surface polaritons.

SURFACE PHONONS AND POLARITIONS

A great deal can be accomplished on the theoretical side using a macroscopic approach based on Maxwell's equations and the electromagnetic boundary conditions. One need only have a knowledge of the frequency-dependent dielectric constant or magnetic permeability of the bulk crystal. However, a complete understanding of certain experiments seems to be possible only if more detailed information is available concerning the properties of the crystal in the immediate vicinity of the surface.

REFERENCES

1. M. Born and Th. von Karman, Phys. Z., 13, 297 (1912); 14, 15 (1913).

2. Lord Rayleigh, Proc. London Math. Soc., 17, 4 (1887).

3. L. N. Rosenzweig, Tr. Fiz. Otdel. Fiz.-Mat. Fakul'teta Khark. Gos. Univ., 2, 19 (1950).

4. I. M. Lifshitz and L. N. Rosenzweig, Zh. Eksp. Teor. Fiz., 18, 1012 (1948).

5. D. C. Gazis, R. Herman, and R. F. Wallis, Phys. Rev., 119, 533 (1960).

6. See, for example: S. G. Kalashnikov and O. I. Zamsha, Zh. Eksp. Teor. Fiz., 9, 1408 (1939); A. U. MacRae and L. H. Germer, Phys. Rev. Lett., 8, 489 (1962).

7. R. V. Pound and G. A. Rebka, Phys. Rev. Lett., 4, 274 (1960); B. D. Josephson, ibid., 4, 341 (1960).

8. A. A. Maradudin, E. W. Montroll, G. H. Weiss, and I. P. Ipatova, Theory of Lattice Dynamics in the Harmonic Approximation (Academic Press, New York, 1971), p. 471.

9. J. Zenneck, Ann. Phys. (Leipz.), 23, 846 (1907).

10. A. Sommerfeld, Ann. Phys. (Leipz.), 28, 665 (1909).

Chapter 2

DYNAMICAL THEORY OF CRYSTAL SURFACES

INTRODUCTION

The modern theory of lattice dynamics, due to Born and his collaborators (1), was based on the assumption of an infinitely extended crystal. Calculations of vibrational properties of such a crystal were made tractable, and were normalized to a finite volume, by imposing cyclic boundary conditions on the atomic displacements. That is, it was postulated that the displacements are periodic functions of position with the periodicity volume being a macrocrystal containing N primitive unit cells, which can be regarded as the crystal of physical interest, whose repetition fills all space. The essential simplification introduced into lattice dynamics by the adoption of these boundary conditions is that the equations of motion of any atom are the same as those of any other atom separated from it by a lattice translation vector.

The justification for the use of these boundary conditions in calculations of extensive vibrational properties of a crystal was provided in 1944 by W. Ledermann (2) in the form of the following theorem: If in a Hermitian matrix the elements of r rows and their corresponding columns are modified in any way whatever, provided that the matrix remains Hermitian, the number of eigenvalues which lie in any given interval cannot increase or decrease by more than 2r.

The relevance of this theorem to the justification of the use of periodic boundary conditions can be seen from the following discussion. Let us first consider an infinitely extended crystal in which the displacements of the atoms from their equilibrium positions are periodic functions of position, with the periodicity of a space-filling macrocrystal containing N atoms. In the harmonic approximation, the independent equations of motion of this crystal can be written schematically in the form

$$M_m \ddot{u}_\alpha(m) = - \sum_{m'\beta} \Phi_{\alpha\beta}^{(o)}(mm')u_\beta(m') \qquad (1)$$

where $\alpha,\beta = x,y,z$; $m,m' = 1,2,\ldots,N$; $u_\alpha(m)$ is the α Cartesian component of the displacement of the mth atom from its equilibrium position; M_m is its mass; and the coefficients $\left\{\Phi_{\alpha\beta}^{(o)}(mm') = \Phi_{\beta\alpha}^{(o)}(m'm)\right\}$, called atomic force constants, are the second partial derivatives of the potential energy of the crystal with respect to the atomic displacements $u_\alpha(m)$ and $u_\beta(m')$, evaluated with all the atoms at their equilibrium positions. Because of the infinite, perfect periodicity of the crystal which results from the assumption of cyclic boundary conditions, the interaction of an atom inside a given macrocrystal with an atom outside it is identical with the interaction between any two atoms inside the macrocrystal separated by the same lattice translation vector. It is this fact that leads to the identity of the equations of motion of all atoms which are related by the translation vectors of the infinitely extended crystal.

Let us now create a true, finite crystal of N atoms from this infinitely extended one by equating to zero all interactions between the atoms inside a given macrocrystal and all the atoms outside it. The equations of motion of the resulting finite crystal can be written schematically as

$$M_m \ddot{u}_\alpha(m) = - \sum_{m'\beta} \Phi_{\alpha\beta}(mm')u_\beta(m') \tag{2}$$

where $\alpha, \beta = x, y, z$ and $m, m' = 1, 2, \ldots, N$. The atomic force constants $\{\Phi_{\alpha\beta}(mm') = \Phi_{\beta\alpha}(m'm)\}$ differ from the force constants $\left\{\Phi^{(o)}_{\alpha\beta}(mm')\right\}$ in at least two ways. First, all constants $\Phi_{\alpha\beta}(mm')$ for which the atom m is inside the macrocrystal while the atom m' is outside it, or vice versa, vanish, while the corresponding constants $\Phi^{(o)}_{\alpha\beta}(mm')$ need not vanish, depending on the range of the interatomic forces. Second, even if both of the atoms m and m' are in the finite crystal, but one or the other, or both, is closer to the surface than a distance of the order of the range of the interatomic forces, the force constants $\Phi_{\alpha\beta}(mm')$ and $\Phi^{(o)}_{\alpha\beta}(mm')$ will be somewhat different because of the rearrangements of the atomic positions in the surface layers of the crystal which will inevitably accompany the creation of the finite crystal by the annulment of all forces crossing the boundary of the macrocrystal.

Thus, if we make the substitution

$$u_\alpha(m) = \frac{v_\alpha(m)}{M_m^{\frac{1}{2}}} e^{-i\omega t} \tag{3}$$

where $v_\alpha(m)$ is a time-independent amplitude, Eqs. (1) and (2) can be put in the forms:

$$\omega^2 v_\alpha(m) = \sum_{m'\beta} D^{(o)}_{\alpha\beta}(mm')v_\beta(m') \tag{4a}$$

and

$$\omega^2 v_\alpha(m) = \sum_{m'\beta} D_{\alpha\beta}(mm')v_\beta(m') \tag{4b}$$

respectively, where

$$D^{(o)}_{\alpha\beta}(mm') = \frac{\Phi^{(o)}_{\alpha\beta}(mm')}{\left(M_m M_{m'}\right)^{\frac{1}{2}}} = D^{(o)}_{\beta\alpha}(m'm) \tag{5a}$$

$$D_{\alpha\beta}(mm') = \frac{\Phi_{\alpha\beta}(mm')}{\left(M_m M_{m'}\right)^{\frac{1}{2}}} = D_{\beta\alpha}(m'm) \tag{5b}$$

are the so-called dynamical matrices of the infinitely extended (cyclic) and finite crystals, respectively. As they are real and symmetric, they are clearly Hermitian.

From Eqs. (4a) and (4b) we see that the squares of the normal mode frequencies of the infinitely extended, cyclic crystal and of the finite crystal are the eigenvalues of the Hermitian matrices $\left\{D^{(o)}_{\alpha\beta}(mm')\right\}$ and $\{D_{\alpha\beta}(mm')\}$, respectively. We can partition these matrices in the following way

$$\overset{\leftrightarrow}{D}^{(o)} = \left(\begin{array}{c|c} \overset{\leftrightarrow}{D}^{(o)}_{ii} & \overset{\leftrightarrow}{D}^{(o)}_{ib} \\ \hline \overset{\leftrightarrow}{D}^{(o)T}_{ib} & \overset{\leftrightarrow}{D}^{(o)}_{bb} \end{array}\right), \quad \overset{\leftrightarrow}{D} = \left(\begin{array}{c|c} \overset{\leftrightarrow}{D}_{ii} & \overset{\leftrightarrow}{D}_{ib} \\ \hline \overset{\leftrightarrow}{D}^{T}_{ib} & \overset{\leftrightarrow}{D}_{bb} \end{array}\right) \tag{6}$$

In these expressions the subscripts (ii) indicate that portion of the corresponding dynamical matrix in which atom m and atom m' are both farther inside the crystal from the bounding surface than the range of interatomic forces; the subscripts (ib) indicate that portion of the corresponding dynamical matrix in which atom m is farther inside the crystal than the range of the interatomic forces, while atom m' is closer to the surface than this range; the subscripts (bb) denote that portion of the corresponding dynamical matrix for which both atoms m and m' lie closer to the bounding surface than the range of the interatomic forces.

If there were no relaxation of the equilibrium positions of the atoms in forming the finite crystal from the infinite one, the matrices $\overleftrightarrow{D}_{ii}^{(o)}$ and $\overleftrightarrow{D}_{ii}$ would be identical, and the dynamical matrices $\overleftrightarrow{D}^{(o)}$ and $\overleftrightarrow{D}$ would differ only in the rows and columns labeled by atoms closer to the surface than the range of interatomic forces. This is still true in the presence of relaxation of the atomic equilibrium positions on the reasonable assumption that the shifts in these equilibrium positions do not propagate farther into the interior of the crystal from the surface than approximately the range of the interatomic forces. The number of such rows and columns is given closely by the product of the number of atoms in the surface of the crystal, $N^{2/3}$, the range of the interatomic forces in units of the nearest neighbor separation between atoms, n_0, assumed finite (small), and the number of degrees of freedom per atom, 3. Since the dimensionality of $\overleftrightarrow{D}^{(o)}$ and $\overleftrightarrow{D}$ is 3N, we see that in replacing the true dynamical matrix $\overleftrightarrow{D}$ by the approximate one $\overleftrightarrow{D}^{(o)}$ the fraction of eigenvalues ω^2 of $\overleftrightarrow{D}$ which lie in any given interval cannot increase or decrease by more than $2n_0N^{-1/3}$. This is the fractional error which occurs in the evaluation of any physical property of a crystal which can be expressed as a sum over its normal mode frequencies, such as the vibrational contribution to its thermodynamic functions. For any crystal of macroscopic size N is of the order of 10^{24}, so that this is a negligible fractional change, and justifies the use of cyclic boundary conditions in such calculations.

The preceding justification is based on the assumption that the range of the interatomic forces is finite. It is therefore inapplicable to the case of ionic or polar crystals, in which the constituent atoms or ions interact through long-range Coulomb forces. However, it has been shown by Hardy (3) that the use of cyclic boundary conditions in the presence of Coulomb interactions affects only the very small minority of vibrational modes whose wavelengths are comparable with the linear dimensions of a crystal, and which possess a dipole moment. The number of such modes is shown to be of the order of $N^{2/3}$ out of a total number of modes of the order of N, so that the fractional error in the distribution of normal modes according to frequency in this case is also of the order of $N^{-1/3}$, which is negligible when N is large.

However, the adoption of cyclic boundary conditions, while mathematically convenient, eliminates the possibility of studying the dynamical properties of atoms in the neighborhood of a free surface of a real crystal, or of dynamical properties associated with a crystal surface as a whole, rather than with individual atoms. This is a serious shortcoming of the cyclic boundary conditions, since the existence of surfaces bounding a crystal gives rise to a variety of physical phenomena which are absent from the dynamical properties of infinitely extended, or cyclic, crystals.

In this chapter we present the dynamical theory of crystal surfaces. We will first give some preliminary considerations and definitions. The next subsections will be devoted respectively to surface waves; the group theory of phonons in a crystal slab; the Green's function for a semi-infinite crystal; the surface specific heat and other thermodynamic functions; the surface mean square displacements; the interaction of defects with a surface; the indirect interaction of adatoms on a surface; the scattering of phonons by a surface; the thermal expansion at a crystal surface; the anharmonic damping of Rayleigh waves; and a continuum approach to surface lattice dynamics.

Finally we have assumed that the reader of this section has a general knowledge of lattice dynamics at the level of such general references as the books by Born and Huang (1) and Maradudin, Montroll, Weiss, and Ipatova (4). However, if the reader prefers to gain first, on the basis of some simple examples, some qualitative understanding about the dynamical theory of crystal surfaces, he can as well start reading Chapter 4 and then come back to Chapter 2.

PRELIMINARY CONSIDERATIONS AND DEFINITIONS

Specification of Atomic Positions in a Semi-Infinite Crystal

In this section we present some general considerations which underlie the dynamical theory of semi-infinite crystals, and provide the basis for the remainder of the discussion of theory.

We begin by specifying the manner in which the atoms of a semi-infinite solid can be labeled. We first focus our attention on a crystal slab, with parallel surfaces, which exhibits translational periodicity in the directions parallel to the surfaces of the slab. We regard the slab as being composed of a two-dimensional array of lattice points, each point having associated with it a basis of r atoms which extends through the slab from surface to surface. The position vectors of the lattice points are given by

$$\vec{x}(\ell) = \ell_1 \vec{a}_1 + \ell_2 \vec{a}_2 \tag{7}$$

where $\vec{a}_1$ and $\vec{a}_2$ are the primitive translation vectors of the two-dimensional lattice, and ℓ_1 and ℓ_2 are any two integers, positive, negative, or zero, which we write collectively as ℓ. The position of any atom in the crystal slab is then given by

$$\vec{x}(\ell\kappa) \equiv \vec{x}(\ell) + \vec{x}(\kappa) \tag{8}$$

where the basis vector $\vec{x}(\kappa)$ is the position vector of the κth basis atom with respect to its lattice point.

The particular basis associated with each lattice point is arbitrary, subject only to the restrictions that it extend through the slab from surface to surface, and that when translated through all possible vectors $\{\vec{x}(\ell)\}$ it generates the entire crystal slab.

The preceding characterization of the structure of a crystal slab is particularly convenient in the study of the dynamical properties of a crystal surface which has undergone reconstruction, or of a crystal slab in which the interplanar spacings in the vicinity of the surfaces have undergone relaxation. Both of these effects can be incorporated into the definition of the atomic positions by suitable definitions of the primitive translation vectors $\vec{a}_1$ and $\vec{a}_2$, and of the basis vectors $\{\vec{x}(\kappa)\}$. We will refer to this way of characterizing the structure of a crystal slab as definition (A).

DYNAMICAL THEORY OF CRYSTAL SURFACES

For the study of certain kinds of dynamical properties of crystal surfaces it is convenient to treat the surface as a defect in an infinitely extended crystal, which is produced by setting equal to zero all of the interactions between atoms on opposite sides of a hypothetical plane which bisects the crystal, but which contains no atoms itself. Alternatively, particularly in the study of surface waves, it is sometimes desirable to consider a semi-infinite crystal, bounded by a single planar surface and occupying the upper half-space $(x_3 \geq 0)$, instead of a crystal slab bounded by two, parallel plane surfaces. In either of these two cases a way of specifying the atomic positions different from definition (A) is required.

In such cases we start with an infinitely extended crystal composed of a three-dimensional Bravais lattice, in which each lattice point is associated with a basis of atoms. The position vectors of the lattice points are given by

$$\vec{x}(\ell) = \ell_1\vec{a}_1 + \ell_2\vec{a}_2 + \ell_3\vec{a}_3 \qquad (9)$$

where $\vec{a}_1$, $\vec{a}_2$, $\vec{a}_3$ are the three primitive translation vectors of the lattice. It is convenient to choose two of these, say $\vec{a}_1$ and $\vec{a}_2$, to be parallel to the plane of the intended surface, which we take to be parallel to the plane $x_3 = 0$, while $\vec{a}_3$ is then directed out of this plane. If the semi-infinite crystal to be studied occupies the upper half-space, it is convenient to choose $\vec{a}_3$ in such a way that its x_3-component is positive.

If a pair of adjacent free surfaces is to be created by equating to zero all interactions between atoms on opposite sides of a hypothetical plane parallel to the plane $x_3 = 0$, the integers ℓ_1, ℓ_2, ℓ_3 in Eq. (9) can take on all values, positive, negative, and zero, and we refer to them collectively by ℓ. The r basis vectors $\{\vec{x}(\kappa)\}$ giving the positions of the r atoms in the basis with respect to their lattice point $\vec{x}(\ell)$ can be chosen in such a way that all atoms whose positions are given by

$$\vec{x}(\ell\kappa) \equiv \vec{x}(\ell) + \vec{x}(\kappa) \tag{10}$$

with $\ell_3 \geq 0$ are in the half space $x_3 \geq 0$, while all atoms whose positions are given by $\vec{x}(\ell\kappa)$ with $\ell_3 < 0$ are in the half-space $x_3 < 0$. We will refer to this way of characterizing the structure of a crystal as definition (B1).

If we are concerned with a semi-infinite crystal occupying the upper half-space $(x_3 > 0)$, the atomic positions are still given in Eqs. (9) and (10), with the primitive translation vectors $\vec{a}_1$, $\vec{a}_2$, $\vec{a}_3$ and the basis vectors $\{\vec{x}(\kappa)\}$ chosen as in definition (B1). However, in this case the values of the integer ℓ_3 in Eq. (9) are restricted to the range $\ell_3 \geq 0$. We will call this characterization of the atomic positions in a semi-infinite crystal definition (B2).

The atomic positions given by definitions (B1) and (B2) are those of the perfectly periodic, infinitely extended crystal, from which a semi-infinite crystal is created by annuling certain atomic interactions, or an entire half-space of atoms, respectively. Surface reconstruction and the relaxation of interplanar spacings in the vicinity of a crystal surface represent departures of the atomic positions from those given by definitions (B1) and (B2), and are therefore not described by definitions (B1) and (B2). The effects of surface reconstruction and the relaxation of interplanar spacings at crystal surfaces have to be determined as perturbations of the results obtained on the basis of definitions (B1) and (B2). Consequently, definitions (B1) and (B2) are most useful in calculations of surface properties which are not expected to be affected significantly by these two effects. Surface reconstruction and the relaxation of interplanar spacings are confined to within a few interatomic spacings of the surface of a crystal, and we will see in the section on Surface Modes that the penetration depth of (acoustic) surface vibration modes into the interior of the crystal from the surface along which they propagate is of the order of their wavelength parallel to the surface, which can be many thousands of interatomic spacings. It follows that definitions (B1) and (B2) can be used for the accurate calculation of long-wavelength, acoustic surface mode frequencies, and their associated displacement fields. Results based on their use should become less accurate for surface modes whose wavelengths, and penetration depths, are of the order of a few interatomic spacings. Similarly, since the changes in the vibrational thermodynamic functions of a crystal from their bulk values due to the presence of a surface are determined in the low-temperature limit by the long-wavelength, low-frequency, modes of a semi-infinite crystal, these are also given accurately by the use of definitions (B1) and (B2), but less accurately at higher temperatures, when shorter wavelength modes are excited. However, for the calculation of a surface property which is localized in the vicinity of a surface, such as the mean square displacement or mean square velocity of a surface atom, and which is sensitive to surface reconstruction and the relaxation of interplanar spacings, definition (A) proves to be the most useful.

In what follows we will base most of our discussion on definition (A), but will indicate at appropriate places the changes required in the treatment if definition (B1) or definition (B2) is employed.

Equations of Motion of a Semi-Infinite Crystal

If we adopt definition (A) the kinetic energy of a crystal slab is given by

$$T = \sum_{\ell\kappa\alpha} \frac{p_\alpha^2(\ell\kappa)}{2M_\kappa} = \frac{1}{2} \sum_{\ell\kappa\alpha} M_\kappa \dot{u}_\alpha^2(\ell\kappa) \tag{11}$$

where $u_\alpha(\ell\kappa)$ is the α Cartesian component of the displacement of the atom $(\ell\kappa)$ from its rest position $\vec{x}(\ell\kappa)$; $p_\alpha(\ell\kappa)$ is the conjugate momentum, and M_κ is the mass of the κth atom in the basis. The potential energy of the slab in the harmonic approximation (omitting the potential energy of the static crystal) can be written in the form

$$\Phi = \frac{1}{2} \sum_{\ell\kappa\alpha} \sum_{\ell'\kappa'\beta} \Phi_{\alpha\beta}(\ell\kappa;\ell'\kappa') u_\alpha(\ell\kappa) u_\beta(\ell'\kappa') \tag{12}$$

where the $\{\Phi_{\alpha\beta}(\ell\kappa;\ell'\kappa')\}$ are the (second-order) atomic force constants.

From Eq. (12) we see that $\Phi_{\alpha\beta}(\ell\kappa;\ell'\kappa')$ is expressible as

$$\Phi_{\alpha\beta}(\ell\kappa;\ell'\kappa') = \left. \frac{\partial^2 \Phi}{\partial u_\alpha(\ell\kappa)\partial u_\beta(\ell'\kappa')} \right|_0 \tag{13}$$

where the subscript zero means that the derivatives are evaluated in the equilibrium configuration, i.e., the configuration in which each atom is situated at its rest position $\vec{x}(\ell\kappa)$. It follows immediately from Eq. (13) that the atomic force constants satisfy the symmetry condition

$$\Phi_{\alpha\beta}(\ell\kappa;\ell'\kappa') = \Phi_{\beta\alpha}(\ell'\kappa';\ell\kappa) \tag{14}$$

because the value of a mixed partial derivative is independent of the order in which the derivatives are taken.

The equations of motion of the crystal slab are therefore given by

$$M_{\kappa}\ddot{u}_{\alpha}(\ell\kappa) = -\frac{\partial\Phi}{\partial u_{\alpha}(\ell\kappa)} = -\sum_{\ell'\kappa'\beta} \Phi_{\alpha\beta}(\ell\kappa;\ell'\kappa')u_{\beta}(\ell'\kappa') \tag{15}$$

Properties of the Atomic Force Constants of a Semi-Infinite Crystal

In addition to the symmetry property of Eq. (14) there are other relations among the atomic force constants that can be stated explicitly, and these fall into two categories. In the first category are the general relations that are imposed by the invariance of the potential energy against rigid body translations and rotations of the crystal slab as a whole. These relations are valid for all collections of atoms, whether they are located at the lattice points of a crystal or not. In the second category are those relations that are imposed by the special structure and symmetry of a particular crystal slab. We discuss these two kinds of relations in turn.

The fact that the force on an atom cannot be changed by a rigid body displacement of the crystal slab as a whole leads to the conditions

$$\sum_{\ell'\kappa'} \Phi_{\alpha\beta}(\ell\kappa;\ell'\kappa') = 0 \tag{16}$$

The transformation properties of the potential energy and its derivatives under an infinitesimal rigid body rotation of the crystal slab lead to the additional conditions

$$\sum_{\ell'\kappa'} \left\{\Phi_{\alpha\beta}(\ell\kappa;\ell'\kappa')x_{\gamma}(\ell'\kappa') - \Phi_{\alpha\gamma}(\ell\kappa;\ell'\kappa')x_{\beta}(\ell'\kappa')\right\} = 0 \tag{17}$$

It has been pointed out by Ludwig and Lengeler (5) (see also the discussion on pp. 523-527 and pp. 577-578 of Ref. 6) that the satisfaction of the conditions expressed by Eq. (17) is a necessary condition for the results of lattice theory to agree with those of continuum (elasticity) theory for the dynamical properties of semi-infinite solids, in the appropriate limit.

We turn now to a discussion of the restrictions on the atomic force constants that are imposed by the structure and symmetry of a particular crystal slab.

The most general symmetry operation that sends a crystal slab into itself is a combination of a rigid body rotation (proper or improper) of the crystal slab about some point, plus a rigid body translation of the crystal slab. Such an operation is represented in the Seitz (7) notation by $\{\overleftrightarrow{S}|\vec{v}(s) + \vec{x}(m)\}$. Applied to the position vector $\vec{x}(\ell\kappa)$ of the rest position of the κth atom in the ℓth unit cell this operation transforms it according to the rule

$$\begin{aligned}\{\overleftrightarrow{S}|\vec{v}(S) + \vec{x}(m)\}\vec{x}(\ell\kappa) &= \overleftrightarrow{S}\vec{x}(\ell\kappa) + \vec{v}(S) + \vec{x}(m)\\ &\equiv \vec{x}(LK) = \vec{x}(L) + \vec{x}(K)\end{aligned} \tag{18}$$

which is to be interpreted in the active sense, i.e., as a rotation of the point $\vec{x}(\ell\kappa)$ with respect to a fixed coordinate system, and not as a rotation of the coordinate system with respect to a crystal slab fixed in space. In Eq. (18) $\overleftrightarrow{S}$ is a 3 x 3 real, orthogonal matrix representation of a proper or improper rotation, $\vec{v}(S)$ is a vector which is smaller than any of the primitive translation vectors $\vec{a}_1, \vec{a}_2$ of the two-dimensional Bravais lattice underlying the crystal slab, and $\vec{x}(m)$ is the translation vector $m_1\vec{a}_1 + m_2\vec{a}_2$. The second equality in Eq. (18) expresses the fact that, because the operation $\{\overleftrightarrow{S}|\vec{v}(S) + \vec{x}(m)\}$ is one that sends the crystal slab into itself, the lattice site $(\ell\kappa)$ must be sent into an equivalent site, which we label by (LK). Here, and where no confusion results from its use, we adopt the convention of labeling by upper case letters the site into which a given site, labeled by the corresponding lower case letters, is transformed by the operation $\{\overleftrightarrow{S}|\vec{v}(S) + \vec{x}(m)\}$.

The totality of all operations $\{\overleftrightarrow{S}|\vec{v}(S) + \vec{x}(m)\}$ that send a crystal slab into itself constitute the space group G of the crystal slab. The space group G is therefore one of the 80 diperiodic groups in three dimensions, which are applicable to three-dimensional arrays that are infinitely periodic in only two dimensions, lacking periodicity in the third. A tabulation of these groups and a discussion of their properties and various applications of them has been given by Wood (8).

The atomic force constants $\{\Phi_{\alpha\beta}(\ell\kappa;\ell'\kappa')$ of a crystal slab obey the transformation law

$$\Phi_{\alpha\beta}(LK;L'K') = \sum_{\mu\nu} S_{\alpha\mu} S_{\beta\nu} \Phi_{\mu\nu}(\ell\kappa;\ell'\kappa') \qquad (19)$$

when the slab is subjected to the symmetry operation $\{\overleftrightarrow{S}|\vec{v}(S) + \vec{x}(m)\}$. It follows from Eq. (19) that if we restrict the operations of the space group G to the subgroup which leaves the sites ($\ell\kappa$) and ($\ell'\kappa'$) fixed or interchanges them, so that (LK) = ($\ell\kappa$) and (L'K') = ($\ell'\kappa'$), or (LK) = ($\ell'\kappa'$) and (L'K') = ($\ell\kappa$), then this equation together with Eq. (14) yields the independent, nonzero elements of the force constants $\{\Phi_{\alpha\beta}(\ell\kappa;\ell'\kappa')\}$.

An important special case of Eq. (19) is obtained when we specialize the space group operation $\{\overleftrightarrow{S}|\vec{v}(S) + \vec{x}(m)\}$, to the operation $\{\overleftrightarrow{E}|\vec{x}(m)\}$, where $\overleftrightarrow{E}$ is the 3 x 3 unit matrix, which is the operation of displacing the crystal slab through the lattice translation vector $\vec{x}(m)$. The effect of this operation on the position vector $\vec{x}(\ell\kappa)$ is

$$\{\overleftrightarrow{E}|\vec{x}(m)\}\vec{x}(\ell\kappa) = \vec{x}(\ell\kappa) + \vec{x}(m) = \vec{x}(\ell + m,\kappa) \qquad (20)$$

If we substitute this result into Eq. (19) we find that $\Phi_{\alpha\beta}(\ell\kappa;\ell'\kappa')$ depends on the indices ℓ and ℓ' only through their difference. For we have that

$$\Phi_{\alpha\beta}(\ell + m\kappa;\ell' + m\kappa') = \Phi_{\alpha\beta}(\ell\kappa;\ell'\kappa') \qquad (21)$$

Setting $m = -\ell'$ and $m = -\ell$ in this equation, we obtain finally

$$\Phi_{\alpha\beta}(\ell\kappa;\ell'\kappa') = \Phi_{\alpha\beta}(\ell - \ell',\kappa;0\kappa') \qquad (22a)$$

$$= \Phi_{\alpha\beta}(0\kappa;\ell' - \ell,\kappa') \qquad (22b)$$

Solution of the Equations of Motion of a Crystal Slab

The result given by Eq. (21) can be expressed alternatively by saying that the atomic force constants

$\Phi_{\alpha\beta}(\ell\kappa;\ell'\kappa')$ commute with the operation of displacing the crystal slab through an arbitrary translation vector $\vec{x}(m)$ of the slab. It follows that the solutions of the equations of motion of the slab, Eqs. (15), must simultaneously be eigenfunctions of this displacement operator. We may then choose as the solution to Eq. (15) a function of the form

$$u_\alpha(\ell\kappa;t) = M_\kappa^{-\frac{1}{2}} v_\alpha(\kappa) e^{i\vec{k}_\parallel \cdot \vec{x}(\ell) - i\omega t} \tag{23}$$

where $v_\alpha(\kappa)$ is independent of ℓ and $\vec{k}_\parallel$ is a two-dimensional wave vector whose components are parallel to the surfaces of the crystal slab, i.e.

$$\vec{k}_\parallel = \hat{x}_1 k_1 + \hat{x}_2 k_2 \tag{24}$$

(where $\hat{x}_1$ and $\hat{x}_2$ are unit vectors in the 1- and 2-directions, respectively). When we substitute Eq. (23) into Eqs. (15) we obtain as the equation for the amplitudes $\{v_\alpha(\kappa)\}$

$$\omega^2 v_\alpha(\kappa) = \sum_{\kappa'\beta} D_{\alpha\beta}(\kappa\kappa'|\vec{k}_\parallel) v_\beta(\kappa') \tag{25}$$

where $\overleftrightarrow{D}(\vec{k}_\parallel)$ is a 3r x 3r Hermitian matrix, the dynamical matrix, given by

$$D_{\alpha\beta}(\kappa\kappa'|\vec{k}_\parallel) = \frac{1}{(M_\kappa M_{\kappa'})^{\frac{1}{2}}} \sum_{\ell'} \Phi_{\alpha\beta}(\ell\kappa;\ell'\kappa') e^{-i\vec{k}_\parallel \cdot (\vec{x}(\ell) - \vec{x}(\ell'))}$$

$$= D^*_{\beta\alpha}(\kappa'\kappa|\vec{k}_\parallel) \tag{26}$$

It should be understood that the summation over ℓ' means summation over all values of ℓ_1' and ℓ_2'.

Since the dynamical matrix has dimensions 3r x 3r, there are 3r solutions of Eq. (25) for each value of the wave vector $\vec{k}_\parallel$, which we label by the index j = 1,2,3,..., 3r. Thus Eq. (25) can be rewritten as

$$\omega_j^2(\vec{k}_\parallel)v_\alpha(\kappa|\vec{k}_\parallel j) = \sum_{\kappa'\beta} D_{\alpha\beta}(\kappa\kappa'|\vec{k}_\parallel)v_\beta(\kappa'|\vec{k}_\parallel j) \qquad (27)$$

where the dependence of the frequency $\omega_j(\vec{k}_\parallel)$ and the corresponding eigenvector $v_\alpha(\kappa|\vec{k}_\parallel j)$ on the wave vector $\vec{k}_\parallel$ has been made explicit.

Equation (27) defines the eigenvector $v_\alpha(\kappa|\vec{k}_\parallel j)$ only to within an arbitrary multiplicative factor. In view of the Hermiticity of the dynamical matrix we can choose this factor in such a way that $v_\alpha(\kappa|\vec{k}_\parallel j)$ satisfies the orthonormality and closure conditions

$$\sum_{\kappa\alpha} v_\alpha^*(\kappa|\vec{k}_\parallel j)v_\alpha(\kappa|\vec{k}_\parallel j') = \delta_{jj'} \qquad (28a)$$

$$\sum_{j} v_\alpha^*(\kappa|\vec{k}_\parallel j)v_\beta(\kappa'|\vec{k}_\parallel j) = \delta_{\kappa\kappa'}\delta_{\alpha\beta} \qquad (28b)$$

In practice the solution of the set of 3r equations (27) in 3r unknowns to obtain the frequencies $\{\omega_j(\vec{k}_\parallel)\}$ and the corresponding unit eigenvectors $\{\vec{v}(\vec{k}_\parallel j)\}$ has to be carried out on a high-speed computer. Some examples of the results of such calculations will be presented in the following section.

The displacement field corresponding to a particular solution of Eq. (27) defined by given values of $\vec{k}_\parallel$ and j, is called a <u>normal mode</u> of the crystal slab

$$u_\alpha(\ell\kappa;t) = M_\kappa^{-\frac{1}{2}} v_\alpha(\kappa|\vec{k}_\parallel j) e^{i\vec{k}_\parallel \cdot \vec{x}(\ell) - i\omega_j(\vec{k}_\parallel)t}$$

We note that the set of equations (27) at $\vec{k}_\parallel = 0$ always has three solutions with vanishing frequency. For if we set $\vec{k}_\parallel = 0$ in Eq. (27) we have

$$\omega_j^2(\vec{o})\frac{v_\alpha(\kappa|\vec{o}j)}{M_\kappa^{\frac{1}{2}}} = \sum_\beta \frac{1}{M_\kappa} \sum_{\ell'\kappa'} \Phi_{\alpha\beta}(\ell\kappa;\ell'\kappa')\frac{v_\beta(\kappa'|\vec{o}j)}{M_{\kappa'}^{\frac{1}{2}}} \tag{29}$$

If, for each of the three values of α, $M_\kappa^{-\frac{1}{2}} v_\alpha(\kappa|\vec{o}j)$ is independent of κ, then in view of Eq. (16) the right hand side of Eq. (29) vanishes, implying the vanishing of $\omega_j^2(\vec{o})$. Thus we have three solutions, one for each value of α, which vanish with vanishing $\vec{k}_\parallel$. From Eq. (23) we find that these three solutions describe rigid body displacements of the crystal slab as a whole.

A complete specification of the eigenvalue problem defined by Eq. (27) requires that we define the values that can be assumed by the wave vector $\vec{k}_\parallel$ introduced in Eq. (23). It is clear that they will be determined by the boundary conditions imposed on the components of the displacement vectors $\{\vec{u}(\ell\kappa;t)\}$.

In the present discussion we will adopt cyclic boundary conditions on the atomic displacements. We consider a crystal slab that is infinitely extended in the 1- and 2-directions. We subdivide the infinite, two-dimensional Bravais lattice upon which the slab is constructed into "macrocells," each of which contains L x L = N lattice points. These macrocells cover the plane of the two-dimensional Bravais lattice without cracks or overlaps, and are parallelograms whose edges are defined by the vectors $L\vec{a}_1$ and $L\vec{a}_2$. The atoms of the basis associated with each lattice point in any one of these macrocells can be regarded as constituting the physical crystal slab of L^2r atoms whose dynamical properties we are studying. The cyclic boundary conditions postulate that the atomic displacements be periodic with the periodicity of the macrocells, that is

$$\vec{u}(\ell_1 + L, \ell_2, \kappa) = \vec{u}(\ell_1, \ell_2 + L, \kappa) = \vec{u}(\ell_1, \ell_2, \kappa) \quad (30)$$

It must be recognized that the cyclic boundary conditions are a purely mathematical fiction, whose adoption, however, does not affect the results of any calculations we will be discussing in this chapter, but simplifies them considerably. They also provide a convenient way of normalizing the kinetic and potential energies of a crystal slab to a finite volume, containing L^2r atoms.

Combined with Eq. (23), Eq. (30) requires that

$$e^{i\vec{k}_\parallel \cdot L\vec{a}_1} = e^{i\vec{k}_\parallel \cdot L\vec{a}_2} = 1 \quad (31)$$

The values of the wave vector $\vec{k}_\parallel$ allowed by these conditions are conveniently expressed in terms of the two-dimensional lattice reciprocal to the direct two-dimensional Bravais lattice defined by the primitive translation vectors $\vec{a}_1$ and $\vec{a}_2$. The primitive translation vectors of the reciprocal lattice, $\vec{b}_1$ and $\vec{b}_2$, are defined by the equations

$$\vec{a}_i \cdot \vec{b}_j = 2\pi\delta_{ij}, \text{ where } i,j = 1,2 \quad (32)$$

The solutions of these equations are

$$\vec{b}_1 = \frac{2\pi}{a_c}(a_{22}, -a_{21}) \tag{33a}$$

$$\vec{b}_2 = \frac{2\pi}{a_c}(-a_{12}, a_{11}) \tag{33b}$$

where a_c is the area of the primitive unit cell of the direct lattice

$$a_c = a_{11}a_{22} - a_{12}a_{21} = a_1a_2\sin\theta \tag{33c}$$

In writing Eqs. (33) we have used the convention that in $a_{i\alpha}$ the first subscript labels the primitive translation vector while the second labels its Cartesian component. In Eq. (33c) θ is the angle between the vectors $\vec{a}_1$ and $\vec{a}_2$, measured counterclockwise from $\vec{a}_1$ to $\vec{a}_2$.

A lattice vector of the reciprocal lattice is given by

$$\vec{\tau}(h) = h_1\vec{b}_1 + h_2\vec{b}_2 \tag{34}$$

where h_1 and h_2 are any two integers, positive, negative, or zero, to which we refer collectively as h. In view of the fact that the scalar product between a vector of the direct lattice and a vector of the reciprocal lattice is 2π times an integer, i.e.

$$\vec{x}(\ell) \cdot \vec{\tau}(h) = 2\pi(\ell_1h_1 + \ell_2h_2) = 2\pi(\text{integer}) \tag{35}$$

the expression for the wave vector $\vec{k}_\parallel$ that satisfies Eqs. (31) is

$$\vec{k}_{\parallel} = \frac{1}{L}\,\vec{\tau}(h) = \left(\frac{h_1}{L}\right)\vec{b}_1 + \left(\frac{h_2}{L}\right)\vec{b}_2 \tag{36}$$

The values of the integers h_1 and h_2 now are not unrestricted, however. From Eqs. (23) and (35) we see that the addition of any reciprocal lattice vector $\vec{\tau}(h)$ to $\vec{k}_{\parallel}$ leaves the value of the displacement amplitude $u_{\alpha}(\ell\kappa;t)$ unaffected. This means that we obtain all the distinct solutions to the equations of motion of a crystal slab if we restrict the allowed values of $\vec{k}_{\parallel}$ to lie in one unit cell of the two-dimensional reciprocal lattice

$$\vec{k}_{\parallel} = \left(\frac{h_1}{L}\right)\vec{b}_1 + \left(\frac{h_2}{L}\right)\vec{b}_2, \quad h_1, h_2 = 1, 2, \ldots, L \tag{37}$$

Thus there are L^2 allowed values of $\vec{k}_{\parallel}$. As there are 3r solutions of the equations of motion for each value of $\vec{k}_{\parallel}$, we see that there are as many distinct solutions, $3rL^2$, as there are degrees of freedom in the crystal slab.

In most calculations of the dynamical properties of a crystal slab, it is not the particular values that $\vec{k}_{\parallel}$ can assume that is important. Rather, it is the fact that they are uniformly and densely distributed throughout the area of the primitive unit cell of the reciprocal lattice. An important consequence of this result is that, where convenient, summation over the allowed values of $\vec{k}_{\parallel}$ can be replaced by integration, according to the rule

$$\sum_{\vec{k}_{\parallel}} \rightarrow \frac{L^2}{b_c}\int_{b_c} d^2\vec{k}_{\parallel} \tag{38}$$

where b_c is the area of a unit cell of the reciprocal

lattice. However, it is readily shown that $b_c = (2\pi)^2 \cdot a_c^{-1}$, so that Eq. (38) can be rewritten as

$$\sum_{\vec{k}_\parallel} \rightarrow \frac{S}{(2\pi)^2} \int_{b_c} d^2\vec{k}_\parallel \tag{39}$$

where S is the area of either surface of the slab.

The choice of a primitive unit cell of the reciprocal lattice as the region in which the allowed values of $\vec{k}_\parallel$ are restricted to lie does not, in general, reflect the symmetry properties of the reciprocal lattice. A choice for the area in reciprocal space in which we can restrict the allowed values of $\vec{k}_\parallel$ to lie and which explicitly displays the symmetry of the reciprocal lattice can be made as follows. We draw vectors from the origin of the reciprocal lattice to all lattice points and then draw the lines which are the perpendicular bisectors of these vectors. The smallest area containing the origin enclosed by these perpendicular bisectors can be shown to be completely equivalent to the unit cell in that every allowed value of $\vec{k}_\parallel$ in the unit cell differs from a corresponding point in the symmetric polygon only by a translation vector of the reciprocal lattice, and hence the two values of $\vec{k}_\parallel$ are equivalent. The symmetric polygon constructed in this way containing all allowed values of $\vec{k}_\parallel$ is called the two-dimensional first Brillouin zone of the reciprocal lattice. The first Brillouin zone has the property that it is invariant under all of the operations of the point group which sends the two-dimensional Bravais lattice underlying the crystal slab into itself. This point group is also the point group of the two-dimensional reciprocal lattice.

Equation (39) can now be rewritten equivalently as

$$\sum_{\vec{k}_\parallel} \to \frac{S}{(2\pi)^2} \int_{BZ} d^2\vec{k}_\parallel \tag{40}$$

where the integral over $\vec{k}_\parallel$ now extends throughout the area of the two-dimensional first Brillouin zone.

We conclude this section by establishing some general properties of the dynamical matrix and of its eigenvectors. We note first from Eq. (26) that

$$D_{\alpha\beta}(\kappa\kappa'|-\vec{k}_\parallel) = D^*_{\alpha\beta}(\kappa\kappa'|\vec{k}_\parallel) \tag{41}$$

If we replace $\vec{k}_\parallel$ by $-\vec{k}_\parallel$ in Eq. (27), take the complex conjugate of the resulting equation, and make use of Eq. (41), we obtain

$$\sum_{\kappa'\beta} D_{\alpha\beta}(\kappa\kappa'|\vec{k}_\parallel)v^*_\beta(\kappa'|-\vec{k}_\parallel j) = \omega^2_j(-\vec{k}_\parallel)v^*_\alpha(\kappa|-\vec{k}_\parallel j) \tag{42}$$

where we have used the fact that because $\overleftrightarrow{D}(\vec{k}_\parallel)$ is a Hermitian matrix, $\omega^2_j(\vec{k}_\parallel)$ is real. From Eq. (42) we see that the squared frequencies $\{\omega^2_j(-\vec{k}_\parallel)\}$ and the squared frequencies $\{\omega^2_j(\vec{k}_\parallel)\}$ are eigenvalues of the same matrix $\overleftrightarrow{D}(\vec{k}_\parallel)$. If $\vec{k}_\parallel$ is not a point in the first Brillouin zone at which $\overleftrightarrow{D}(\vec{k}_\parallel)$ has degenerate eigenvalues, we therefore obtain the result that

$$\omega^2_j(-\vec{k}_\parallel) = \omega^2_j(\vec{k}_\parallel) \tag{43}$$

By continuity, we extend this result to points of degeneracy, since in such cases Eq. (43) merely gives us a prescription for labeling the modes at $-\vec{k}_\parallel$ in terms of those at $\vec{k}_\parallel$.

Because the vector $\vec{v}^*(-\vec{k}_\parallel j)$ satisfies the same equation as the eigenvector $\vec{v}(\vec{k}_\parallel j)$, then as long as $\vec{k}_\parallel$ is not a point of degeneracy, the two vectors can differ at most by an arbitrary factor of modulus unity (to preserve normalization)

$$\vec{v}^*(-\vec{k}_\parallel j) = e^{i\phi}\vec{v}(\vec{k}_\parallel j) \tag{44}$$

Since a particular choice of the phase factor cannot affect the result of a calculation of any physical property of the crystal slab or bar, we adopt the convenient convention that $e^{i\phi} = 1$. Thus, in what follows we assume that

$$v_\alpha^*(\kappa|-\vec{k}_\parallel j) = v_\alpha(\kappa|\vec{k}_\parallel j) \tag{45}$$

Equation (45) was derived on the assumption that $\vec{k}_\parallel$ is not a point of degeneracy. When $\vec{k}_\parallel$ is a point of degeneracy, the most we can infer from Eq. (42) is that $v_\alpha^*(\kappa|\vec{k}_\parallel j)$ is an arbitrary linear combination of the eigenvectors $\{v_\alpha(\kappa|\vec{k}_\parallel j')\}$ for which $\omega_{j'}^2(\vec{k}_\parallel) = \omega_j^2(\vec{k}_\parallel)$. However, it has become conventional to choose this linear combination in such a way that Eq. (45) remains valid at such points as well if the degeneracy is due to spatial symmetry only. For these points, Eq. (45) gives us a convention for labeling the normal modes at $-\vec{k}_\parallel$ in terms of those at $\vec{k}_\parallel$, which is consistent with that given by Eq. (43).

Finally, we note that from its definition, Eq. (26) and from the fact that

$$\vec{x}(\ell) \cdot \vec{\tau}(h) = 2\pi \times \text{integer}$$

we have that $D_{\alpha\beta}(\kappa\kappa'|\vec{k}_{\parallel})$ is a periodic function of $\vec{k}_{\parallel}$ with the periodicity of the reciprocal lattice

$$D_{\alpha\beta}(\kappa\kappa'|\vec{k}_{\parallel} + \vec{\tau}) = D_{\alpha\beta}(\kappa\kappa'|\vec{k}_{\parallel}) \tag{46}$$

Thus, the normal mode frequencies and polarization vectors can be assumed to possess the periodicity of the two-dimensional reciprocal lattice

$$\omega_j(\vec{k}_{\parallel} + \vec{\tau}) = \omega_j(\vec{k}_{\parallel}) \tag{47}$$

$$v_\alpha(\kappa|\vec{k}_{\parallel} + \vec{\tau}j) = v_\alpha(\kappa|\vec{k}_{\parallel}j) \tag{48}$$

Normal Coordinates of a Crystal Slab

The eigenvectors $\{v_\alpha(\kappa|\vec{k}_{\parallel}j)$ of the dynamical matrix $D_{\alpha\beta}(\kappa\kappa'|\vec{k}_{\parallel})$ can be used as the basis for a normal coordinate transformation which diagonalizes the vibrational Hamiltonian of the crystal slab

$$H = \sum_{\ell\kappa\alpha} \frac{p_\alpha^2(\ell\kappa)}{2M_\kappa} + \tfrac{1}{2}\sum_{\ell\kappa\alpha}\sum_{\ell'\kappa'\beta} \Phi_{\alpha\beta}(\ell\kappa;\ell'\kappa')u_\alpha(\ell\kappa)u_\beta(\ell'\kappa') \tag{49}$$

If we introduce the decompositions

$$u_\alpha(\ell\kappa) = \left(\frac{\hbar}{2L^2M_\kappa}\right)^{\frac{1}{2}} \sum_{\vec{k}_{\parallel}j} \frac{v_\alpha(\kappa|\vec{k}_{\parallel}j)}{[\omega_j(\vec{k}_{\parallel})]^{\frac{1}{2}}} e^{i\vec{k}_{\parallel}\cdot\vec{x}(\ell)} A_{\vec{k}_{\parallel}j} \tag{50a}$$

$$p_\alpha(\ell\kappa) = \frac{1}{i}\left(\frac{\hbar M_\kappa}{2L^2}\right)^{\frac{1}{2}} \sum_{\vec{k}_{\parallel}j} [\omega_j(\vec{k}_{\parallel})]^{\frac{1}{2}} v_\alpha(\kappa|\vec{k}_{\parallel}j) e^{i\vec{k}_{\parallel}\cdot\vec{x}(\ell)} B_{\vec{k}_{\parallel}j} \tag{50b}$$

into the Hamiltonian (49), we find that the latter takes the form

$$H = \frac{1}{4} \sum_{\vec{k}_\parallel j} \hbar\omega_j(\vec{k}_\parallel) \left[B^+_{\vec{k}_\parallel j} B_{\vec{k}_\parallel j} + A^+_{\vec{k}_\parallel j} A_{\vec{k}_\parallel j} \right] \tag{51}$$

In Eqs. (50) $A_{\vec{k}_\parallel j}$ and $B_{\vec{k}_\parallel j}$ are phonon field and momentum operators, which are defined in terms of the usual phonon creation and destruction operators $b^+_{\vec{k}_\parallel j}$ and $b_{\vec{k}_\parallel j}$, respectively, by

$$A_{\vec{k}_\parallel j} = b_{\vec{k}_\parallel j} + b^+_{-\vec{k}_\parallel j} = A^+_{-\vec{k}_\parallel j}$$

(52)

$$B_{\vec{k}_\parallel j} = b_{\vec{k}_\parallel j} - b^+_{-\vec{k}_\parallel j} = - B^+_{-\vec{k}_\parallel j}$$

The operators $\{A_{\vec{k}_\parallel j}\}$ and $\{B_{\vec{k}_\parallel j}\}$ obey the commutation relations

$$[A_{\vec{k}_\parallel j}, B_{\vec{k}'_\parallel j'}] = -2\delta_{jj'}\Delta(\vec{k}_\parallel + \vec{k}'_\parallel)$$

(53)

$$[A_{\vec{k}_\parallel j}, A_{\vec{k}'_\parallel j'}] = [B_{\vec{k}_\parallel j}, B_{\vec{k}'_\parallel j'}] = 0$$

where $\Delta(\vec{k}_\parallel)$ equals unity if $\vec{k}_\parallel$ is a translation vector of the two-dimensional reciprocal lattice and vanishes otherwise. In the case of the function $\Delta(\vec{k}_\parallel - \vec{k}'_\parallel)$, since both $\vec{k}_\parallel$ and $\vec{k}'_\parallel$ are confined to the two-dimensional first Brillouin zone, there is no nonzero reciprocal lattice vector by which they can differ, so that $\Delta(\vec{k}_\parallel - \vec{k}'_\parallel)$ is equivalent to the Kronecker symbol $\delta_{\vec{k}_\parallel, \vec{k}'_\parallel}$. The phonon creation and destruction operators obey the commutation relations

$$[b_{\vec{k}_\parallel j}, b^{+}_{\vec{k}'_\parallel j'}] = \delta_{jj'}\Delta(\vec{k}_\parallel - \vec{k}'_\parallel)$$
$$[b_{\vec{k}_\parallel j}, b_{\vec{k}'_\parallel j'}] = [b^{+}_{\vec{k}_\parallel j}, b^{+}_{\vec{k}'_\parallel j'}] = 0 \tag{54}$$

In terms of the phonon creation and destruction operators the vibrational Hamiltonian of the crystal slab takes the form

$$H = \sum_{\vec{k}_\parallel j} \hbar\omega_j(\vec{k}_\parallel)[b^{+}_{\vec{k}_\parallel j} b_{\vec{k}_\parallel j} + \tfrac{1}{2}] \tag{55}$$

From the equations of motion

$$i\hbar\dot{b}_{\vec{k}_\parallel j} = [b_{\vec{k}_\parallel j}, H] = \hbar\omega_j(\vec{k}_\parallel) b_{\vec{k}_\parallel j} \tag{56a}$$

$$i\hbar\dot{b}^{+}_{\vec{k}_\parallel j} = [b^{+}_{\vec{k}_\parallel j}, H] = -\hbar\omega_j(\vec{k}_\parallel) b^{+}_{\vec{k}_\parallel j} \tag{56b}$$

we find that the Heisenberg representation operators $b_{\vec{k}_\parallel j}(t)$ and $b^{+}_{\vec{k}_\parallel j}(t)$ are given by

$$b_{\vec{k}_\parallel j}(t) = e^{i\frac{t}{\hbar}H} b_{\vec{k}_\parallel j} e^{-i\frac{t}{\hbar}H} = b_{\vec{k}_\parallel j} e^{-i\omega_j(\vec{k}_\parallel)t} \tag{57a}$$

$$b^{+}_{\vec{k}_\parallel j}(t) = e^{i\frac{t}{\hbar}H} b^{+}_{\vec{k}_\parallel j} e^{-i\frac{t}{\hbar}H} = b^{+}_{\vec{k}_\parallel j} e^{i\omega_j(\vec{k}_\parallel)t} \tag{57b}$$

Finally, from the form of the Hamiltonian (55) and the commutation relations (54) (i.e., from the fact that the operators $b^{+}_{\vec{k}_\parallel j} b_{\vec{k}_\parallel j}$ and $b^{+}_{\vec{k}'_\parallel j'} b_{\vec{k}'_\parallel j'}$ commute for all $(\vec{k}_\parallel j)$ and $(\vec{k}'_\parallel j')$, so that the partial Hamiltonians corresponding to the individual normal modes can be simultaneously diagonalized), it follows by standard methods that the thermodynamic averages of the products of creation and destruction operators are given by

$$\langle b^{+}_{\vec{k}_{\parallel} j} b_{\vec{k}'_{\parallel} j'} \rangle = \delta_{jj'} \Delta(\vec{k}_{\parallel} - \vec{k}'_{\parallel}) n_{\vec{k}_{\parallel} j} \tag{58a}$$

$$\langle b_{\vec{k}_{\parallel} j} b^{+}_{\vec{k}'_{\parallel} j'} \rangle = \delta_{jj'} \Delta(\vec{k}_{\parallel} - \vec{k}'_{\parallel}) \left(n_{\vec{k}_{\parallel} j} + 1\right) \tag{58b}$$

$$\langle b_{\vec{k}_{\parallel} j} b_{\vec{k}'_{\parallel} j'} \rangle = \langle b^{+}_{\vec{k}_{\parallel} j} b^{+}_{\vec{k}'_{\parallel} j'} \rangle = 0 \tag{58c}$$

where

$$n_{\vec{k}_{\parallel} j} = \left(\exp \beta\hbar\omega_j(\vec{k}_{\parallel}) - 1\right)^{-1}, \quad \beta = \left(k_B T\right)^{-1} \tag{58d}$$

and where k_B is Boltzmann's constant and T is the absolute temperature. The angular brackets in Eqs. (58) are defined for any operator O by

$$\langle O \rangle = \frac{\mathrm{Tr}\ \exp(-\beta H) O}{\mathrm{Tr}\ \exp(-\beta H)} = \frac{\sum_n \exp\left(-\beta E_n\right) \langle n|O|n\rangle}{\sum_n \exp\left(-\beta E_n\right)} \tag{59}$$

where E_n is the energy of the eigenstate $|n\rangle$ of the Hamiltonian H

$$H|n\rangle = E_n|n\rangle \tag{60}$$

The preceding results enable us to obtain formal expressions for several useful time-dependent correlation functions of atomic displacements and momenta. We have immediately

$$\left\langle u_\alpha(\ell\kappa;t)u_\beta(\ell'\kappa';0)\right\rangle = \frac{\hbar}{2L^2(M_\kappa M_{\kappa'})^{\frac{1}{2}}} \sum_{\vec{k}_\parallel j} \frac{v_\alpha(\kappa|\vec{k}_\parallel j)v_\beta^*(\kappa'|\vec{k}_\parallel j)}{\omega_j(\vec{k}_\parallel)}$$

$$\times\ e^{i\vec{k}_\parallel \cdot (\vec{x}(\ell) - \vec{x}(\ell'))}\left\{(n_{\vec{k}_\parallel j} + 1)e^{-i\omega_j(\vec{k}_\parallel)t} + n_{\vec{k}_\parallel j}e^{i\omega_j(\vec{k}_\parallel)t}\right\}$$

(61a)

$$\left\langle p_\alpha(\ell\kappa;t)p_\beta(\ell'\kappa';0)\right\rangle = \frac{\hbar(M_\kappa M_{\kappa'})^{\frac{1}{2}}}{2L^2} \sum_{\vec{k}_\parallel j} \omega_j(\vec{k}_\parallel)v_\alpha(\kappa|\vec{k}_\parallel j)v_\beta^*(\kappa'|k_\parallel j)$$

$$\times\ e^{ik_\parallel \cdot (\vec{x}(\ell) - \vec{x}(\ell'))}\left\{(n_{\vec{k}_\parallel j} + 1)e^{-i\omega_j(\vec{k}_\parallel)t} + n_{\vec{k}_\parallel j}e^{i\omega_j(\vec{k}_\parallel)t}\right\}$$

(61b)

$$\left\langle u_\alpha(\ell\kappa;t)p_\beta(\ell'\kappa';0)\right\rangle = \frac{1}{i}\frac{\hbar}{2L^2}\left(\frac{M_{\kappa'}}{M_\kappa}\right)^{\frac{1}{2}}\sum_{\vec{k}_\parallel j} v_\alpha(\kappa|\vec{k}_\parallel j)v^*_\beta(\kappa'|\vec{k}_\parallel j)$$

$$\times\, e^{ik_\parallel\cdot(\vec{x}(\ell)-\vec{x}(\ell'))}\left\{n_{\vec{k}_\parallel j}e^{i\omega_j(\vec{k}_\parallel)t} - \left(n_{\vec{k}_\parallel j}+1\right)e^{-i\omega_j(\vec{k}_\parallel)t}\right\}$$

(61c)

The equal time correlation functions, obtained by setting t = 0 in Eqs. (61), are given by

$$\left\langle u_\alpha(\ell\kappa)u_\beta(\ell'\kappa')\right\rangle = \frac{\hbar}{2L^2\left(M_\kappa M_{\kappa'}\right)^{\frac{1}{2}}}\sum_{\vec{k}_\parallel j}\frac{v_\alpha(\kappa|\vec{k}_\parallel j)v^*_\beta(\kappa'|\vec{k}_\parallel j)}{\omega_j(\vec{k}_\parallel)}$$

$$\times\, e^{ik_\parallel\cdot(\vec{x}(\ell)-\vec{x}(\ell'))}\left[2n_{\vec{k}_\parallel j}+1\right] \qquad (62a)$$

$$\left\langle p_\alpha(\ell\kappa)p_\beta(\ell'\kappa')\right\rangle = \frac{\hbar\left(M_\kappa M_{\kappa'}\right)^{\frac{1}{2}}}{2L^2}\sum_{\vec{k}_\parallel j}\omega_j(\vec{k}_\parallel)v_\alpha(\kappa|\vec{k}_\parallel j)$$

$$\times\, v^*_\beta(\kappa'|\vec{k}_\parallel j)e^{i\vec{k}_\parallel\cdot(\vec{x}(\ell)-\vec{x}(\ell'))}\left[2n_{\vec{k}_\parallel j}+1\right] \qquad (62b)$$

$$\left\langle u_\alpha(\ell\kappa)p_\beta(\ell'\kappa')\right\rangle = i\,\frac{\hbar}{2L^2}\left(\frac{M_{\kappa'}}{M_\kappa}\right)^{\frac{1}{2}} \sum_{\vec{k}_\parallel j}$$

$$\cdot\; v_\alpha(\kappa|\vec{k}_\parallel j)v^*_\beta(\kappa'|\vec{k}_\parallel j)e^{i\vec{k}_\parallel\,\left(\vec{x}(\ell) - \vec{x}(\ell')\right)}$$

$$- i\,\frac{\hbar}{2}\,\delta_{\alpha\beta}\delta_{\kappa\kappa'}\delta_{\ell\ell'} \qquad (62c)$$

For future reference we present alternative expressions for the first two of these equal time-correlation functions. To obtain them we use the following representations for $2n_{\vec{k}_\parallel j} + 1$

$$2n_{\vec{k}_\parallel j} + 1 = \frac{2\omega_j(\vec{k}_\parallel)}{\beta\hbar} \sum_{n=-\infty}^{\infty} \frac{1}{\Omega_n^2 + \omega_j^2(\vec{k}_\parallel)}$$

$$\Omega_n = \frac{2\pi n}{\beta\hbar} \qquad (63)$$

and obtain

$$\left\langle u_\alpha(\ell\kappa)u_\beta(\ell'\kappa')\right\rangle = \frac{1}{\beta L^2(M_\kappa M_{\kappa'})^{\frac{1}{2}}} \sum_{n=-\infty}^{\infty} \sum_{\vec{k}_\parallel j} \frac{v_\alpha(\kappa|\vec{k}_\parallel j)v^*_\beta(\kappa'|\vec{k}_\parallel j)}{\Omega_n^2 + \omega_j^2(\vec{k}_\parallel)}$$

$$\times\; e^{i\vec{k}_\parallel\,\cdot\,\left(\vec{x}(\ell) - \vec{x}(\ell')\right)} \qquad (64a)$$

$$\left\langle p_\alpha(\ell\kappa)p_\beta(\ell'\kappa')\right\rangle =$$

$$\frac{(M_\kappa M_{\kappa'})^{\frac{1}{2}}}{\beta L^2}\sum_{n=-\infty}^{\infty}\sum_{\vec{k}_\parallel j}\frac{\omega_j^2(\vec{k}\ )v_\alpha(\kappa|\vec{k}_\parallel j)v_\beta^*(\kappa'|\vec{k}_\parallel j)}{\Omega_n^2 + \omega_j^2(\vec{k}_\parallel\ )}$$

$$\times\ e^{i\vec{k}_\parallel\ \cdot\ (\vec{x}(\ell) - \vec{x}(\ell'))} \tag{64b}$$

For many applications the Fourier transform with respect to time of one of these correlation functions is required. We find that

$$\int_{-\infty}^{\infty} dt\ e^{i\omega t}\left\langle u_\alpha(\ell\kappa;t)u_\beta(\ell'\kappa';0)\right\rangle$$

$$= \frac{2\pi\hbar[n(\omega) + 1]\text{sgn}\omega}{L^2(M_\kappa M_{\kappa'})^{\frac{1}{2}}}\sum_{\vec{k}_\parallel j} v_\alpha(\kappa|\vec{k}_\parallel j)v_\beta^*(\kappa'|\vec{k}_\parallel j)$$

$$\times\ e^{i\vec{k}_\parallel\ \cdot\ (\vec{x}(\ell) - \vec{x}(\ell'))}\,\delta(\omega^2 - \omega_j^2(\vec{k}_\parallel)) \tag{65a}$$

$$\int_{-\infty}^{\infty} dt\ e^{i\omega t}\left\langle p_\alpha(\ell\kappa;t)p_\beta(\ell'\kappa';0)\right\rangle$$

$$= 2\pi\hbar\omega^2[n(\omega) + 1]\text{sgn}\omega\ \frac{(M_\kappa M_{\kappa'})^{\frac{1}{2}}}{L^2}$$

$$\times \sum_{\vec{k}_\parallel j} v_\alpha(\kappa|\vec{k}_\parallel j)v_\beta^*(\kappa'|\vec{k}_\parallel j)e^{i\vec{k}_\parallel\ \cdot\ (\vec{x}(\ell) - \vec{x}(\ell'))}$$

$$\cdot\ \delta(\omega^2 - \omega_j^2(\vec{k}_\parallel)) \tag{65b}$$

$$\int_{-\infty}^{\infty} dt\ e^{i\omega t} \left\langle u_\alpha(\ell\kappa;t)p_\beta(\ell'\kappa';0)\right\rangle = i2\pi\hbar\omega[n(\omega) + 1]\operatorname{sgn}\omega$$

$$\cdot \frac{1}{L^2}\left(\frac{M_{\kappa'}}{M_\kappa}\right)^{\frac{1}{2}} \sum_{\vec{k}_\parallel j} v_\alpha(\kappa|\vec{k}_\parallel j) \times v^*_\beta(\kappa'|\vec{k}_\parallel j)e^{i\vec{k}_\parallel \cdot (\vec{x}(\ell) - \vec{x}(\ell'))}$$

$$\cdot\ \delta\left(\omega^2 - \omega_j^2(\vec{k}_\parallel)\right) \tag{65c}$$

where $n(\omega) = (\exp \beta\hbar\omega - 1)^{-1}$ and $n(-\omega) = -n(\omega) - 1$.

If we use the fact that

$$\frac{1}{(\omega + i\varepsilon)^2 - \omega_j^2(\vec{k}_\parallel)} - \frac{1}{(\omega - i\varepsilon)^2 - \omega_j^2(\vec{k}_\parallel)}$$

$$= -2\pi i\delta\left(\omega^2 - \omega_j^2(\vec{k}_\parallel)\right)\operatorname{sgn}\omega \tag{66}$$

where ε is a positive infinitesimal, we can rewrite these results in a form that will be useful later

$$\int_{-\infty}^{\infty} dt\ e^{i\omega t} \left\langle u_\alpha(\ell\kappa;t)u_\beta(\ell'\kappa';0)\right\rangle$$

$$= \frac{i\hbar[n(\omega) + 1]}{L^2\left(M_\kappa M_{\kappa'}\right)^{\frac{1}{2}}} \sum_{\vec{k}_\parallel j} e^{i\vec{k}_\parallel \cdot (\vec{x}(\ell) - \vec{x}(\ell'))} v_\alpha(\kappa'|\vec{k}_\parallel j)v^*_\beta(\kappa'|\vec{k}_\parallel j)$$

$$\times \left\{\frac{1}{(\omega + i\varepsilon)^2 - \omega_j^2(\vec{k}_\parallel)} - \frac{1}{(\omega - i\varepsilon)^2 - \omega_j^2(\vec{k}_\parallel)}\right\} \tag{67a}$$

$$\int_{-\infty}^{\infty} dt\; e^{i\omega t} \left\langle p_\alpha(\ell\kappa;t)p_\beta(\ell'\kappa';0)\right\rangle = i\hbar\omega^2[n(\omega)+1]$$

$$\cdot \frac{(M_\kappa M_{\kappa'})^{\frac{1}{2}}}{L^2} \sum_{\vec{k}_\parallel j} e^{i\vec{k}_\parallel \cdot (\vec{x}(\ell) - \vec{x}(\ell'))} v_\alpha(\kappa|\vec{k}_\parallel j) v_\beta(\kappa'|\vec{k}_\parallel j)$$

$$\times \left\{ \frac{1}{(\omega+i\varepsilon)^2 - \omega_j^2(\vec{k}_\parallel)} - \frac{1}{(\omega-i\varepsilon)^2 - \omega_j^2(\vec{k}_\parallel)} \right\} \tag{67b}$$

$$\int_{-\infty}^{\infty} dt\; e^{i\omega t} \left\langle u_\alpha(\ell\kappa;t)p_\beta(\ell'\kappa';0)\right\rangle = -\hbar\omega[n(\omega)+1]$$

$$\cdot \frac{1}{L^2}\left(\frac{M_{\kappa'}}{M_\kappa}\right)^{\frac{1}{2}} \sum_{\vec{k}_\parallel j} e^{i\vec{k}_\parallel \cdot (\vec{x}(\ell) - \vec{x}(\ell'))} v_\alpha(\kappa|\vec{k}_\parallel j) v_\beta^*(\kappa'|\vec{k}_\parallel j)$$

$$\times \left\{ \frac{1}{(\omega+i\varepsilon)^2 - \omega_j^2(\vec{k}_\parallel)} - \frac{1}{(\omega-i\varepsilon)^2 - \omega_j^2(\vec{k}_\parallel)} \right\} \tag{67c}$$

Applications of several of these results will be discussed in succeeding sections of this chapter.

SURFACE VIBRATION MODES

We have already mentioned, in the introduction to this chapter, that a free surface of a crystal has the property that it can bind vibrational modes of the crystal to it in the form of surface vibration modes. These are modes in which the displacement amplitudes are relatively large for atoms in the vicinity of the surface and decay with increasing distance into the crystal in an essentially exponential fashion. Because the crystal retains translational periodicity in the directions parallel to the free surface the atomic displacement amplitudes are wavelike in directions parallel to the surface, and the two-dimensional continuum of surface modes is characterized by a two-dimensional wave vector, whose components are also parallel to the surface.

Surface modes can be viewed as arising in the following way. Starting from an infinitely extended crystal, one can create a pair of adjacent free surfaces by setting equal to zero the interactions coupling atoms on opposite sides of a plane lying between two adjacent planes of atoms. Equating to zero the atomic force constants coupling atoms in the two halves of a crystal is the limit of softening these force constants until they vanish. Therefore, by Rayleigh's theorem's (9) the frequencies of the normal modes of the infinitely extended crystal are lowered by the introduction of the free surfaces. In general, this lowering is of the order of the separation between consecutive eigenfrequencies of the infinitely extended crystal. If, however, there are gaps of finite width in the distribution of normal mode frequencies of the infinitely extended crystal corresponding to a given value of the two-dimensional wave vector whose components are parallel to the surface, then the creation of a free surface can displace the frequencies of the normal modes corresponding to the high-frequency ends of the gaps at a finite distance down into the gaps. Since the frequencies of these modes now lie in ranges (gaps) forbidden the normal modes of the infinitely extended crystal, they are nonpropagating, and hence localized.

Surface modes can be characterized as either acoustic or optical, depending on whether or not their frequencies vanish with the components of the two-dimensional wave vector characterizing their propagation parallel to the surface. In the former case, the surface modes split off from the bottom of the lowest band of acoustic bulk mode frequencies having the same wave vector components parallel to the surface. In the latter case, the surface modes split off from the bottom of a band of optical mode frequencies having the same wave vector components parallel to the surface.

However, as we will see later in this section, such a classification is too crude. If one studies the spectrum of acoustic and optical vibration modes of an infinitely extended crystal, corresponding to nonzero values of the components of the wave vector parallel to the crystal surface, it is often the case that gaps are observed, in which no normal mode frequencies of the infinitely extended crystal can lie. These gaps have a finite extent with respect to the values of the components of the wave vector parallel to the surface (see Fig. 2.1). It is possible that surface modes can exist whose frequencies lie in these gaps and which are split off from the upper boundary of each gap. They can exist, consequently, for only a finite range of the (nonzero) values of the components of the wave vector parallel to the surface. They cannot be classified simply as either acoustic or optical surface modes, according to the definitions of such modes given above. They are, nevertheless, genuine surface modes, in the sense that the atomic displacement amplitudes in these modes are largest at the surface, and

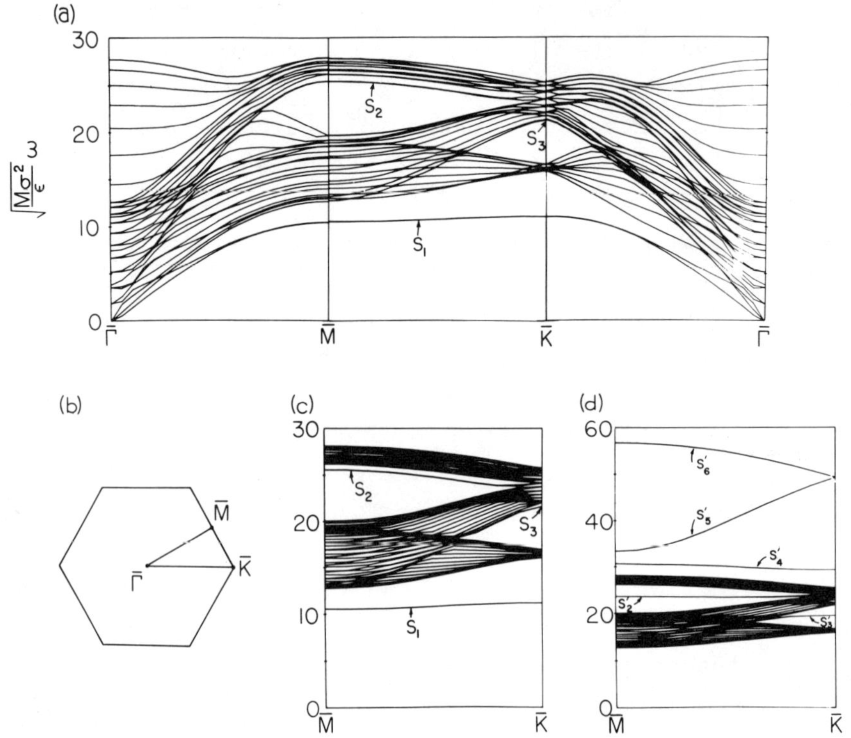

Fig. 2.1.--Variation of frequency with wave vector for a monatomic fcc crystal with two (111) surfaces. The surface waves are denoted by S_i: (a) 11-layer crystal; (b) Brillouin zone; (c) 21-layer crystal; (d) 21-layer crystal with outermost layers consisting of light adsorbed atoms, where the adatom-host atom mass ratio is 1:5. After Allen et al. (28).

decrease in an essentially exponential fashion with increasing distance into the crystal.

In this section, we present a discussion of the various kinds of surface modes that are possible, and of methods for their calculation. We begin this discussion by considering the continuum theory of the surface vibrational modes which are split off from the bottom of the lowest acoustical branch of vibrations of the corresponding infinitely extended crystal, the so-called Rayleigh waves (10). In subsequent parts of this section these and the other kinds of surface modes described above will be discussed from the discrete or lattice point of view.

Rayleigh Waves

Historically, the earliest surface waves studied are those which propagate along the stress-free, planar surface of an isotropic elastic medium. These are called Rayleigh waves, after their discoverer. Although many articles and books have been devoted to various theoretical and experimental aspects of Rayleigh waves, and their generalization to anistropic elastic media (11-13), for completeness we present here a brief discussion of these waves. The results obtained will be useful in later parts of this section and in subsequent sections of this chapter.

We consider an elastic medium occupying the half-space $x_3 \geq 0$. Within the framework of the linear theory of elasticity the equations of motion of the medium are given by

$$\rho \frac{\partial^2 u_\alpha}{\partial t^2} = \sum_\beta \frac{\partial T_{\alpha\beta}}{\partial x_\beta}, \quad \alpha = 1,2,3 \tag{68}$$

where ρ is the mass density of the medium, $u_\alpha(\vec{x},t)$ is the α Cartesian component of the displacement of the medium at the point $\vec{x}$ and time t, and $T_{\alpha\beta}(\vec{x},t)$ is the stress tensor. The latter can be expressed as

$$T_{\alpha\beta} = \sum_{\mu\nu} C_{\alpha\beta\mu\nu} \frac{\partial u_\mu}{\partial x_\nu}, \quad \alpha,\beta = 1,2,3 \tag{69}$$

where the $\{C_{\alpha\beta\mu\nu}\}$ are the elements of the elastic modulus tensor. Eq. (69) is one way of writing Hooke's law relating the stress in an elastic medium to the strain. The elements of the tensor $C_{\alpha\beta\mu\nu}$ are symmetric in α and β; in μ and ν; and in the interchange of the pairs $\alpha\beta$ and $\mu\nu$. Combining Eqs. (68) and (69) we obtain the equations of motion of the medium in the form

$$\rho \frac{\partial^2 u_\alpha}{\partial t^2} = \sum_\beta C_{\alpha\beta\mu\nu} \frac{\partial^2 u_\mu}{\partial x_\beta \partial x_\nu} \tag{70}$$

These equations have to be supplemented by the conditions expressing the fact that the stresses acting on the surface $x_3 = 0$ vanish

$$T_{\alpha 3} = 0 \quad \text{at } x_3 = 0; \; \alpha = 1,2,3 \tag{71}$$

The elastic modulus tensor for an isotropic medium has only two, independent, nonzero components (12)

$$C_{\alpha\beta\mu\nu} = \lambda\delta_{\alpha\beta}\delta_{\mu\nu} + \mu\left(\delta_{\alpha\mu}\delta_{\beta\nu} + \delta_{\alpha\nu}\delta_{\beta\mu}\right) \tag{72}$$

where λ and μ are the two Lamé constants. It is more convenient, however, to express $C_{\alpha\beta\mu\nu}$ in terms of the speeds of sound for longitudinal and transverse waves in an isotropic medium, which are given by (14)

$$c_\ell^2 = \frac{\lambda + 2\mu}{\rho}, \quad c_t^2 = \frac{\mu}{\rho} \tag{73}$$

Thus Eq. (72) can be rewritten in the alternative form

$$C_{\alpha\beta\mu\nu} = \rho\left(c_\ell^2 - 2c_t^2\right)\delta_{\alpha\beta}\delta_{\mu\nu} + \rho c_t^2\left(\delta_{\alpha\mu}\delta_{\beta\nu} + \delta_{\alpha\nu}\delta_{\beta\mu}\right) \tag{74}$$

Since all directions are equivalent in an isotropic medium, with no loss of generality we can restrict our attention to waves propagating in the x_1-direction. The displacement field can be expressed in the form

$$u_\alpha(\vec{x},t) = \bar{u}_\alpha(x_3)e^{ikx_1 - i\omega t} \tag{75}$$

where $\alpha = 1,2,3$. The equations satisfied by the amplitude functions $\{\bar{u}_\alpha(x_3)\}$ obtained by substituting Eqs. (74) and (75) into Eq. (70) are

$$\begin{pmatrix} c_t^2D^2 + \omega^2 - c_\ell^2k^2 & 0 & ik\left(c_\ell^2 - c_t^2\right)D \\ 0 & c_t^2D^2 + \omega^2 - c_t^2k^2 & 0 \\ ik\left(c_\ell^2 - c_t^2\right)D & 0 & c_\ell^2D^2 + \omega^2 - c_t^2k^2 \end{pmatrix} \times \begin{pmatrix} \bar{u}_1 \\ \bar{u}_2 \\ \bar{u}_3 \end{pmatrix} = 0 \tag{76}$$

where we have written $D \equiv d/dx_3$. The boundary conditions (71) take the form

$$\left.\begin{aligned} [ik\bar{u}_3 + D\bar{u}_1]_{x_3=0} &= 0 \end{aligned}\right\} \tag{77}$$

$$\left.\begin{aligned} &[D\bar{u}_2]_{x_3=0} = 0 \\ &\left[ik\left(c_\ell^2 - 2c_t^2\right)\bar{u}_1 + c_\ell^2 D\bar{u}_3\right]_{x_3=0} = 0 \end{aligned}\right\} \quad (77)$$

Since we are interested in solutions which are localized in the vicinity of the surface $x_3 = 0$, we seek them in the form

$$\bar{u}_\alpha(x_3) = \text{const.}\ e^{-\alpha x_3} \qquad (78)$$

It follows immediately that $\bar{u}_2(x_3)$ must vanish identically because a function of the form given by Eq. (78) cannot satisfy simultaneously the equation of motion and the boundary condition obeyed by $\bar{u}_2(x_3)$. The equations of motion for $\bar{u}_1(x_3)$ and $\bar{u}_3(x_3)$ are satisfied by the functions

$$\bar{u}_1(x_3) = Ae^{-\alpha_\ell x_3} + Be^{-\alpha_t x_3} \qquad (79a)$$

$$\bar{u}_3(x_3) = i\left(\frac{\alpha_\ell}{k}\ Ae^{-\alpha_\ell x_3} + \frac{k}{\alpha_t}\ Be^{-\alpha_t x_3}\right) \qquad (79b)$$

where A and B are arbitrary constants, and

$$\alpha_\ell = \left(k^2 - \frac{\omega^2}{c_\ell^2}\right)^{\frac{1}{2}}, \quad \alpha_t = \left(k^2 - \frac{\omega^2}{c_t^2}\right)^{\frac{1}{2}} \qquad (80)$$

Both α_ℓ and α_t must be real for Eqs. (79) to represent waves localized at the surface $x_3 = 0$. Since c_ℓ^2 is always larger than c_t^2 (11), we see that such localized waves can exist in only that part of the (ω, k) plane for which

$$\omega^2 < c_t^2 k^2 \tag{81}$$

Substitution of Eqs. (79) into the boundary conditions (77) yields a pair of homogeneous linear equations for the constants A and B. The solvability condition for this system yields

$$\left(k^2 + \alpha_t^2\right)^2 - 4\alpha_t \alpha_\ell k^2 = 0 \tag{82}$$

Equation (82) relates the frequency ω of the wave and the wave number k. The form of this equation indicates that this relation must be of the form

$$\omega = c_R k \tag{83}$$

where c_R, the speed of Rayleigh surface waves, is the solution of

$$4\left(1 - \frac{c_R^2}{c_\ell^2}\right)^{\frac{1}{2}}\left(1 - \frac{c_R^2}{c_t^2}\right)^{\frac{1}{2}} = \left(2 - \frac{c_R^2}{c_t^2}\right)^2 \tag{84}$$

It follows that Rayleigh waves are acoustic surface waves. Equation (84) can be rearranged into the form

$$\left(\frac{c_R^2}{c_t^2}\right)^3 - 8\left(\frac{c_R^2}{c_t^2}\right)^2 + 8\left(3 - 2\,\frac{c_t^2}{c_\ell^2}\right) - 16\left(1 - \frac{c_t^2}{c_\ell^2}\right) = 0 \tag{85}$$

Of the three solutions of this cubic equation for $\left(c_R^2/c_t^2\right)$ only one, the smallest, satisfies the condition $\left(c_R/c_t\right) < 1$, obtained by combining Eqs. (81) and (83). This is the condition for the existence of a localized wave.

Physical values of the ratio $\left(c_R/c_t\right)$ range from 0.96 for $\left(c_\ell^2/c_t^2\right) = \infty$, corresponding to an incompressible solid, to 0.69 for $\left(c_\ell^2/c_t^2\right) = 4/3$, the smallest value consistent with the stability of the elastic medium (i.e., with a positive definite strain energy density).

When Eq. (82) is satisfied, we have

$$B/A = -\frac{1}{2}\left(2 - \left(c_R^2/c_t^2\right)\right)$$

The displacement field of a Rayleigh wave is therefore given by

$$u_1(\vec{x},t) = A\left\{e^{-\alpha_\ell x_3} - \left(1 - \frac{c_R^2}{2c_t^2}\right)e^{-\alpha_t x_3}\right\}e^{ikx - i\omega t} \tag{86a}$$

$$u_2(\vec{x},t) = 0 \tag{86b}$$

$$u_3(\vec{x},t) = iA\left(1 - \frac{c_R^2}{c_\ell^2}\right)^{\frac{1}{2}}\left\{e^{-\alpha_\ell x_3} - \frac{e^{-\alpha_t x_3}}{\left(1 - \frac{c_R^2}{2c_t^2}\right)}\right\}e^{ikx - i\omega t} \tag{86c}$$

where

$$\alpha_\ell = k\left(1 - \frac{c_R^2}{c_\ell^2}\right)^{\frac{1}{2}}, \quad \alpha_t = k\left(1 - \frac{c_R^2}{c_t^2}\right)^{\frac{1}{2}} \tag{86d}$$

We see from Eqs. (86) that the particle displacements in a Rayleigh wave execute ellipses in the sagittal plane, i.e., the plane containing both the normal to the surface and the direction of propagation of the wave. In addition, we find that the attenuation lengths of the wave into the elastic medium, α_ℓ^{-1} and α_t^{-1}, are of the order of the wavelength $\lambda = (2\pi/k)$ of the wave along the surface. For example, in the typical case in which $c_\ell^2 = 3c_t^2$ (the so-called Poisson case, for which the two Lamé constants are equal), we find that $(c_R/c_t) = [2 - \frac{2}{3}\sqrt{3}]^{\frac{1}{2}} = 0.9194$. It follows that

$$\alpha_\ell^{-1} = \frac{1.18}{2\pi}\lambda, \quad \alpha_t^{-1} = \frac{2.54}{2\pi}\lambda \tag{87}$$

The analysis of Rayleigh waves propagating in an arbitrary direction along a planar, stress-free surface of an anisotropic elastic medium has many features in common with the discussion just presented, although the calculations are more complicated. There are qualitative differences, however. In the general case, a Rayleigh wave is characterized by three decay constants instead of the two, that is, α_ℓ and α_t, obtained for an isotropic medium. The particle displacements still trace out ellipses, but the latter, in general, are now inclined at some nonzero angle to the sagittal plane. The speed of Rayleigh waves now depends on the direction of their propagation. The three attenuation constants are the roots of a cubic equation. It is found that, depending on the elastic moduli of the solid, all three can be real, or one can be real while the remaining two are complex conjugates of each other, with positive real parts. Surface waves characterized by attenuation constants that are all real are called ordinary Rayleigh waves, while those characterized by attenuation constants which include a complex conjugate pair are called generalized Rayleigh waves (15). Pseudosurface waves are also possible (16,17). These are surface waves for which one of the three attenuation constants is pure imaginary rather than real or complex. This means that there is a component of the displacement field whose amplitude does not decay exponentially with increasing distance into the solid from the surface; hence the name pseudosurface wave. The nondecaying component of the displacement field gives rise to a flow of energy away from the surface into the solid. This in turn requires that the components of the wave vector parallel to the surface $\vec{k}_{\parallel}$ have imaginary parts, so that the wave attenuates as it propagates along the surface. However, in many cases, the attenuation is so small that the pseudosurface wave is readily observable.

Elastic surface waves have also been studied in slab-shaped solids bounded by a pair of parallel, planar, stress-free surfaces, and of infinite extent in the directions parallel to the surfaces (18). In such slabs, in addition to Rayleigh waves, perturbed by the existence of a pair of surfaces, a new kind of surface wave can exist when they are sufficiently thin, which has no counterpart for a semi-infinite solid. It involves the vibration of the slab as a whole, as if it were a two-dimensional membrane, and has a dispersion relation of the form $\omega = c_L^2 k_{\parallel}^2$, rather than of the linear form given by Eq. (83) (19).

Elastic surface waves have also been studied at the planar interface of two different elastic media (20) or, if the medium is anisotropic, at the interface between two pieces of the same medium which are oriented differently with respect to each other (21). Elastic surface waves have also been studied at the interface between a solid and a liquid (20,22), and at the surface of a solid which is not planar but curved (23).

Finally, elastic surface waves have been studied at the surface of a piezoelectric material (24,25). In this case the mechanical displacements of the medium give rise to an accompanying electric field, in the medium and in the vacuum outside it, which is localized in the vicinity of the surface, and which can have interesting and useful technological consequences. In piezoelectric crystals, moreover, a type of surface wave is possible that has no counterpart in ordinary, non-piezoelectric crystals. This wave, discovered independently by Bleustein (26) and Gulyaev (27), consists of a transverse wave whose mechanical displacement is parallel to the surface. In an ordinary surface wave, on the other hand, the mechanical displacement is not parallel to the surface.

Space does not permit a more extensive discussion of the various kinds of acoustic surface waves that have been enumerated above. The interested reader is referred to the papers cited for detailed discussions of their properties, as well as to the review articles by Farnell (11) and by Dieulesaint and Royer (13), and the books by Auld (12).

Surface Waves in a Slab of a Nonionic Crystal

We first consider the vibrations of a nonionic crystal slab, described by definition (A). The extension of this discussion to a slab of an ionic crystal will be described in the next section. The solutions of Eq. (27) give the eigenfrequencies $\omega_j^2(\vec{k}_{\parallel})$ and the eigenvectors $v_\alpha(\kappa|\vec{k}_{\parallel}j)$. To distinguish perturbed bulk modes from localized surface modes one has to plot the eigenvectors $v_\alpha(\kappa|\vec{k}_{\parallel}j)$ for a given mode as functions of the penetration κ into the slab. If these eigenvectors are localized near the surfaces of the slab, one has a localized phonon.

Surface modes of vibration in hexagonal close-packed crystals and in monatomic fcc lattices with (111), (100), and (110) surfaces have been calculated by this method by Allen et al. (28). They assumed central interaction of the Lennard-Jones 6-12 type between all pairs of atoms in plates up to 21 atomic layers thick. The model may be considered appropriate for a slab of a rare-gas solid. They found the expected dispersion of the Rayleigh modes and also additional surface modes of non-Rayleigh type, whose frequencies fall in the gaps in the one-dimensional density of states corresponding to fixed values of the components of the wave vector $\vec{k}_{\parallel}$. The existence of such surface modes had been predicted by Feuchtwang (29). Allen et al. discuss the circumstances under which the displacement ellipse of the particles lies in the saggital plane (surfaces with "axial-inversion symmetry"), and when surface modes can have frequencies lying within bulk-mode sub-bands. The number of modes they found are probably dependent, to some extent, on the model employed, that is, one in which there are interactions between all pairs of atoms through a Lennard-Jones potential. Some of these surface modes are primarily localized in the second layer from the surface, or even a deeper layer, rather than in the surface layer itself. These various surface modes can be regarded as "peeling off" in succession from the bulk branches.

A study of surface phonons at the (001) surface of some fcc and bcc transition metals was made by Castiel et al. (30) by the slab method and at the same time by calculating directly the surface modes in a semi-infinite crystal (see the discussion in a later section). A model consisting of first- and second-neighbor central interactions together with angle bending forces was used. Specific results have been obtained for surface phonons on Pt, Pd, Cu, and Ag as well as on Fe and W.

The effects of surface force-constant changes were considered in these two works (28,30), as well as by Ludwig (32) and by Trullinger et al. (33). Castiel et al. (30,31) showed the possibility of the existence of soft surface modes on (001) Pt when the surface forces were decreased by about 35 to 40% from their bulk values. This may account for surface superstructures. Ludwig (32) and Trullinger et al. (33) obtained soft-surface phonons for the (111) surface of Si. The last authors (33) introduced an effective dynamical charge on the surface atoms.

Some calculations emphasizing the long-wavelength region have been carried out by Alldredge et al. (34) for the (100) surface of a fcc crystal. In addition to the usual Rayleigh waves and pseudosurface waves, they find a surface wave propagating in the (110) direction, which is polarized shear horizontal. Alldredge (35) has shown, however, that the attenuation constant for this wave is proportional to $k_{\parallel}^2$, and hence, this mode is not a Rayleigh-type surface wave.

Surface Waves in a Slab of an Ionic Crystal

The range of the interatomic forces has not been specified in the preceding discussion. Consequently, it would seem that the results just obtained apply equally to crystal slabs in which the constituent atoms interact through short-range forces, and to ionic crystal slabs in which the ions interact through long-range Coulomb forces. In fact, the method of the preceding section has been used for the determination of the normal vibration modes of a slab of an ionic crystal, using the approximation that the ions are rigid, or point charges. In this case the forces between the ions, and consequently their equations of motion, can be derived from a potential function: the Coulomb potential plus the potential describing the short-range repulsive interactions. However, ions are not point charges. The electronic charge distribution surrounding each nucleus is of finite extent. It can deform and be displaced with respect to its nucleus as a neighboring ion is brought up to it and the overlap of the two electronic charge distributions increases. This is a consequence of the Pauli exclusion principle. This ionic deformability alters the dipole moment of an ion from the value obtained by displacing the static charge of the ion from its equilibrium position (36), and can even induce dipole moments on atoms which bear no static charge, such as those comprising tellurium and graphite (37). The finite extent of the electronic charge distribution about an ion also renders the ion polarizable. It can deform in response to an electric field acting on it in such a way as to acquire a dipole moment in addition to the one it posseses as a consequence of its deformability in response to overlap and its static charge.

Because they are essentially electronic in origin, the effects of ionic deformability and polarizability cannot be incorporated into a dynamical theory of ionic crystal slabs by starting from an expansion of the crystal potential energy in powers of the atomic displacements alone: the electronic coordinates have to be taken into account either explicitly, as in the "shell model" of ionic crystals (38), or implicitly, through their consequences, as in the "deformation dipole" model (39). At the phenomenological level the results of either approach can be reproduced by supplementing the ionic displacements with the components of the electric field acting on each ion as additional, independent dynamical variables in writing the equations of motion of the crystal slab and the dipole moment at each site. These electric field components are ultimately related to the ionic displacements through the equations of electromagnetic theory, so that the final equations of motion from which the normal mode frequencies are obtained are again expressed in terms of the ionic displacements alone.

The positions of the ions constituting the crystal slab under study here are specified by definition (A), which was discussed at the beginning of this section. The index κ in Eq. (8) assumes the values 0, 1, 2,..., r-1, where r is the number of ions in the basis. (The number r is not necessarily the number of ionic planes in the slab, as it is possible for a given plane to contain more than one ion from the basis.) The case of a semi-infinite crystal is obtained by letting r become infinite.

Thus the essential physical phenomena which play a role in determining the vibrational properties of a semi-infinite crystal slab made up of polarizable and deformable ions can be taken into account phenomenologically by basing the dynamical theory of such a system on the following expression for their potential energy (40):

$$\Phi = \Phi_0 + \tfrac{1}{2} \sum_{\ell\kappa\alpha} \sum_{\ell'\kappa'\beta} \Phi_{\alpha\beta}(\ell\kappa;\ell'\kappa')u_\alpha(\ell\kappa)u_\beta(\ell'\kappa')$$

$$- \sum_{\ell\kappa\alpha} \sum_{\ell'\kappa'\beta} M_{\alpha\beta}(\ell\kappa;\ell'\kappa')E_\alpha(\ell\kappa)u_\beta(\ell'\kappa')$$

$$- \tfrac{1}{2} \sum_{\ell\kappa\alpha} \sum_{\ell'\kappa'\beta} P_{\alpha\beta}(\ell\kappa;\ell'\kappa')E_\alpha(\ell\kappa)E_\beta(\ell'\kappa') \qquad (88)$$

The $\{\Phi_{\alpha\beta}(\ell\kappa;\ell'\kappa')\}$ are the force constants associated with all the short-range interactions in the crystal slab, including those due to the Lorentz field, but contain no contribution associated with the macroscopic field; the $\{M_{\alpha\beta}(\ell\kappa;\ell'\kappa')\}$ are the components of the (nonlocal) transverse effective charge tensor; the $\{E_\alpha(\ell\kappa)\}$ are the components of the macroscopic electric field acting on each ion; and the $\{P_{\alpha\beta}(\ell\kappa;\ell'\kappa')\}$ are the components of the (nonlocal) electronic polarizability tensor. General properties of these $\{\Phi_{\alpha\beta}(\ell\kappa;\ell'\kappa')\}$ are given by Eqs. (14), (16), (17), (19), and (21). Similar properties are obtained for the same reasons for the $\{P_{\alpha\beta}(\ell\kappa;\ell'\kappa')\}$ and $\{M_{\alpha\beta}(\ell\kappa;\ell'\kappa')\}$

$$P_{\alpha\beta}(\ell\kappa;\ell'\kappa') = P_{\beta\alpha}(\ell'\kappa';\ell\kappa) \qquad (89)$$

$$\sum_{\ell'\kappa'} M_{\alpha\beta}(\ell\kappa;\ell'\kappa') = 0 \qquad (90)$$

$$\sum_{\ell'\kappa'} \{M_{\alpha\beta}(\ell\kappa;\ell'\kappa')x_\gamma(\ell'\kappa') - M_{\alpha\gamma}(\ell\kappa;\ell'\kappa')x_\beta(\ell'\kappa')\} = 0 \qquad (91)$$

$$M_{\alpha\beta}(LK;L'K') = \sum_{\mu\nu} S_{\alpha\mu}S_{\beta\nu}M_{\mu\nu}(\ell\kappa;\ell'\kappa') \qquad (92a)$$

$$P_{\alpha\beta}(LK;L'K') = \sum_{\mu\nu} S_{\alpha\mu}S_{\beta\nu}P_{\mu\nu}(\ell\kappa;\ell'\kappa') \tag{92b}$$

It is found also, as in Eq. (22), that the coefficients $M_{\alpha\beta}(\ell\kappa;\ell'\kappa')$ and $P_{\alpha\beta}(\ell\kappa;\ell'\kappa')$ depend on the cell indices ℓ and ℓ' only through their difference.

The equations of motion of the crystal slab, and the dipole moment of the ion $(\ell\kappa)$, are given by

$$M_\kappa \ddot{u}_\alpha(\ell\kappa) = -\frac{\partial\Phi}{\partial u_\alpha(\ell\kappa)} \tag{93a}$$

$$p_\alpha(\ell\kappa) = -\frac{\partial\Phi}{\partial E_\alpha(\ell\kappa)} \tag{93b}$$

where M_κ is the mass of the κth ion in the basis. Substituting the following

$$u_\alpha(\ell\kappa) = \frac{v_\alpha(\kappa)}{M_\kappa^{\frac{1}{2}}} e^{i\vec{k}_\parallel \cdot \vec{x}(\ell) - i\omega t} \tag{94a}$$

$$p_\alpha(\ell\kappa) = p_\alpha(\kappa)e^{i\vec{k}_\parallel \cdot \vec{x}(\ell) - i\omega t} \tag{94b}$$

$$E_\alpha(\ell\kappa) = E_\alpha(\kappa)e^{i\vec{k}_\parallel \cdot \vec{x}(\ell) - i\omega t} \tag{94c}$$

into Eqs. (93), where $\vec{k}_\parallel$ is a two-dimensional wave vector, $\vec{k}_\parallel = \hat{x}_1k_1 + \hat{x}_2k_2$, yields the equations

$$\omega^2 v_\alpha(\kappa) = \sum_{\kappa'\beta} D_{\alpha\beta}(\kappa\kappa'|\vec{k}_\parallel)v_\beta(k') - \frac{1}{M_\kappa^{\frac{1}{2}}}\sum_{\kappa'\beta} M^*_{\beta\alpha}(\kappa'\kappa|\vec{k}_\parallel)E_\beta(\kappa') \tag{95a}$$

$$p_\alpha(\kappa) = \sum_{\kappa'\beta} M_{\alpha\beta}(\kappa\kappa'|\vec{k}_\parallel)\frac{v_\beta(\kappa')}{M_{\kappa'}^{\frac{1}{2}}} + \sum_{\kappa'\beta} P_{\alpha\beta}(\kappa\kappa'|\vec{k}_\parallel)E_\beta(\kappa') \tag{95b}$$

where

$$D_{\alpha\beta}(\kappa\kappa'|\vec{k}_\parallel) = \frac{1}{(M_\kappa M_{\kappa'})^{\frac{1}{2}}} \sum_{\ell'} \Phi_{\alpha\beta}(\ell\kappa;\ell'\kappa')e^{-i\vec{k}_\parallel \cdot (\vec{x}(\ell) - \vec{x}(\ell'))} \tag{96a}$$

$$M_{\alpha\beta}(\kappa\kappa'|\vec{k}_\parallel) = \sum_{\ell'} M_{\alpha\beta}(\ell\kappa;\ell'\kappa')e^{-i\vec{k}_\parallel \cdot (\vec{x}(\ell) - \vec{x}(\ell'))} \tag{96b}$$

$$P_{\alpha\beta}(\kappa\kappa'|\vec{k}_\parallel) = \sum_{\ell'} P_{\alpha\beta}(\ell\kappa;\ell'\kappa')e^{-ik_\parallel \cdot (\vec{x}(\ell) - \vec{x}(\ell'))} \tag{96c}$$

The macroscopic field $E_\alpha(\ell\kappa)$ is a consequence of the presence of a dipole with moment $p_\alpha(\ell\kappa)$ at each ion in the slab. The relation between the amplitudes $E_\alpha(\kappa)$ and $p_\alpha(\kappa)$ is found by Ewald's method to be (see Ref. 40)

$$E_\alpha(\kappa) = -\sum_{\kappa'\beta} e^{i\vec{k}_\parallel \cdot \vec{x}_\parallel(\kappa)} T_{\alpha\beta}(\kappa\kappa'|\vec{k}_\parallel) e^{-i\vec{k}_\parallel \cdot \vec{x}_\parallel(\kappa')} p_\beta(\kappa') \tag{97}$$

where

$$\overleftrightarrow{T}(\kappa\kappa'|\vec{k}_{\parallel}) = \frac{2\pi}{a_c} e^{-k_{\parallel}|x_3(\kappa\kappa')|}$$

$$\times \begin{bmatrix} \frac{k_1^2}{k_{\parallel}} & \frac{k_1 k_2}{k_{\parallel}} & ik_1 \operatorname{sgn} x_3(\kappa\kappa') \\ \frac{k_1 k_2}{k_{\parallel}} & \frac{k_2^2}{k_{\parallel}} & ik_2 \operatorname{sgn} x_3(\kappa\kappa') \\ ik_1 \operatorname{sgn} x_3(\kappa\kappa') & ik_2 \operatorname{sgn} x_3(\kappa\kappa') & -k_{\parallel} \end{bmatrix} \tag{98}$$

provided that we use the notation $x_3(\kappa\kappa') = x_3(\kappa) - x_3(\kappa')$ and the interpretation $\operatorname{sgn}(o) \equiv 0$; In Eq. (98), a_c is the area of a primitive unit cell of the two-dimensional Bravais lattice.

Let us, for simplicity, make the definition

$$S_{\alpha\beta}(\kappa\kappa'|\vec{k}_{\parallel}) = e^{i\vec{k}_{\parallel}\cdot\vec{x}_{\parallel}(\kappa)} T_{\alpha\beta}(\kappa\kappa'|\vec{k}_{\parallel}) e^{-i\vec{k}_{\parallel}\cdot\vec{x}_{\parallel}(\kappa')} \tag{99}$$

Then, on combining Eqs. (95b), (97), and (99), we obtain the equation relating the amplitude of the macroscopic electric field to the reduced displacement amplitude

$$E_{\alpha}(\kappa) = -\sum_{\kappa'\beta} S_{\alpha\beta}(\kappa\kappa'|\vec{k}_{\parallel}) \left\{ \sum_{\kappa''\gamma} M_{\beta\gamma}(\kappa'\kappa''|\vec{k}_{\parallel}) \frac{v_{\gamma}(\kappa'')}{(M_{\kappa''})^{\frac{1}{2}}} + \sum_{\kappa''\gamma} P_{\beta\gamma}(\kappa'\kappa''|\vec{k}_{\parallel}) E_{\gamma}(\kappa'') \right\} \tag{100}$$

We write the solution of this equation formally as

$$E_{\alpha}(\kappa) = -\sum_{\kappa'\beta}\sum_{\kappa''\gamma} U_{\alpha\beta}(\kappa\kappa'|\vec{k}_{\parallel})M_{\beta\gamma}(\kappa'\kappa''|\vec{k}_{\parallel})\frac{v_{\gamma}(\kappa')}{(M_{\kappa''})^{\frac{1}{2}}} \tag{101}$$

where $U_{\alpha\beta}(\kappa\kappa'|\vec{k}_{\parallel})$ is the solution of the equation

$$U_{\alpha\beta}(\kappa\kappa'|\vec{k}_{\parallel}) = S_{\alpha\beta}(\kappa\kappa'|\vec{k}_{\parallel}) - \sum_{\kappa''\gamma}\sum_{\kappa'''\delta} S_{\alpha\gamma}(\kappa\kappa''|\vec{k}_{\parallel})P_{\gamma\delta}(\kappa''\kappa'''|\vec{k}_{\parallel})U_{\delta\beta}(\kappa'''\kappa'|\vec{k}_{\parallel}) \tag{102}$$

When we substitute Eq. (101) into Eq. (95a) we obtain the time-independent equations of motion of the crystal slab in the form

$$\omega^2 v_{\alpha}(\kappa) = \sum_{\kappa'\beta} D_{\alpha\beta}(\kappa\kappa'|\vec{k}_{\parallel})v_{\beta}(\kappa') + \sum_{\kappa'\beta}\sum_{\kappa''\gamma}\sum_{\kappa'''\delta}\frac{M^{*}_{\gamma\alpha}(\kappa''\kappa|\vec{k}_{\parallel})}{(M_{\kappa})^{\frac{1}{2}}} U_{\gamma\delta}(\kappa''\kappa'''|\vec{k}_{\parallel}) \cdot \frac{M_{\delta\beta}(\kappa'''\kappa'|\vec{k}_{\parallel})}{(M_{\kappa'})^{\frac{1}{2}}} v_{\beta}(\kappa') \tag{103}$$

Let us now turn to a discussion of the matrix $U_{\alpha\beta}(\kappa\kappa'|\vec{k}_{\parallel})$. If we write

$$U_{\alpha\beta}(\kappa\kappa'|\vec{k}_\parallel) = e^{i\vec{k}_\parallel \cdot \vec{x}_\parallel(\kappa)} V_{\alpha\beta}(\kappa\kappa'|\vec{k}_\parallel) e^{-i\vec{k} \cdot \vec{x}(\kappa')} \tag{104}$$

the equation satisfied by the matrix $V_{\alpha\beta}(\kappa\kappa'|\vec{k}_\parallel)$ is

$$V_{\alpha\beta}(\kappa\kappa'|\vec{k}_\parallel) = T_{\alpha\beta}(\kappa\kappa'|\vec{k}_\parallel) - \sum_{\kappa''\gamma}\sum_{\kappa'''\delta} T_{\alpha\gamma}(\kappa\kappa''|\vec{k}_\parallel) p_{\gamma\delta}(\kappa''\kappa'''|\vec{k}_\parallel) V_{\delta\beta}(\kappa'''\kappa'|\vec{k}_\parallel) \tag{105}$$

where

$$\rho_{\alpha\beta}(\kappa\kappa'|\vec{k}_\parallel) = e^{-i\vec{k}_\parallel \cdot \vec{x}_\parallel(\kappa)} P_{\alpha\beta}(\kappa\kappa'|\vec{k}_\parallel) e^{i\vec{k}_\parallel \cdot \vec{x}_\parallel(\kappa')} \tag{106}$$

We wish to show that there is a qualitative difference between the form of the matrix $V_{\alpha\beta}(\kappa\kappa'|\vec{k})$ in the long-wavelength limit in the case of a crystal slab and its form in the long-wavelength limit in the case of a semi-infinite crystal.

For this purpose it is convenient to define a matrix $t_{\alpha\beta}(\kappa\kappa'|\vec{k}_\parallel)$ by

$$T_{\alpha\beta}(\kappa\kappa'|\vec{k}_\parallel) = \frac{2\pi k_\parallel}{a_c} t_{\alpha\beta}(\kappa\kappa'|\vec{k}_\parallel) \tag{107}$$

From a comparison of Eqs. (98) and (107) we see that, as a function of $\vec{k}_\parallel$, $t_{\alpha\beta}(\kappa\kappa'|\vec{k}_\parallel)$ is of $O(1)$ as $\vec{k}_\parallel \to 0$.

The first few terms in the iterative solution of Eq. (105) are given by

$$V_{\alpha\beta}(\kappa\kappa'|\vec{k}_{\|}) = \left(\frac{2\pi k_{\|}}{a_c}\right) t_{\alpha\beta}(\kappa\kappa'|\vec{k}_{\|})$$

$$- \left(\frac{2\pi k_{\|}}{a_c}\right)^2 \sum_{\kappa_2\gamma_2} \sum_{\kappa_3\gamma_3} t_{\alpha\gamma_2}(\kappa\kappa_2|\vec{k}_{\|})$$

$$\cdot \rho_{\gamma_2\gamma_3}(\kappa_2\kappa_3|\vec{k}_{\|}) t_{\gamma_3\beta}(\kappa_3\kappa'|\vec{k}_{\|})$$

$$+ \left(\frac{2\pi k_{\|}}{a_c}\right)^3 \sum_{\kappa_2\gamma_2} \sum_{\kappa_3\gamma_3} \sum_{\kappa_4\gamma_4} \sum_{\kappa_5\gamma_5}$$

$$\cdot t_{\alpha\gamma_2}(\kappa\kappa_2|\vec{k}_{\|}) \rho_{\gamma_2\gamma_3}(\kappa_2\kappa_3|\vec{k}_{\|})$$

$$\cdot t_{\gamma_3\gamma_4}(\kappa_3\kappa_4|\vec{k}_{\|})$$

$$\times \rho_{\gamma_4\gamma_5}(\kappa_4\kappa_5|\vec{k}_{\|}) t_{\gamma_5\beta}(\kappa_5\kappa'|\vec{k}_{\|}) - \cdots \quad (108)$$

In general $\rho_{\alpha\beta}(\kappa\kappa'|\vec{k}_{\|})$ tends to a finite, nonzero limit as $\vec{k}_{\|} \to 0$. Consequently, for a crystal slab of finite thickness ($\kappa_p = 0,1,2,\ldots, r-1$), each of the sums in the expansion in Eq. (108) is of O(1) in the limit as $\vec{k}_{\|} \to 0$. However, the nth term in this expansion contains an explicit factor of $(2\pi k_{\|}/a_c)^n$. It follows, therefore, that the leading term in the small $\vec{k}_{\|}$ limit of the matrix $V_{\alpha\beta}(\kappa\kappa'|\vec{k}_{\|})$ for a crystal slab of finite thickness can be obtained as follows

$$V_{\alpha\beta}(\kappa\kappa'|\vec{k}_{\|}) = T_{\alpha\beta}(\kappa\kappa'|\vec{k}_{\|}) \quad \text{(crystal slab)} \tag{109}$$
$$\vec{k}_{\|} \to 0$$

The conclusion that follows from Eq. (109) is that the long-range Coulomb interaction in a crystal slab is not screened in the long-wavelength limit.

In the case of a semi-infinite crystal ($r \to \infty$) the situation is different. In this case, because $t_{\alpha\beta}(\kappa\kappa'|\vec{k}_{\|})$ and $\rho_{\alpha\beta}(\kappa\kappa'|\vec{k}_{\|})$ have finite, nonzero limits as $\vec{k}_{\|} \to 0$, it is clear that each of the sums in Eq. (108) diverges. It is not difficult to show that if the nonlocal electronic polarizability tensor $P_{\alpha\beta}(\ell\kappa;\ell'\kappa')$ is nonzero only when the ions $(\ell\kappa)$ and $(\ell'\kappa')$ are within some finite distance of each other, which is the usual case, the sum in the second term in Eq. (108) diverges as $k_{\|}^{-1}$ when $k_{\|} \to 0$, the sum in the third term as $k_{\|}^{-2}$, and in general the sum in the nth term diverges as $k_{\|}^{n-1}$ in this limit. Consequently every term in the series in Eq. 108 is of $O(k_{\|})$ in the small $k_{\|}$ limit for a semi-infinite crystal, and this means that the entire series must be summed to obtain the leading term in the matrix $V_{\alpha\beta}(\kappa\kappa'|\vec{k}_{\|})$ in this limit.

The physical content of the result that $V_{\alpha\beta}(\kappa\kappa'|\vec{k}_{\|})$ differs from $T_{\alpha\beta}(\kappa\kappa'|\vec{k}_{\|})$ in the case of a semi-infinite crystal is that in such a crystal the Coulomb interaction is screened in the long-wavelength limit.

A large number of studies have been devoted to the microscopic calculations of surface waves in a slab of an ionic crystal. Extensive reviews of these works have appeared recently (41-44). However, the optical surface modes can couple to photons and the corresponding polaritons can be obtained, by a simpler procedure, as pointed out by Kliewer and Fuchs (45, 46), by treating the problem macroscopically in terms of Maxwell's equations and the appropriate boundary conditions. These surface polaritons are discussed in Chapter 3. This section contains a short summary of the extensive literature devoted to the microscopic computations of surface waves in a slab of an ionic crystal.

Lucas (47) investigated optical surface modes in a thin slab of an NaCl-type crystal with a (001) surface. Nearest neighbor, central forces were assumed for the short-range interactions. Relaxation of the ions near the surfaces was neglected. The ions were assumed to be rigid, and the appropriate Coulomb sums were evaluated in plane-wise fashion (48). Lucas carried out specific calculations only for $\vec{k}_{\parallel} = 0$. He found a transverse optical (TO) surface mode (displacements parallel to the surface) with a frequency very slightly below that of the bulk transverse optical mode for $\vec{k} = 0$ and no longitudinal optical (LO) surface modes (displacements perpendicular to the surface) for $\vec{k}_{\parallel} = 0$.

Tong and Maradudin (49) took into account the possibility of changes in the interplanar spacing near the (001) surfaces of an NaCl slab. Their calculations were carried out for $\vec{k}_{\parallel}$ covering the entire two-dimensional first Brillouin zone. For propagation in the [100] direction, they found four surface branches. One corresponds to Rayleigh waves and three to optical surface modes. A high-frequency branch lies between ω_L and ω_T and approaches a frequency somewhat below ω_L as $\vec{k}_{\parallel} \to 0$. A low-frequency branch lies below the bulk optical mode frequencies for small $\vec{k}_{\parallel}$, approaches a frequency just below ω_T as $\vec{k}_{\parallel} \to 0$, and is polarized in the sagittal plane. An intermediate-frequency branch has the same frequency at $\vec{k}_{\parallel} = 0$ as the low-frequency branch, but rises into the region between ω_L and ω_T, and is polarized parallel to the surface.

The problem of reconciling the two sets of results just described has been undertaken by Jones and Fuchs (50) and Chen et al. (51, 52). As shown in Fig. 2.2 the following picture appears. The high-frequency

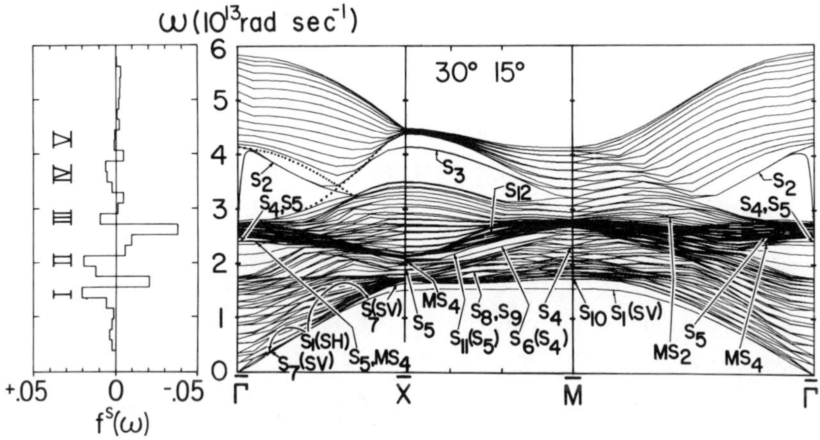

Fig. 2.2.--Variation of frequencies with wave vector for a 15-layer crystal of NaCl with two (001) surfaces. The surface modes are labeled by S_i and mixed modes by MS_i. After Chen et al. (51, 52).

branch of Tong et al. (49) splits as $\vec{k}_{\parallel} \to 0$ into two branches, one approaching ω_L and the other ω_T. These are the Fuchs-Kliewer (45) modes. Near $\vec{k}_{\parallel} = 0$, these modes enter the bulk LO and TO continua and become "mixed" or pseudosurface waves. The low-frequency branch enters the acoustical bulk continuum. The intermediate branch does not couple to bulk modes, and remains a pure surface branch.

Several other surfaces and crystals have been investigated more recently (42,53) and a variety of surface modes were obtained, especially in gaps in the bulk-frequency spectrum.

The Surface of a Crystal Considered as a Defect

The overwhelming majority of the calculations of the normal mode frequencies of a semi-infinite crystal are carried out purely numerically today, on the basis of the representation of the crystal by a slab of a finite number of atomic layers, as described in the preceding two subsections. There are alternative approaches to such calculations, however, which can be usefully employed in specific kinds of calculations, e.g., when only surface modes are required rather than the complete spectrum of vibrational modes of a semi-infinite crystal; with simple models of crystal lattices; or for establishing certain kinds of formal results. In this and the following two subsections we describe three such approaches.

In the present subsection we discuss the determination of the frequencies of those normal modes of a semi-infinite crystal which are affected by the introduction of surfaces into an initially infinitely extended crystal. That is, we explore the consequences of regarding the surface of a crystal as an extended defect in an otherwise perfect, infinitely extended crystal.

We will accordingly use definition (B1) for the specification of the atomic positions in the crystal. However, to simplify the notation somewhat we will consider only the case of a Bravais crystal. The extension of the present treatment to the case of a nonprimitive crystal requires only some inessential generalizations of the notation used here.

We begin by considering an infinitely extended Bravais crystal in which the rest positions of the atoms are given by the vectors

$$\vec{x}(\ell) = \ell_1\vec{a}_1 + \ell_2\vec{a}_2 + \ell_3\vec{a}_3 \tag{110}$$

The two primitive translation vectors $\vec{a}_1$ and $\vec{a}_2$ are assumed to be in the same plane, which we take to be the x_1x_2 plane; the third primitive translation vector $\vec{a}_3$ is assumed to have a positive projection on the x_3 axis. As usual, ℓ_1, ℓ_2, ℓ_3 are three integers, positive, negative, or zero, to which we refer collectively as ℓ.

We now create a pair of adjacent free surfaces at the planes $\ell_3 = 0$ and $\ell_3 = 1$ by equating to zero all interatomic forces between atoms on opposite sides of a fictitious plane between these two planes which contains no atoms, say the plane specified by $\ell_3 = \frac{1}{2}$. This can be effected mathematically by subtracting from the atomic force constants of the infinitely extended crystal the interactions which correspond to bonds between atoms on opposite sides of the fictitious surface. However, this must be done carefully. If we denote by $\{\Phi^{(o)}_{\alpha\beta}(\ell\ell')\}$ the atomic force constants of the infinitely extended crystal, those of the crystal with the interactions crossing the fictitious plane subtracted out can be written as

$$\Phi_{\alpha\beta}(\ell\ell') = \Phi^{(o)}_{\alpha\beta}(\ell\ell') - \Delta\Phi_{\alpha\beta}(\ell\ell') \tag{111}$$

It is tempting to think that $\Delta\Phi_{\alpha\beta}(\ell\ell')$ is given simply by

$$\Delta\Phi_{\alpha\beta}(\ell\ell') = \Phi^{(o)}_{\alpha\beta}(\ell\ell')[\theta(\ell_3 - 1)\theta(-\ell'_3) + \theta(\ell'_3 - 1)\theta(-\ell_3)] \tag{112}$$

where $\theta(\ell_3) = 1$ if $\ell_3 = 0,1,2,\ldots$, and $\theta(\ell_3) = 0$ if $\ell_3 = -1,-2,-3,\ldots$. In fact Eq. (112) applies only to crystals in which the atoms interact with two-body forces. For more complicated interactions the subtraction of the interactions between atoms on opposite sides of the fictitious plane must be carried out more carefully. For example, if the atoms interact through three-body forces, e.g., angle-bending forces, so that the force constant $\Phi^{(o)}_{\alpha\beta}(\ell\ell')$ can be written as

$$\Phi_{\alpha\beta}(\ell\ell') = \sum_{\ell''} \Phi_{\alpha\beta}(\ell\ell',\ell'') \tag{113}$$

then in subtracting from the potential energy the interactions between atoms on opposite sides of the fictitious plane we must include in what we subtract off not only the terms in which the atoms ℓ and ℓ' are on opposite sides, as in Eq. (112), but also the terms in which the atoms ℓ and ℓ' are in the same half-crystal while the atom ℓ'' is in the other half-crystal. It is only in this way that all interactions between the two halves of the crystal are removed.

The force constants $\{\Phi^{(o)}_{\alpha\beta}(\ell\ell')\}$ and their perturbations $\{\Delta\Phi_{\alpha\beta}(\ell\ell')\}$ must satisfy the general conditions

$$\Phi^{(o)}_{\alpha\beta}(\ell\ell') = \Phi^{(o)}_{\beta\alpha}(\ell'\ell); \quad \Delta\Phi_{\alpha\beta}(\ell\ell') = \Delta\Phi_{\beta\alpha}(\ell'\ell) \tag{114}$$

$$\sum_{\ell} \Phi^{(o)}_{\alpha\beta}(\ell\ell') = \sum_{\ell'} \Phi^{(o)}_{\alpha\beta}(\ell\ell') = 0 \tag{115a}$$

$$\sum_{\ell} \Delta\Phi_{\alpha\beta}(\ell\ell') = \sum_{\ell'} \Delta\Phi_{\alpha\beta}(\ell\ell') = 0 \qquad (115b)$$

$$\sum_{\ell'} \Phi^{(o)}_{\alpha\beta}(\ell\ell')x_{\gamma}(\ell') = \sum_{\ell'} \Phi^{(o)}_{\alpha\gamma}(\ell\ell')x_{\beta}(\ell') \qquad (116a)$$

$$\sum_{\ell'} \Delta\Phi_{\alpha\beta}(\ell\ell')x_{\gamma}(\ell') = \sum_{\ell'} \Delta\Phi_{\alpha\gamma}(\ell\ell')x_{\beta}(\ell') \qquad (116b)$$

The force constants $\{\Phi_{\alpha\beta}(\ell\ell')\}$ of the perturbed crystal must satisfy these conditions, as must the force constants of the initially infinitely extended crystal, $\{\Phi^{(o)}_{\alpha\beta}(\ell\ell')\}$.

In addition, the following symmetry-imposed conditions must be satisfied by both the $\{\Phi^{(o)}_{\alpha\beta}(\ell\ell')\}$ and the $\{\Delta\Phi^{(o)}_{\alpha\beta}(\ell\ell')\}$

$$\Phi^{(o)}_{\alpha\beta}(LL') = \sum_{\mu\nu} S_{\alpha\mu}\, S_{\beta\nu}\, \Phi^{(o)}_{\mu\nu}(\ell\ell') \qquad (117a)$$

$$\Delta\Phi_{\alpha\beta}(LL') = \sum_{\mu\nu} S_{\alpha\mu}\, S_{\beta\nu}\, \Delta\Phi_{\mu\nu}(\ell\ell') \qquad (117b)$$

However, the space group operations entering Eqs. (117a) and (117b) are different, in general. In the former they belong to the space group of the infinitely extended crystal, one of the 230 triply periodic space groups in three dimensions. In the latter they belong to the space group of the semi-infinite crystal, one of the 80 diperiodic groups in three dimensions.

The equations of motion of the perturbed crystal can be written in the form

$$M\ddot{u}_\alpha(\ell) = -\sum_{\ell'\beta}\left\{\Phi^{(o)}_{\alpha\beta}(\ell\ell') - \Delta\Phi_{\alpha\beta}(\ell\ell')\right\}u_\beta(\ell') \qquad (118)$$

where M is the mass of each atom. Assuming a harmonic time dependence proportional to $\exp(-i\omega t)$, the resulting time-independent equations of motion can be rewritten as

$$\sum_{\ell'\beta}\left\{M\omega^2\delta_{\ell\ell'}\delta_{\alpha\beta} - \Phi^{(o)}_{\alpha\beta}(\ell\ell')\right\}u_\beta(\ell') = -\sum_{\ell'\beta}\Delta\Phi_{\alpha\beta}(\ell\ell')u_\beta(\ell') \qquad (119)$$

One of the consequences of Eqs. (115) is that both $\Phi^{(o)}_{\alpha\beta}(\ell\ell')$ and $\Delta\Phi_{\alpha\beta}(\ell\ell')$ depend on the indices ℓ_1, ℓ_2, ℓ_1', ℓ_2' only through the differences $(\ell_1-\ell_1')$ and $(\ell_2-\ell_2')$. ($\Phi^{(o)}_{\alpha\beta}(\ell\ell')$ of course also depends on the indices ℓ_3 and ℓ_3' only through their difference.) We exploit this fact to seek $u_\alpha(\ell)$ in the form

$$u_\alpha(\ell) = u_\alpha(\vec{k}_\parallel|\ell_3)\, e^{i\vec{k}_\parallel \cdot \vec{x}_\parallel(\ell)} \qquad (120)$$

where

$$\vec{x}_\parallel(\ell) = \ell_1\vec{a}_1 + \ell_2\vec{a}_2 \qquad (121)$$

The wave vector $\vec{k}_\parallel$ as before is given by $\vec{k}_\parallel = \hat{x}_1k_1 + \hat{x}_2k_2$. If we impose periodic boundary conditions on the atomic displacements in the directions parallel to the surface, as in Eqs. (30)-(37), the allowed values of $\vec{k}_\parallel$ are densely distributed throughout the corresponding two-dimensional first Brillouin zone.

The substitution of Eq. (120) into Eq. (119) yields the following equation for $u_\alpha(\vec{k}\,|\ell_3)$

$$\sum_{\ell_3'\beta}\left\{M\omega^2\delta_{\ell_3\ell_3'}\delta_{\alpha\beta} - \phi^{(o)}_{\alpha\beta}(\vec{k}_\parallel|\ell_3\ell_3')\right\}u_\beta(\vec{k}_\parallel|\ell_3')$$

$$= -\sum_{\ell_3'\beta}\Delta\Phi_{\alpha\beta}(\vec{k}_\parallel|\ell_3\ell_3')u_\beta(\vec{k}_\parallel|\ell_3') \tag{122}$$

where

$$\Phi^{(o)}_{\alpha\beta}(\vec{k}_\parallel|\ell_3\ell_3') = \sum_{\ell_1'\ell_2'}\Phi^{(o)}_{\alpha\beta}(\ell\ell')\, e^{-i\vec{k}_\parallel\cdot\left(\vec{x}_\parallel(\ell) - \vec{x}_\parallel(\ell')\right)} \tag{123a}$$

and

$$\Delta\Phi_{\alpha\beta}(\vec{k}_\parallel|\ell_3\ell_3') = \sum_{\ell_1'\ell_2'}\Delta\Phi_{\alpha\beta}(\ell\ell')\, e^{-i\vec{k}_\parallel\cdot\left(\vec{x}_\parallel(\ell) - \vec{x}_\parallel(\ell')\right)} \tag{123b}$$

The perturbation on the equations of motion of an infinitely extended crystal provided by the matrix $\Delta\Phi_{\alpha\beta}(\vec{k}_\parallel|\ell_3\ell_3')$ is localized, if the interatomic forces are of short range. The dimensionality of this matrix is 3n x 3n, where n is the number of atomic planes directly touched by the cutting of the bonds crossing the fictitious plane between the planes $\ell_3 = 0$ and $\ell_3 = 1$. For example, n = 2 in the case of a simple cubic crystal with nearest and next nearest neighbor interactions and also in the case of a face-centered cubic crystal with nearest neighbor interactions, if the planes $\ell_3 = 0$ and $\ell_3 = 1$ are (001) surfaces.

We exploit this fact to solve Eq. (122) in the following way. We introduce a Green's function for the infinitely extended crystal, $G_{\alpha\beta}(\vec{k}_{\parallel}\omega|\ell_3\ell_3')$, as the solution of the equation

$$\sum_{\ell_3'\beta}\left\{M\omega^2\delta_{\ell_3\ell_3'}\delta_{\alpha\beta} - \Phi^{(o)}_{\alpha\beta}(\vec{k}_{\parallel}|\ell_3\ell_3')\right\}G_{\beta\gamma}(\vec{k}_{\parallel}\omega|\ell_3\ell_3')$$

$$= \delta_{\ell_3\ell_3'}\delta_{\alpha\gamma} \tag{124}$$

Equation (122) can be rewritten as

$$u_{\alpha}(\vec{k}_{\parallel}|\ell_3) = -\sum_{\ell_3'\beta}\sum_{\ell_3''\gamma} G_{\alpha\beta}(\vec{k}_{\parallel}\omega|\ell_3\ell_3')\ \Delta\Phi_{\beta\gamma}(\vec{k}_{\parallel}|\ell_3'\ell_3'')u_{\gamma}(\vec{k}_{\parallel}|\ell_3'') \tag{125}$$

One can use the representation

$$\delta_{\ell_3\ell_3'} = \frac{1}{2\pi}\int_{-\pi}^{\pi} d\theta\, e^{i(\ell_3 - \ell_3')\theta} \tag{126}$$

together with the fact that $\Phi^{(o)}_{\alpha\beta}(\vec{k}_{\parallel}|\ell_3\ell_3')$ depends on ℓ_3 and ℓ_3' only through their difference, to represent $G_{\alpha\beta}(\vec{k}_{\parallel}\omega|\ell_3\ell_3')$ in the form

$$G_{\alpha\beta}(\vec{k}_{\parallel}\omega|\ell_3\ell_3') = \frac{1}{2\pi}\int_{-\pi}^{\pi} d\theta G_{\alpha\beta}(\vec{k}_{\parallel}\omega|\theta)e^{i(\ell_3 - \ell_3')\theta} \tag{127}$$

For sufficiently simple crystal models, or for $\vec{k}_{\parallel}$ lying along high-symmetry directions in the two-dimensional first Brillouin zone, it is often possible to obtain $G_{\alpha\beta}(\vec{k}_{\parallel}\omega|\theta)$ in a simple enough form that the integral over θ in Eq. (127) can be carried out analytically. In carrying out this calculation it is convenient to give ω^2 a small positive imaginary part, $\omega^2 \to \omega^2 + i\varepsilon$, to define the way poles in the denominator of $G_{\alpha\beta}(\vec{k}_{\parallel}\omega|\theta)$ are to be treated when ω falls in the range of bulk crystal band mode frequencies corresponding to the given value of $\vec{k}_{\parallel}$.

If we denote the values of $(\ell_3\alpha)$ for which the matrix $\Delta\Phi_{\alpha\beta}(\vec{k}_{\parallel}|\ell_3\ell_3')$ has nonzero elements as constituting the "space of the surface," we can reduce the problem of solving Eq. (125) for the frequencies of those normal modes of the semi-infinite crystal perturbed by the creation of the pair of adjacent surfaces to a problem confined to this space. We begin by writing the matrix $\Delta\Phi_{\alpha\beta}(\vec{k}_{\parallel}|\ell_3\ell_3')$ in partitioned form

$$\Delta\overleftrightarrow{\Phi}(\vec{k}_{\parallel}) = \left(\begin{array}{c|c} \Delta\overleftrightarrow{\phi}(\vec{k}_{\parallel}) & \overleftrightarrow{0} \\ \hline \overleftrightarrow{0} & \overleftrightarrow{0} \end{array}\right) \tag{128}$$

The rows and columns defining the upper left-hand submatrix are those constituting the space of the surface. The remaining rows and columns are labeled by the values of $(\ell_3\alpha)$ and $(\ell_3'\beta)$ constituting the complementary space.

We partition the matrix $\overleftrightarrow{G}(\vec{k}_{\parallel}\omega)$ and the column vector $\vec{u}(\vec{k}_{\parallel})$ in the same way

$$\overleftrightarrow{G}(\vec{k}_{\parallel}\omega) = \left(\begin{array}{c|c} \overleftrightarrow{g}(\vec{k}_{\parallel}\omega) & \overleftrightarrow{G}_{12}(\vec{k}_{\parallel}\omega) \\ \hline \overleftrightarrow{G}_{21}(\vec{k}_{\parallel}\omega) & \overleftrightarrow{G}_{22}(\vec{k}_{\parallel}\omega) \end{array}\right) \tag{129a}$$

$$\vec{u}(\vec{k}_{\parallel}) = \left(\begin{array}{c} \vec{u}_1(\vec{k}_{\parallel}) \\ \hline \vec{u}_2(\vec{k}_{\parallel}) \end{array}\right) \tag{129b}$$

The values of $(\ell_3\alpha)$ defining the components of $\vec{u}(\vec{k}_{\parallel})$ are confined to the space of the surface, for example.

When the decompositions (128) and (129) are substituted into Eq. (125), the latter breaks up into the pair of matrix equations

$$\vec{u}_1(\vec{k}_{\parallel}) = -\overleftrightarrow{g}(\vec{k}_{\parallel}\omega)\Delta\overleftrightarrow{\phi}(\vec{k}_{\parallel})\vec{u}_1(\vec{k}_{\parallel}) \tag{130a}$$

$$\vec{u}_2(\vec{k}_{\parallel}) = -\overleftrightarrow{G}_{21}(\vec{k}_{\parallel}\omega)\Delta\overleftrightarrow{\phi}(\vec{k}_{\parallel})\vec{u}_1(\vec{k}_{\parallel}) \tag{130b}$$

The solvability condition for the set of homogeneous equations (130a) is that the determinant of the coefficients vanish

$$\left|\overleftrightarrow{I} + \overleftrightarrow{g}(\vec{k}_{\parallel}\omega)\Delta\overleftrightarrow{\phi}(\vec{k}_{\parallel})\right| = 0 \tag{131}$$

The roots of this equation, $\omega_j(\vec{k}_{\parallel})$, are the frequencies of the normal modes of the crystal perturbed by the introduction of the surfaces into it. These include surface vibration modes if they exist. Note that each of the matrices $\overleftrightarrow{I}$, $\overleftrightarrow{g}(\vec{k}_{\parallel}\omega)$, $\Delta\overleftrightarrow{\phi}(\vec{k}_{\parallel})$ entering Eq. (131) is a 3n x 3n matrix. The use of symmetry and group theory arguments can often lead to a block diagonalization of the matrix $\overleftrightarrow{I} + \overleftrightarrow{g}(\vec{k}_{\parallel}\omega)\Delta\overleftrightarrow{\phi}(\vec{k}_{\parallel})$, which further reduces the dimensionality of the matrices that have to be dealt with.

This method is particularly convenient for the study of surface modes, whose frequencies lie outside the continuum of bulk band modes corresponding to the given value of $\vec{k}_{\parallel}$. To obtain the perturbed bulk-mode frequencies from Eq. (125), ω^2 must be treated as purely real in obtaining $G_{\alpha\beta}(\vec{k}_{\parallel}\omega|\ell_3\ell_3')$, and the integral representation (127) must be replaced by a sum of the form

$$G_{\alpha\beta}(\vec{k}_{\parallel}|\ell_3\ell_3') = \sum_j \frac{v_\alpha^{(j)}(\vec{k}_{\parallel}|\ell_3)v_\beta^{(j)}(\vec{k}_{\parallel}|\ell_3')^*}{\omega^2 - \omega_j^2(\vec{k}_{\parallel})} \tag{132}$$

In Eq. (132), the $\{\omega_j^2(\vec{k}_{\parallel})\}$ and $\{v_\alpha^{(j)}(\vec{k}_{\parallel}|\ell_3\}$ are the eigenvalues and suitably normalized eigenvectors of the matrix $M^{-1}\Phi_{\alpha\beta}^{(o)}(\vec{k}_{\parallel}|\ell_3\ell_3')$. This in general is very difficult to do. However, when ω lies outside the range of frequencies covered by the $\{\omega_j(\vec{k}_{\parallel})\}$, the summation in Eq. (132) can be replaced by integration, since $G_{\alpha\beta}(\vec{k}_{\parallel}\omega|\ell_3\ell_3')$ has no poles for ω in the indicated ranges.

Once the solutions of Eq. (131) are known, the values of the components of the vector $\vec{u}_1(\vec{k}_{\parallel})$ corresponding to each solution can be obtained to within a constant multiplicative factor, which can be made definite by the use of some suitable normalization condition. Equation (130b) then gives the displacement amplitudes of all the remaining atoms in the crystal for each of the solutions of Eq. (130a) in terms of the displacement amplitudes in the localized space of the surface.

The approach outlined here has not been used a great deal (54-57) in the study of surface vibration modes. Offering as it does a possibility of obtaining analytic results in interesting cases, for example, interfaces (58-60), we believe it deserves further development and application.

Surface Modes in Semi-Infinite Crystals With Short-Range Interatomic Forces

The preceding approaches to the determination of the normal modes of vibration of a semi-infinite crystal are characterized by the fact that they yield all the normal mode frequencies and the corresponding unit eigenvectors of the crystal, those of the wavelike modes as well as those of the localized surface modes.

If it is desired to study only the localized surface modes, more direct methods are available which do not involve obtaining the non-surface modes at the same time as well. In this section we outline one such approach that is well suited for the study of surface vibrational modes in crystals with short-range interatomic forces, i.e., non-ionic crystals.

As before, in order to simplify the notation, we consider a Bravais crystal. The extension of the present discussion to the case of nonprimitive crystals can be found in Refs. 61 and 62.

We will use Definition (B2) to define the positions of the atoms in a semi-infinite crystal occupying the upper half space $x_3 \geq 0$. Accordingly, the position vectors giving the rest positions of the atoms are

$$\vec{x}(\ell) = \ell_1\vec{a}_1 + \ell_2\vec{a}_2 + \ell_3\vec{a}_3 \tag{133}$$

The two vectors $\vec{a}_1$ and $\vec{a}_2$ are parallel to the plane $x_3 = 0$, while the third vector $\vec{a}_3$ is directed out of this plane in such a way that it has a nonzero projection on the positive x_3 axis. The index ℓ_3 is restricted to the integer values $\ell_3 = 0,1,2,\ldots.$ We will assume that the atomic displacements are periodic in the directions parallel to the surface, with the periodicity element being the rhombus whose edges are defined by the vectors $L\vec{a}_1$ and $L\vec{a}_2$.

The equations of motion for the atoms in a semi-infinite crystal that occupies the upper half-space $x_3 \geq 0$ differ from those of the atoms of an infinite crystal for the atoms closer to the surface than those in the range of the interatomic forces. For these atoms the interactions with the atoms in the half-space $x_3 < 0$ are absent, while they are present for the corresponding atoms in the infinitely extended crystal.

If, for the moment, we neglect these differences, the equations of motion of the atoms in the semi-infinite crystal are the same as those of the atoms in the infinite crystal

$$-M\ddot{u}_\alpha(\ell) = \sum_{\ell'\beta} \Phi^{(o)}_{\alpha\beta}(\ell\ell')u_\beta(\ell') \tag{134}$$

where the atom ℓ is assumed to be in the half-space $x_3 \geq 0$.

We now assume a solution of Eq. (134) of the form

$$u_\alpha(\ell) = \frac{w_\alpha}{m^{\frac{1}{2}}} e^{i\vec{k}_\parallel \cdot \vec{x}_\parallel(\ell) - q\ell_3 - i\omega t} \tag{135}$$

where $x_{\parallel}(\ell) = \ell_1\vec{a}_1 + \ell_2\vec{a}_2$, and the allowed values of the two-dimensional wave vector $\vec{k}_{\parallel}$ are determined by our periodic boundary conditions and are given explicitly in Eqs. (36) and (37).

When we substitute Eq. (135) into Eq. (134) we obtain the set of equations

$$\omega^2 w_\alpha = \sum_\beta D_{\alpha\beta}(\vec{k}_{\parallel};q)w_\beta \tag{136a}$$

where

$$D_{\alpha\beta}(\vec{k}_{\parallel};q) = \frac{1}{M}\sum_{\ell'} \Phi^{(o)}_{\alpha\beta}(\ell\ell')e^{-i\vec{k}_{\parallel}\cdot\left(\vec{x}_{\parallel}(\ell) - \vec{x}_{\parallel}(\ell')\right)} \cdot e^{q\left(\ell_3 - \ell_3'\right)} \tag{136b}$$

The result that the matrix $\overleftrightarrow{D}(\vec{k}_{\parallel};q)$ is independent of the index ℓ is due to the fact that the atomic force constants of the infinite crystal, $\{\Phi^{(o)}_{\alpha\beta}(\ell\ell')\}$, depend on ℓ and ℓ' only through their difference, and we have so far neglected the differences between the equations of motion of the semi-infinite crystal and those of the infinite crystal.

The condition that the set of equations (136) have a nontrivial solution is that the determinant of the coefficients vanish

$$\left|\omega^2\delta_{\alpha\beta} - D_{\alpha\beta}(\vec{k}_{\parallel};q)\right| \equiv \left|\overleftrightarrow{M}(\vec{k}_{\parallel}\,\omega;q)\right| = 0 \tag{137}$$

For a given value of the two-dimensional wave vector $\vec{k}_{\parallel}$, Eq. (137) constitutes a relationship between the frequency ω and the attenuation constant q. To a given value of ω correspond several values of q, which we label by the index j. The precise number of these values of q depends on the range of the interatomic forces. A surface wave results only if all the $q_j(\vec{k}_\parallel \omega)$ corresponding to a given value of $\vec{k}_\parallel$ and a given value of ω have positive real parts. More precisely, the solutions of Eqs. (137) correspond to pure surface waves if all q_j are real and positive, and to generalized surface waves if the q_j are complex with positive real parts.

The amplitude $\{w_\alpha(j)\}$ associated with a particular q_j is determined by

$$\frac{w_\alpha(j)}{c_\alpha(j)} = K_j, \quad \alpha = 1,2,3 \tag{138}$$

The term c_α is the cofactor of the 3 x 3 matrix $\overleftrightarrow{M}(\vec{k}_\parallel \omega; q_j)$ obtained by evaluating the determinant of the matrix formed by striking out the first row of $\overleftrightarrow{M}(\vec{k}_\parallel \omega; q_j)$ and the column corresponding to w_α, and giving it a sign in accordance with the rule for determining the sign of a cofactor. The $\{K_j\}$ are new amplitudes.

The general solution of Eqs. (137) corresponding to surface waves is therefore given by

$$u_\alpha(\ell) = \frac{1}{M^{\frac{1}{2}}} e^{i\vec{k}_\parallel \cdot \vec{x}_\parallel(\ell) - i\omega t} \sum_j K_j c_\alpha(j) e^{-q_j \ell_3} \tag{139}$$

(Recall that both q_j and $c_\alpha(j)$ are functions of $\vec{k}_\parallel$ and ω. For notational simplicity we do not indicate this dependence explicitly.)

The new amplitudes $\{K_j\}$ are to be determined from the boundary conditions. We have remarked above that the equations of motion for atoms in the surface layers of a semi-infinite crystal differ from those for the corresponding atoms of an infinite crystal by the absence of forces acting on these atoms from the half-space $x_3 < 0$. The equations of motion (134) and the determinantal equation (137) were obtained from them by ignoring this difference, and we must now take account of it.

The force exerted in the α direction on the atom ℓ in the half-space $x_3 \geq 0(\ell_3 \geq 0)$ by the atoms in the half-space $x_3 < 0(\ell_3 < 0)$, must be equated to zero, and this condition can be expressed as

$$\sum_{\ell'\beta}\left\{\theta(\ell_3)\Phi_{\alpha\beta}^{(o)}(\ell\ell')\theta(-\ell_3' - 1)\right\}u_\beta(\ell') = 0 \qquad (140)$$

[As it stands, this equation is rigorously correct only for crystals in which the atoms interact with two-body forces. The modification of Eq. (140) required in the case of many body forces has already been discussed.] When we substitute Eq. (139) into this equation we obtain a set of homogeneous equations for the coefficients $\{K_j\}$

$$\sum_j\left\{\sum_{\ell'\beta}\left[\theta(\ell_3)\Phi_{\alpha\beta}^{(o)}(\ell\ell')\theta(-\ell_3' - 1)\right]e^{-i\vec{k}_\parallel\cdot\left(\vec{x}_\parallel(\ell) - \vec{x}_\parallel(\ell')\right)}\right.$$

$$\left.\times\, e^{-q_j\ell_3}c_\beta(j)\right\}K_j = 0 \qquad (141)$$

The left-hand side of this equation is independent of ℓ_1 and ℓ_2 because $\theta(\ell_3)\Phi_{\alpha\beta}^{(o)}(\ell\ell')\theta(-\ell_3' - 1)$ is a function of ℓ_1, ℓ_2, ℓ_1', ℓ_2' only through the differences $\ell_1 - \ell_1'$ and $\ell_2 - \ell_2'$. It is, consequently, a function of ℓ_3 and α, and the number of equations (141) is equal to three times the number of values of ℓ_3 that label atomic planes which lie closer to the surface than the range of interatomic forces. This is just the number of solutions for q obtained from Eq. (137) for fixed $\vec{k}_{\parallel}$ and ω.

The simultaneous solution of Eqs. (139) and (141) gives the attenuation constants $\{q_j\}$ and the frequencies of surface modes as functions of $\vec{k}_{\parallel}$. Such calculations ordinarily have to be carried out on a high-speed computer.

This method has been used for the determination of dispersion relations of acoustical and optical surface vibration modes (61-65). In Fig. 2.3 is plotted the dispersion curve obtained in this way for acoustic waves propagating in the [100] direction on a (001) surface of a simple cubic crystal representing KCℓ (63).

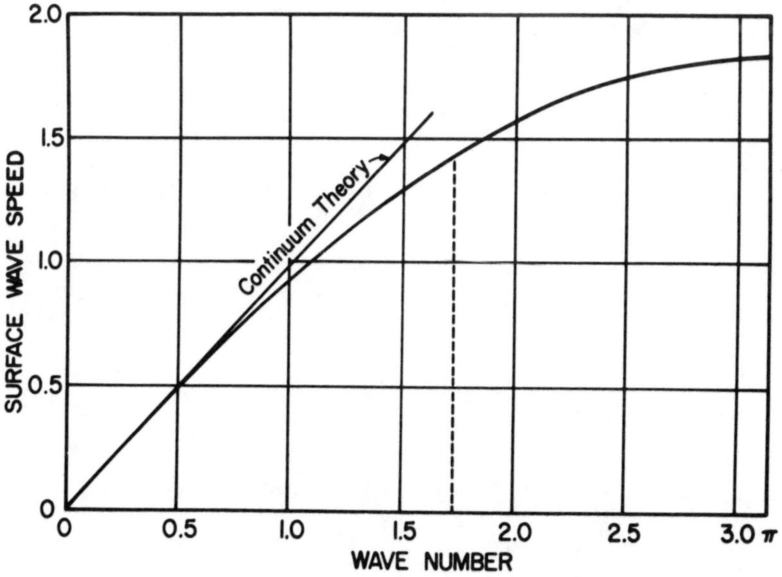

Fig. 2.3.--The dispersion relation for acoustic waves propagating in the [100] direction on a (001) surface of a simple cubic crystal representing KCℓ. After Gazis et al. (63).

We conclude this section by illustrating the method through its application to a semi-infinite, alternating diatomic linear chain. We label the atoms by an index $n = 0,1,2,\ldots$. The even-numbered atoms have a mass m, while the odd-numbered atoms have a mass M, and we assume that $m < M$. The atoms are equally spaced (this model possesses the one-dimensional analogue of the rocksalt structure), and interaction with nearest neighbor forces is characterized by a force constant γ. The equations of motion of this crystal are

$$m\ddot{u}_0 = \gamma(u_1 - u_0) \tag{142a}$$

$$M\ddot{u}_1 = \gamma(u_2 - 2u_1 + u_0) \tag{142b}$$

$$m\ddot{u}_2 = \gamma(u_3 - 2u_2 + u_1) \tag{142c}$$

$$m\ddot{u}_{2n} = \gamma(u_{2n+1} - 2u_{2n} + u_{2n-1}) \qquad n > 0 \tag{142d}$$

$$Mu_{2n+1} = \gamma(u_{2n+2} - 2u_{2n+1} + u_{2n}) \qquad n \geq 0 \tag{142e}$$

We seek the solutions of Eqs. (142d) and (142e) in the forms

$$u_{2n} = A\ e^{-2nq\ -\ i\omega t} \tag{143a}$$

$$u_{2n+1} = B\ e^{-(2n+1)q\ -\ i\omega t} \tag{143b}$$

and are led to the following determinantal equation for the amplitudes A and B

$$\begin{pmatrix} 2 - \frac{m\omega^2}{\gamma} & -2\cosh q \\ -2\cosh q & 2 - \frac{M\omega^2}{\gamma} \end{pmatrix} \begin{pmatrix} A \\ B \end{pmatrix} = 0 \qquad (144)$$

From the solvability condition for this system of equations we find

$$\cosh^2 q = -\left(1 - \frac{m\omega^2}{2\gamma}\right)\left(\frac{M\omega^2}{2\gamma} - 1\right) < 0, \quad \frac{2\gamma}{M} \le \omega^2 \le \frac{2\gamma}{m} \qquad (145)$$

$$\frac{B}{A} = \frac{\left(1 - \frac{m\omega^2}{2\gamma}\right)}{\cosh q} \qquad (146)$$

Equation (145) takes this form because the frequency spectrum of the infinitely extended diatomic linear chain has a gap for frequencies satisfying $(2\gamma/M) \le \omega^2 \le (2\gamma/m)$, i.e., between the acoustic and optical branches. Since the normal mode frequencies of a semi-infinite crystal are expected to be lower than those of the corresponding infinite crystal, and because the frequencies of surface vibration modes must lie in regions forbidden to the infinite crystal, we expect the frequency of a surface mode in the present case to lie in the indicated gap. For frequencies in the gap, the right-hand side of Eq. (145) is negative definite. We must therefore have

$$q = q_0 + i\,\frac{\pi}{2} \qquad (147)$$

where q_0 is real, and find that

$$\sinh^2 q_0 = \left(1 - \frac{m\omega^2}{2\gamma}\right)\left(\frac{M\omega^2}{2\gamma} - 1\right) > 0 \tag{148}$$

$$\frac{B}{A} = \frac{\left(1 - \frac{m\omega^2}{2\gamma}\right)}{i \sinh q_0} \tag{149}$$

We now turn to the boundary condition at the surface satisfied by the atomic displacements. We see from Eq. (142a) that the equation of motion of the atom $n = 0$ is different from that of every other even-numbered atom in the chain. We can write it in the same form

$$m\ddot{u}_0 = \gamma\left(u_1 - 2u_0 + u_{-1}\right) \tag{150a}$$

This is satisfied by Eqs. (142a) and (142d), provided that

$$u_0 - u_{-1} = 0 \tag{150b}$$

This is the needed boundary condition. When we substitute Eqs. (143) into Eqs. (150) we obtain a second relation between the coefficients B and A

$$\frac{B}{A} = e^{-q} = -ie^{-q_0} \tag{151}$$

When we eliminate (B/A) between Eqs. (149) and (151), and make use of Eq. (148), we finally obtain the following simple results

$$\omega^2 = \gamma\left(\frac{1}{m} + \frac{1}{M}\right) \tag{152a}$$

$$q_0 = \sinh^{-1}\left\{\tfrac{1}{2}\left[\left(\frac{M}{m}\right)^{\frac{1}{2}} - \left(\frac{m}{M}\right)^{\frac{1}{2}}\right]\right\} \tag{152b}$$

Since the maximum frequency of the infinite crystal is given by $\omega_L^2 = 2\gamma(m^{-1} + M^{-1})$, we see that the squared frequency of the surface mode is half the maximum squared frequency, and lies precisely in the middle of the gap between the squared acoustic and squared optical frequency branches.

This simple example exhibits most of the features found in calculations based on more realistic, three-dimensional crystal models, except for the occurrence of more than one decay constant q; however, these calculations cannot ordinarily be carried out analytically.

Simple and Approximate Methods for Calculation of Surface Modes

The methods discussed above for the calculation of surface modes are exact. It is often possible, and useful, to make some approximations in such calculations to simplify them, and yet obtain the surface mode frequencies with sufficiently good precision. The kinds of approximations that can be made depend on the nature of the surface mode being studied. We will distinguish among surface modes having a short penetration (a few atomic layers) into the bulk of the crystal and those having a large penetration. In the first case we can use either a density-of-states moments method (66) or a frozen substrate approximation (64,67,68). In the case of deep penetration, an effective modulus method (69,70) will be appropriate.

The Moment Method

We start by creating a pair of adjacent free surfaces by equating to zero all interatomic forces between atoms on opposite sides of a fictitious plane which contains no atoms. Taking due account of the translation symmetry parallel to the free surfaces, we can rewrite Eqs. (27) in the following matrix form

$$\sum_{\ell_3'\beta} \left[D_{\alpha\beta}(\vec{k}_\parallel | \ell_3 \ell_3') - \omega^2 \delta_{\alpha\beta} \delta_{\ell_3 \ell_3'} \right] v_\beta(\vec{k}_\parallel | \ell_3') = 0 \tag{153}$$

The dynamical matrix can be written as

$$\overleftrightarrow{D}(\vec{k}_\parallel) = \overleftrightarrow{D}^o(\vec{k}_\parallel) + \Delta\overleftrightarrow{D}(\vec{k}_\parallel) \tag{154}$$

where $\overleftrightarrow{D}^o(\vec{k}_\parallel)$ is the dynamical matrix of the infinitely extended crystal, and $\Delta\overleftrightarrow{D}(\vec{k}_\parallel)$ is the perturbation introduced by the cutting procedure which created the two free surfaces.

We define the moments μ_{2n}^o and μ_{2n} as the respective traces of the matrices $[\overleftrightarrow{D}^o(\vec{k}_\parallel)]^n$ and $[\overleftrightarrow{D}(\vec{k}_\parallel)]^n$. Their variations due to the creation of the two free surfaces are

$$\delta\mu_{2n} = \mu_{2n} - \mu_{2n}^o = \mathrm{Tr}[\overleftrightarrow{D}(\vec{k}_\parallel)]^n - [\overleftrightarrow{D}^o(\vec{k}_\parallel)]^n \tag{155}$$

and can be easily calculated.

Once one knows how many states are pulled out of the bulk bands when the two free surfaces are created, one can approximate at a given $\vec{k}_\parallel$ the density of states lost in the bulk bands by one or several Dirac δ-

functions, and then obtain a simple relation for the localized modes frequencies (66).

The precision obtained by this method increases with increasing distance of the localized mode frequency from the bulk bands, or equivalently, with decreasing localized mode penetration into the interior of the crystal. An application of this method is given in the chapter on Examples.

The Frozen-Substrate Method

For localized modes with a shorter penetration into the interior of the crystal an even simpler method for the determination of their frequencies is the frozen substrate method. This method consists of setting $v_\beta(\kappa|\vec{k}_\parallel j) = 0$ in Eq. (27) for all the planes κ' after a given distance κ^o from the free surface at $\kappa' = 1$. On taking $\kappa^o = 1,2,\ldots,n$, one can check the convergence of the localized modes solutions.

The simplicity of this method makes it well suited for investigating the effects of force constant and mass changes, especially for well-localized optical surface modes (64) and soft-surface phonons (71). An application of this method is given in the chapter on Examples.

The Effective-Modulus Method

For long decay lengths, the preceding methods cannot be used. However, another method has been proposed (69,70) for dealing with this case. This effective-modulus approximation was developed for obtaining the dispersion relation and displacement field of acoustic or optical surface modes associated with external points in the surfaces of constant frequency of the corresponding, infinitely extended crystal at some extremal point of wave vector $\vec{k}_o$ on the surfaces of constant frequency.

Using the dynamical matrix of the infinitely extended crystal [definition (B2)]

$$C_{\alpha\beta}(\kappa\kappa'|\vec{k}) = \exp\left(-i\vec{k}\cdot\vec{x}(\kappa)\right)D_{\alpha\beta}(\kappa\kappa'|\vec{k})\exp\left(i\vec{k}\cdot\vec{x}(\kappa')\right)$$

and using normal mode frequencies $\omega_j^2(\vec{k})$ and eigenvectors $\{w_\alpha(\kappa|\vec{k}j)\}$, one assumes a solution for the semi-infinite crystal of the form

$$u_\alpha(\ell\kappa) = e^{i\vec{k}_o\cdot\vec{x}(\ell\kappa)}\sum_{j=1}^{3r}\frac{w_\alpha(\kappa|\vec{k}_o j)}{M_\kappa^{\frac{1}{2}}}\,\xi^{(j)}(\ell) \qquad (156)$$

where the sum over j indicates that the solutions for the finite crystal are linear combinations of the 3r solutions for the infinite crystal corresponding to the wave vector $\vec{k}_o$. The displacement field $\xi^{(j)}(\ell)$ incorporates the effects of the surface into the solution, and equals a constant for all values of ℓ in an infinitely extended crystal. Substitution of Eq. (156) into the equations of motion for a semi-infinite crystal and use of the orthonormality condition on the bulk eigenvectors yields as the equation for the amplitudes $\{\xi^{(j)}(\ell)\}$

$$\omega^2\xi^{(j)}(\ell) = \sum_{j'\ell'} C^{(jj')}(\ell\ell')\xi^{(j')}(\ell') \qquad (157a)$$

The coefficient $C^{(jj')}(\ell\ell')$ is defined as

$$C^{(jj')}(\ell\ell') \equiv \sum_{\kappa\alpha}\sum_{\kappa'\beta} w_\alpha^*(\kappa|\vec{k}_o j)e^{-i\vec{k}_o\cdot\vec{x}(\ell\kappa)}$$

$$\times \frac{\Phi_{\alpha\beta}(\ell\kappa;\ell'\kappa')}{(M_\kappa M_{\kappa'})^{\frac{1}{2}}} w_\beta(\kappa'|\vec{k}_o j') e^{i\vec{k}_o \cdot \vec{x}(\ell'\kappa')} \tag{157b}$$

Here, the $\{\Phi_{\alpha\beta}(\ell\kappa;\ell'\kappa')\}$ are atomic force constants of a crystal with a free surface. Although the coefficient $C^{(jj')}(\ell\ell')$ is a function of the wave vector $\vec{k}_o$, we suppress explicit reference to this dependence to simplify the notation in what follows.

By assuming a solution of the form given in Eq. (156), we have factored out the rapidly varying part of the displacement field, namely $\exp i\vec{k}_o \cdot \vec{x}(\ell\kappa)$. Hence we are left with a set of equations, Eqs. (157), for the displacement amplitudes $\{\xi^{(j)}(\ell)\}$, which vary slowly from one lattice site to another. Because of this slow variation, we can expand $\xi^{(j')}(\ell')$ in a Taylor series about its value at the site ℓ and replace the discrete variable $\vec{x}(\ell)$ by a continuous variable $\vec{x}$

$$\begin{aligned}\xi^{(j')}(\ell') &= \xi^{(j')}[\vec{x}(\ell) + \vec{x}(\ell') - \vec{x}(\ell)] \\ &= \xi^{(j')}(\vec{x}) + \sum_\mu x_\mu(\ell'\ell)\frac{\partial}{\partial x_\mu} \xi^{(j')}(\vec{x}) \\ &+ \tfrac{1}{2}\sum_{\mu\nu} x_\mu(\ell'\ell)x_\nu(\ell'\ell)\frac{\partial^2}{\partial x_\mu \partial x_\nu} \xi^{(j')}(\vec{x}) + \ldots\end{aligned} \tag{158}$$

where we denote by $x_\mu(\ell'\ell)$ the μth Cartesian component of the difference between the vectors $\vec{x}(\ell')$ and $\vec{x}(\ell)$

$$x_\mu(\ell'\ell) = x_\mu(\ell') = x_\mu(\ell) \tag{159}$$

Substitution of the Taylor series expansion for the

displacement field into Eq. (157) yields the following set of 3r coupled, second-order partial differential equations for the functions $\xi^{(j)}(\vec{x})$

$$\omega^2\xi^{(j)}(\vec{x}) = \sum_{j'} c^{(jj')}(\vec{x})\xi^{(j')}(\vec{x})$$

$$+ \sum_{j'\mu} c_{\mu}^{(jj')}(\vec{x})\frac{\partial}{\partial x_\mu} \xi^{(j')}(\vec{x})$$

$$+ \frac{1}{2} \sum_{j'\mu\nu} c_{\mu\nu}^{(jj')}(\vec{x})\frac{\partial^2}{\partial x_\mu \partial x_\nu} \xi^{(j')}(\vec{x}) \qquad (160)$$

where the coefficient functions are the continuous analogues of the lattice point functions

$$c^{(jj')}(\ell) = \sum_{\ell'} c^{(jj')}(\ell\ell') \qquad (161)$$

$$c_{\mu}^{(jj')}(\ell) = \sum_{\ell'} c^{(jj')}(\ell\ell')x_{\mu}(\ell\ell) \qquad (162)$$

$$c_{\mu\nu}^{(jj')}(\ell) = \sum_{\ell'} c^{(jj')}(\ell\ell')x_{\mu}(\ell'\ell)x_{\nu}(\ell'\ell) \qquad (163)$$

The surface enters the present approach through the restrictions that its presence imposes on the values that ℓ' can take in the sums in Eqs. (161)-(163). These restrictions ultimately lead to the boundary conditions imposed on the equations of motion obeyed by $\vec{\xi}^{(j)}(\vec{x})$.

This method has been illustrated (70) by application to one- and three-dimensional semi-infinite crystal models, and can be useful where exact methods become too cumbersome.

GROUP THEORY OF PHONONS IN A SEMI-INFINITE CRYSTAL SLAB

Just as group theory has proved to be a powerful tool for simplifying certain types of lattice dynamical problems for crystals of infinite extent, it is also a powerful tool for simplifying certain types of lattice dynamical problems for semi-infinite crystals in the forms of films or slabs of a finite number of atomic layers [definition (A)]. These include the determination of the form of the dynamical matrix of the slab and of its eigenvectors, which describe the vibrational displacement pattern of the atoms in the slab, and the determination of possible degeneracies among the eigenvalues of the dynamical matrix (the squares of the normal mode frequencies of the slab), for values of the two-dimensional wave vector $\vec{k}_{\parallel}$ lying at points of symmetry inside or on the boundary of the two-dimensional first Brillouin zone for the slab.

In this section we outline the way in which group theory can be applied to these ends. Our approach is based on the paper by Trullinger and Maradudin (72), who modified the earlier work of Maradudin and Vosko (73), and the subsequent discussion of it by Maradudin, Montroll, Weiss, and Ipatova (74), for infinitely extended crystals.

Symmetry Properties of the Dynamical Matrix and its Eigenvectors

The starting point for our analysis is the equations of motion for a crystal slab in the form

$$\omega_j^2(\vec{k}_\parallel)v_\alpha(\kappa|\vec{k}_\parallel j) = \sum_{\kappa'\beta} D_{\alpha\beta}(\kappa\kappa'|\vec{k}_\parallel)v_\beta(\kappa'|\vec{k}_\parallel j) \quad (164)$$

The dynamical matrix $D_{\alpha\beta}(\kappa\kappa'|\vec{k}_\parallel)$ is given by

$$D_{\alpha\beta}(\kappa\kappa'|\vec{k}_\parallel) = \frac{1}{(M_\kappa M_{\kappa'})^{\frac{1}{2}}} \sum_{\ell'} \Phi_{\alpha\beta}(\ell\kappa;\ell'\kappa') \cdot e^{-i\vec{k}_\parallel \cdot (\vec{x}(\ell) - \vec{x}(\ell'))} \quad (165)$$

We first determine the transformation properties of the dynamical matrix and its eigenvectors and eigenvalues when the crystal slab is subjected to an operation from its space group G.

Since for a symmetry operation the atoms labeled K and κ must be of the same kind, we have $M_K = M_\kappa$, and the matrix element $D_{\alpha\beta}(KK'|\overleftrightarrow{S}\vec{k}_\parallel)$ may be written with the aid of Eqs. (18) and (19) as the result of a unitary transformation

$$\overleftrightarrow{D}(\overleftrightarrow{S}\vec{k}_\parallel) = \overleftrightarrow{\Gamma}(\vec{k}_\parallel;\{\overleftrightarrow{S}|\vec{v}(S) + \vec{x}(m)\})\overleftrightarrow{D}(\vec{k}_\parallel)\overleftrightarrow{\Gamma}^\dagger(\vec{k}_\parallel;\{\overleftrightarrow{S}|\vec{v}(S) + \vec{x}(m)\}) \quad (166)$$

where

$$\Gamma_{\alpha\beta}(\kappa\kappa'|\vec{k}_{\parallel};\{\overleftrightarrow{S}|\vec{v}(S) + \vec{x}(m)\}) = S_{\alpha\beta}\delta(F_0^{-1}(\kappa;S),\kappa') \cdot \exp\{i\vec{k}_{\parallel} \cdot [\{\overleftrightarrow{S}|\vec{v}(S) + \vec{x}(m)\}^{-1} \cdot \{\vec{x}(\kappa) - \vec{x}(\kappa')]\} \quad (167)$$

In writing Eq. (167), we have used the Kronecker symbol and the definition, $F_0^{-1}(K;S) = \kappa$ or its equivalent, $K = F_0(\kappa;S)$.

To each element $\{\overleftrightarrow{S}|\vec{v}(S) + \vec{x}(m)\}$ of the space group G there corresponds a matrix $\overleftrightarrow{\Gamma}(\vec{k}_{\parallel};\{\overleftrightarrow{S}|\vec{v}(S) + \vec{x}(m)\})$. However, the latter do not obey the same multiplication law as the group elements, so that they do not provide a representation of the space group G of the crystal slab.

We can, however, define a set of matrices in terms of the matrices $\overleftrightarrow{\Gamma}(\vec{k}_{\parallel};\{\overleftrightarrow{S}|\vec{v}(S) + \vec{x}(m)\})$ which on the one hand commute with the dynamical matrix $\overleftrightarrow{D}(\vec{k}_{\parallel})$, and on the other hand provide a representation, not of the space group G itself, but of a related group.

For this purpose we restrict ourselves to those operations of the group G which comprise the space group $G_{\vec{k}_{\parallel}}$ of the wave vector. These are the operations $\{\overleftrightarrow{R}|\vec{v}(R) + \vec{x}(m)\}$ which take the crystal slab into itself and whose purely rotational elements, $\{\overleftrightarrow{R}\}$, have the property

$$\overleftrightarrow{R}\vec{k}_{\parallel} = \vec{k}_{\parallel} - \vec{\tau}(\vec{k}_{\parallel};\overleftrightarrow{R}) \quad (168)$$

where $\vec{\tau}(\vec{k}_{\parallel};\overleftrightarrow{R})$, defined by Eq. (168), is a translation vector of the reciprocal lattice of the slab. The purely rotational elements $\{\overleftrightarrow{R}\}$ in the space group $G_{\vec{k}_{\parallel}}$ taken by themselves constitute the point group of the wave vector $\vec{k}_{\parallel}$, $G_0(\vec{k}_{\parallel})$. With each element R of the group $G_0(\vec{k}_{\parallel})$ we associate a matrix $\overleftrightarrow{T}(\vec{k}_\parallel;\overleftrightarrow{R})$ which is defined in terms of the matrices $\overleftrightarrow{T}(\vec{k}_\parallel;\{\overleftrightarrow{R}|\vec{v}(R) + \vec{x}(m)\})$ by

$$\overleftrightarrow{T}(\vec{k}_\parallel;\overleftrightarrow{R}) = \exp\left(i\vec{k}_\parallel \cdot [\vec{v}(R) + \vec{x}(m)]\right)\overleftrightarrow{T}\left(\vec{k}_\parallel;\{R|\vec{v}(R) + \vec{x}(m)\}\right) \tag{169a}$$

or

$$T_{\alpha\beta}(\kappa\kappa'|\vec{k}_\parallel;\overleftrightarrow{R}) = R_{\alpha\beta}\delta\left(\kappa,F_0(\kappa';R)\right)\exp\{i\vec{k}_\parallel[\vec{x}(\kappa) - \overleftrightarrow{R}\vec{x}(\kappa')]\} \tag{169b}$$

The multiplication rule which the $\overleftrightarrow{T}(\vec{k};\overleftrightarrow{R})$ matrices obey is given by

$$\overleftrightarrow{T}(\vec{k}_\parallel;\overleftrightarrow{R}_i)\overleftrightarrow{T}(\vec{k}_\parallel;\overleftrightarrow{R}_j) = \phi(\vec{k}_\parallel;\overleftrightarrow{R}_i,\overleftrightarrow{R}_j)\overleftrightarrow{T}(\vec{k}_\parallel;\overleftrightarrow{R}_i\overleftrightarrow{R}_j) \tag{170}$$

where

$$\phi(\vec{k}_\parallel;\overleftrightarrow{R}_i,\overleftrightarrow{R}_j) = \exp[i\vec{\tau}(\vec{k}_\parallel;\overleftrightarrow{R}_i^{-1}) \cdot \vec{v}(R_j)] \tag{171}$$

The matrices $\{\overleftrightarrow{T}(\vec{k}_\parallel;\overleftrightarrow{R})\}$ are thus seen to provide a 3r-dimensional unitary multiplier-representation of the point group $G_0(\vec{k}_\parallel)$ of the wave vector $\vec{k}_\parallel$, and they also commute with the partially Fourier-transformed dynamical matrix $\overleftrightarrow{D}(\vec{k}_\parallel)$

$$\overleftrightarrow{D}(\vec{k}_\parallel) = \overleftrightarrow{T}^{-1}(\vec{k}_\parallel;\overleftrightarrow{R})\overleftrightarrow{D}(\vec{k}_\parallel)\overleftrightarrow{T}(\vec{k}_\parallel;\overleftrightarrow{R}) \tag{172}$$

It is easily shown (see Ref. 72) that the normal mode frequencies have the symmetry of the point group of the crystal slab, i.e.:

$$\omega_j^2(\overleftrightarrow{S}\vec{k}_\parallel) = \omega_j^2(\vec{k}_\parallel) \tag{173}$$

Furthermore, the transformation law for the eigenvector $\vec{v}(\vec{k}_\parallel j)$ can be written as

$$\vec{v}(\overleftrightarrow{S}\vec{k}_\parallel j) = \overleftrightarrow{\Gamma}\left(\vec{k}_\parallel;\{\overleftrightarrow{S}|\vec{v}(S)\}\right)\vec{v}(\vec{k}_\parallel j) \tag{174}$$

if $\vec{k}_\parallel$ is a point at which none of the eigenvalues of $\overleftrightarrow{D}(\vec{k})$ are degenerate, and may be assumed to be valid at points of degeneracy as well, using exactly the same arguments as those given by Maradudin, et al. (74).

Multiplying both sides of Eq. (164) from the left by the matrix $\overleftrightarrow{T}(\vec{k}_\parallel;\overleftrightarrow{R})$, and using the fact that $\overleftrightarrow{T}(\vec{k}_\parallel;\overleftrightarrow{R})$ commutes with $\overleftrightarrow{D}(\vec{k}_\parallel)$, we obtain

$$\overleftrightarrow{D}(\vec{k}_\parallel)\{\overleftrightarrow{T}(\vec{k}_\parallel;\overleftrightarrow{R})\vec{v}(\vec{k}_\parallel j)\} = \omega_j^2(\vec{k}_\parallel)\{\overleftrightarrow{T}(\vec{k}_\parallel;\overleftrightarrow{R})\vec{v}(\vec{k}_\parallel j)\} \tag{175}$$

Equation (175) implies that if $\vec{v}(\vec{k}_\parallel j)$ is an eigenvector of $\overleftrightarrow{D}(\vec{k}_\parallel)$ with an eigenvalue $\omega_j^2(\vec{k}_\parallel)$, then so is $\overleftrightarrow{T}(\vec{k}_\parallel;\overleftrightarrow{R})\vec{v}(\vec{k}_\parallel j)$, for every operation $\overleftrightarrow{R}$ in the point group $G_0(\vec{k}_\parallel)$. Thus, $\overleftrightarrow{T}(\vec{k}_\parallel;\overleftrightarrow{R})\vec{v}(\vec{k}_\parallel j)$ must be a linear combination of the eigenvectors of $\overleftrightarrow{D}(\vec{k}_\parallel)$ whose eigenvalues are equal to $\omega_j^2(\vec{k}_\parallel)$. This result is conveniently expressed if we replace the single index j by a double index $\sigma\lambda$, where σ labels the distinct values of $\omega_j^2(\vec{k}_\parallel)$ for a given wave vector $\vec{k}_\parallel$, and $\lambda(=1,2,\ldots,f_\sigma)$ labels the linearly independent eigenvectors associated with the eigenvalue $\omega_\sigma^2(\vec{k}_\parallel)$. The number f_σ is then the degeneracy of the normal mode whose frequency is $\omega_\sigma^2(\vec{k}_\parallel)$. In this notation, the eigenvalue equation (164) takes the form

$$\overleftrightarrow{D}(\vec{k}_{\parallel})\vec{v}(\vec{k}_{\parallel}\sigma\lambda) = \omega_{\sigma}^{2}(\vec{k}_{\parallel})\vec{v}(\vec{k}_{\parallel}\sigma\lambda), \quad \lambda = 1,2,\ldots,f_{\sigma} \tag{176}$$

From Eq. (175) and the discussion following it, we can write

$$\overleftrightarrow{T}(\vec{k}_{\parallel};\overleftrightarrow{R})\vec{v}(\vec{k}_{\parallel}\sigma\lambda) = \sum_{\lambda'=1}^{f_{\sigma}} \vec{\tau}_{\lambda\lambda'}^{(\sigma)}(\vec{k}_{\parallel};\overleftrightarrow{R})\vec{v}(\vec{k}_{\parallel}\sigma\lambda') \tag{177}$$

for every operation $\overleftrightarrow{R}$ of the point group $G_0(\vec{k}_{\parallel})$. The f_{σ}-dimensional matrices $\{\overleftrightarrow{\tau}^{(\sigma)}(\vec{k}_{\parallel};\overleftrightarrow{R})\}$ provide a multiplier representation of $G_0(\vec{k}_{\parallel})$, with the same factor system (multipliers) as occur in the representation provided by the matrices $\{\overleftrightarrow{T}(\vec{k}_{\parallel};\overleftrightarrow{R})\}$. It is straightforward (73) to show that $\overleftrightarrow{\tau}^{(\sigma)}(\vec{k}_{\parallel};\overleftrightarrow{R})$ is a unitary matrix. In the absence of accidental degeneracy, the set of matrices $\{\overleftrightarrow{\tau}^{(\sigma)}(\vec{k}_{\parallel};\overleftrightarrow{R})\}$ constitute an f_{σ}-dimensional irreducible multiplier representation of the point group $G_0(\vec{k}_{\parallel})$.

Let us now assume that we know all the eigenvectors $\{\vec{v}(\vec{k}_{\parallel}\sigma\lambda)\}$ of $\overleftrightarrow{D}(\vec{k}_{\parallel})$ for a given value of $\vec{k}_{\parallel}$. From Eq. (177) it follows that if we construct a 3r x 3r matrix $\vec{v}(\vec{k}_{\parallel})$, whose columns are just the vectors $\{\vec{v}(\vec{k}_{\parallel}\sigma\lambda)\}$, so that the $(\alpha\kappa;\sigma\lambda)$ element of this matrix is given by

$$v_{\alpha\kappa;\sigma\lambda}(\vec{k}_{\parallel}) = v_{\alpha}(\kappa|\vec{k}_{\parallel}\sigma\lambda) \tag{178}$$

we obtain as the equation for $\overleftrightarrow{v}(\vec{k})$

$$\overleftrightarrow{T}(\vec{k}_{\parallel};\overleftrightarrow{R})\overleftrightarrow{v}(\vec{k}_{\parallel}) = \overleftrightarrow{v}(\vec{k}_{\parallel})\overleftrightarrow{\Delta}(\vec{k}_{\parallel};\overleftrightarrow{R}) \tag{179}$$

where the 3r x 3r matrix $\overleftrightarrow{\Delta}(\vec{k}_{\parallel};\overleftrightarrow{R})$ has the form

$$\overleftrightarrow{\Delta}(\vec{k}_{\parallel};\overleftrightarrow{R}) = \begin{bmatrix} \overleftrightarrow{\tau}^{(1)}(\vec{k}_{\parallel};\overleftrightarrow{R}) & \cdot & 0 & \cdot \\ 0 & \overleftrightarrow{\tau}^{(2)}(\vec{k}_{\parallel};\overleftrightarrow{R}) & 0 & \cdot \\ 0 & \cdot & \cdot & \cdot \\ \cdot & \cdot & \cdot & \cdot \\ \cdot & \cdot & \cdot & \cdot \\ \cdot & \cdot & \cdot & \cdot \end{bmatrix} \tag{180}$$

In Eq. (180), $\overleftrightarrow{\tau}^{(1)}(\vec{k}_{\parallel};R)$, $\overleftrightarrow{\tau}^{(2)}(\vec{k}_{\parallel};\overleftrightarrow{R})$, . . . are the matrices of the irreducible multiplier representations of $G_0(\vec{k}_{\parallel})$ corresponding to the frequencies $\omega_1^2(\vec{k}_{\parallel})$, $\omega_2^2(\vec{k}_{\parallel})$, . . . , respectively.

From Eq. (179), we see that the matrix $\overleftrightarrow{v}(\vec{k}_{\parallel})$ is the matrix which block-diagonalizes every matrix $\overleftrightarrow{T}(\vec{k}_{\parallel};\overleftrightarrow{R})$ corresponding to one of the operations of $G_o(\vec{k}_{\parallel})$ into the form given in Eq. (180) by a similarity transformation

$$\overleftrightarrow{\Delta}(\vec{k}_{\parallel};\overleftrightarrow{R}) = \overleftrightarrow{v}^{-1}(\vec{k}_{\parallel})\overleftrightarrow{T}(\vec{k}_{\parallel};\overleftrightarrow{R})\overleftrightarrow{v}(\vec{k}_{\parallel}) = \overleftrightarrow{v}^{T}(\vec{k}_{\parallel})^{*}\overleftrightarrow{T}(\vec{k}_{\parallel};\overleftrightarrow{R})\overleftrightarrow{v}(\vec{k}_{\parallel}) \tag{181}$$

i.e., the matrix of the eigenvectors of $\overleftrightarrow{D}(\vec{k}_{\parallel})$ reduces the reducible representation of $G_0(\vec{k}_{\parallel})$ provided by the $3r \times 3r$ matrices $\{\overleftrightarrow{T}(\vec{k}_{\parallel};\overleftrightarrow{R})\}$ into its irreducible representations.

It may be the case that some irreducible representation, i.e., some matrix $\overleftrightarrow{\tau}^{(\sigma)}(\vec{k}_{\parallel};\overleftrightarrow{R})$, appears more than once, say c times, in the reduction of the representation of $G_0(\vec{k}_{\parallel})$ given by the set of matrices $\{\overleftrightarrow{T}(\vec{k}_{\parallel};\overleftrightarrow{R})\}$, as expressed by Eq. (180). This means that there are that many distinct sets of f_σ eigenvectors $\{\vec{v}(\vec{k}_{\parallel}\sigma\lambda)\}$, each of which corresponds to a different value of $\omega^2(\vec{k}_\parallel)$ (since σ labels the distinct eigenvalues), which have the property that the eigenvectors comprising a given set transform into linear combinations of each other under the operations $\{\overleftrightarrow{T}(\vec{k}_\parallel;\overleftrightarrow{R})\}$ in the same way as do the eigenvectors comprising each of the remaining c-1 sets. This circumstance that c sets $(c > 1)$ of matrices $\{\overleftrightarrow{\tau}^{(\sigma)}(\vec{k}_\parallel;\overleftrightarrow{R})\}$ are in fact identical has the consequence that σ is not a unique label for the irreducible representations of $G_o(\vec{k}_\parallel)$ contained in the reducible represenation provided by the matrices $\{\overleftrightarrow{T}(\vec{k}_\parallel;\overleftrightarrow{R})\}$. Thus, it is convenient to generalize our notation a bit more. We replace the single index σ by the double index sa, where s labels the irreducible representations of $G_o(\vec{k}_\parallel)$, and a is a "repetition" index which differentiates among the different eigenvalues whose associated eigenvectors transform according to the same irreducible representation of $G_o(\vec{k}_\parallel)$. The index a takes on the values 1, 2, . . . , c_s, where c_s is the number of times the sth irreducible representation of $G_o(\vec{k}_\parallel)$ is contained in the representation given by the matrices $\{\overleftrightarrow{T}(\vec{k}_\parallel;\overleftrightarrow{R})\}$.

In the new notation the eigenvalue equation (176) takes the form

$$\overleftrightarrow{D}(\vec{k}_\parallel)\vec{v}(\vec{k}_\parallel sa\lambda) = \omega_{sa}^2(\vec{k}_\parallel)\vec{v}(\vec{k}_\parallel sa\lambda) \tag{182}$$

for $\lambda = 1, 2, \ldots, f_s$ and $a = 1, 2, \ldots, c_s$.

Also, Eq. (177) takes the form

$$\overleftrightarrow{T}(\vec{k}_\parallel;\overleftrightarrow{R})\vec{v}(\vec{k}_\parallel sa\lambda) = \sum_{\lambda'=1}^{f_s} \tau_{\lambda'\lambda}^{(s)}(\vec{k}_\parallel;\overleftrightarrow{R})\vec{v}(\vec{k}_\parallel sa\lambda') \tag{183}$$

The reduction of the reducible representation of $G_o(\vec{k}_\parallel)$ provided by the matrices $\{\overleftrightarrow{T}(\vec{k}_\parallel;\overleftrightarrow{R})\}$ into its irreducible representations can be carried out by standard methods (see Refs. 75 and 76). If we denote the characters of the reducible representation $\{\overleftrightarrow{T}(\vec{k}_\parallel;\overleftrightarrow{R})\}$ by $\chi(\vec{k}_\parallel;R)$

$$\chi(\vec{k}_\parallel;\overleftrightarrow{R}) = \mathrm{Tr}\,\overleftrightarrow{T}(\vec{k}_\parallel;\overleftrightarrow{R})$$

$$= \sum_{\kappa\alpha} R_{\alpha\alpha}\delta\left(\kappa, F_o(\kappa;R)\right)\exp\left(i\vec{k}_\parallel \cdot [\vec{x}(\kappa) - \overleftrightarrow{R}\vec{x}(\kappa)]\right) \tag{184}$$

and if we denote the characters of the sth irreducible representation $\{\overleftrightarrow{\tau}^{(s)}(\vec{k}_\parallel;\overleftrightarrow{R})\}$ by $\chi^{(s)}(\vec{k}_\parallel;\overleftrightarrow{R})$

$$\chi^{(s)}(\vec{k}_\parallel;\overleftrightarrow{R}) = \mathrm{Tr}\,\overleftrightarrow{\tau}^{(s)}(\vec{k}_\parallel;\overleftrightarrow{R}) \tag{185}$$

then the number of times the sth irreducible representation is contained in the representation $\{\overleftrightarrow{T}(\vec{k}_\parallel;\overleftrightarrow{R})\}$ is (see Refs. 75 and 76)

$$c_s = \frac{1}{h}\sum_{R} \chi(\vec{k}_\parallel;\overleftrightarrow{R})\chi^{(s)}(\vec{k}_\parallel;\overleftrightarrow{R})^* \tag{186}$$

where h is the order of the group $G_o(\vec{k}_\parallel)$.

The forms of the eigenvectors $\{v(\vec{k}_\parallel sa\lambda)\}$ can be obtained by projection operator techniques. The projection operator in our 3r-dimensional space is the 3r x 3r matrix $\overleftrightarrow{P}^{(s)}_{\lambda\lambda'}(\vec{k}_\parallel)$ defined as

$$\overleftrightarrow{P}^{(s)}_{\lambda\lambda'}(\vec{k}_\parallel) = (f_s/h)\sum_{R} \tau^{(s)}_{\lambda\lambda'}(\vec{k}_\parallel;\overleftrightarrow{R})^{*}\overleftrightarrow{T}(\vec{k}_\parallel;\overleftrightarrow{R}) \qquad (187)$$

If we denote by $\vec{\Psi}$ an arbitrary 3r-component vector whose elements are $\{\Psi_\alpha(\kappa)\}$, then it is straightforward to show (Ref. 73) that the vector

$$\vec{V}(\vec{k}_\parallel;s\lambda) = \overleftrightarrow{P}^{(s)}_{\lambda\lambda'}(\vec{k}_\parallel)\vec{\Psi} \qquad (188)$$

for any fixed λ', transforms under the application of the matrix $\overleftrightarrow{T}(\vec{k}_\parallel;\overleftrightarrow{R})$ in exactly the same way as the eigenvector $\vec{v}(\vec{k}_\parallel sa\lambda)$ does, i.e.

$$\overleftrightarrow{T}(\vec{k}_\parallel;\overleftrightarrow{R}')\vec{V}(\vec{k}_\parallel;s\lambda) = \sum_{\lambda_1} \tau^{(s)}_{\lambda_1\lambda}(\vec{k}_\parallel;\overleftrightarrow{R}')\vec{V}(\vec{k}_\parallel;s\lambda_1) \qquad (189)$$

The fact that $\overleftrightarrow{P}^{(s)}_{\lambda\lambda'}(\vec{k}_\parallel)$ is indeed a projection operator is expressed by the relation

$$\overleftrightarrow{P}^{(s)}_{\lambda\lambda'}(\vec{k}_\parallel)\vec{v}(\vec{k}_\parallel s'a\lambda_1) = \delta_{ss'}\delta_{\lambda'\lambda_1}\vec{v}(\vec{k}_\parallel sa\lambda) \qquad (190)$$

In the special case that the sth irreducible representation appears only once in the reduction of the representation of $G_o(\vec{k}_\parallel)$ provided by the matrices $\{\overleftrightarrow{T}(\vec{k}_\parallel;\overleftrightarrow{R})\}$, so that c_s and the corresponding repetition index, a, both equal unity, the vector $\vec{V}(\vec{k}_\parallel;s\lambda)$ constructed according to Eq. (188) and normalized to unity can be taken to be the eigenvector $\vec{v}(\vec{k}_\parallel;sa\lambda)$ of $\overleftrightarrow{D}(\vec{k}_\parallel)$.

When the sth irreducible representation of $G_o(\vec{k}_\parallel)$ appears more than once, the dynamical matrix can be simplified (block-diagonalized) in the following way.

The vector $\vec{V}(\vec{k}_{\parallel};s\lambda)$ is no longer an (unnormalized) eigenvector of $\overleftrightarrow{D}(\vec{k}_{\parallel})$. It is, in general, some linear combination of the c_s eigenvectors $\vec{v}(\vec{k}_{\parallel}sa\lambda)$ $(a = 1,2,\ldots,c_s)$ corresponding to the distinct eigenfrequencies $\{\omega^2_{sa}(\vec{k}_{\parallel})\}$, and consequently contains c_s arbitrary constants. Since the transformation properties of the vector $\vec{V}(\vec{k}_{\parallel};s\lambda)$ are independent of the numerical values of the c_s-independent constants on which it depends, we can construct c_s linearly independent vectors from $\vec{V}(\vec{k}_{\parallel};s\lambda)$, each of which still belongs to the λth row of the sth irreducible representation of $G_o(\vec{k}_{\parallel})$, by setting each of the c_s parameters on which it depends equal to unity in succession, while setting the remaining parameters equal to zero. If these vectors are not already mutually orthogonal, they can be made so by the Schmidt method, for example. Normalized to unity, these c_s vectors can be denoted by $\vec{V}(\vec{k}_{\parallel};sa\lambda)$, where $a = 1,2,\ldots,c_s$. Each of these unit vectors belongs to the λth row of the sth irreducible representation of $G_o(\vec{k}_{\parallel})$. The vectors $\{\vec{V}(\vec{k}_{\parallel};sa\lambda)\}$ obtained in this manner obey the orthonormality and completeness relations (Ref. 72)

$$\sum_{\kappa\alpha} V_\alpha(\kappa|\vec{k}_{\parallel};sa\lambda)^* V_\alpha(\kappa|\vec{k}_{\parallel};s'a'\lambda') = \delta_{ss'}\delta_{aa'}\delta_{\lambda\lambda'} \tag{191a}$$

$$\sum_{sa\lambda} V_\alpha(\kappa|\vec{k}_{\parallel};sa\lambda)^* V_\beta(\kappa'|\vec{k}_{\parallel};sa\lambda) = \delta_{\kappa\kappa'}\delta_{\alpha\beta} \tag{191b}$$

The usefulness of the vectors $\{\vec{V}(\vec{k}_{\parallel};sa\lambda)\}$ arises from the fact that a matrix element of any matrix $M_{\alpha\beta}(\kappa\kappa'|\vec{k})$, which commutes with all the matrices $\{\overleftrightarrow{T}(\vec{k}_{\parallel};\overleftrightarrow{R})\}$, e.g., the dynamical matrix $D_{\alpha\beta}(\kappa\kappa'|\vec{k}_{\parallel})$, taken between two vectors $\vec{V}(\vec{k}_{\parallel};sa\lambda)$ and $\vec{V}(\vec{k}_{\parallel};s'a'\lambda')$ vanishes unless $s = s'$ and $\lambda = \lambda'$ (Ref. 72).

We thus have available a method for block-diagonalizing the dynamical matrix $\overleftrightarrow{D}(\vec{k}_\parallel)$ into submatrices, the eigenvalues of which are the squares of the normal mode frequencies which belong to a given irreducible representation contained in the representation of $G_o(\vec{k}_\parallel)$ provided by the $\{\overleftrightarrow{T}(\vec{k}_\parallel;\overleftrightarrow{R})\}$. For if we construct a 3r x 3r unitary matrix $\overleftrightarrow{V}(\vec{k}_\parallel)$ whose columns are just the vectors $\{\vec{V}(\vec{k}_\parallel;sa\lambda)\}$, so that the $(\alpha\kappa;sa\lambda)$ element of this matrix is given by

$$V_{\alpha\kappa;sa\lambda}(\vec{k}_\parallel) = V_\alpha(\kappa|\vec{k}_\parallel;sa\lambda) \tag{192}$$

it is then easily shown that (see Ref. 72)

$$\overleftrightarrow{V}^{-1}(\vec{k}_\parallel)\overleftrightarrow{D}(\vec{k}_\parallel)\overleftrightarrow{V}(\vec{k}_\parallel) = \overleftrightarrow{\bar{D}}(\vec{k}_\parallel) \tag{193}$$

where

$$\overleftrightarrow{\bar{D}}(\vec{k}_\parallel) = \begin{bmatrix} \overleftrightarrow{\bar{D}}^{(1)}(\vec{k}_\parallel) & 0 & 0 & \cdot \\ 0 & \overleftrightarrow{\bar{D}}^{(2)}(\vec{k}_\parallel) & 0 & \cdot \\ 0 & \cdot & \cdot & \cdot \\ \cdot & \cdot & \cdot & \cdot \\ \cdot & \cdot & \cdot & \cdot \end{bmatrix} \tag{194}$$

The elements of the c_s x c_s matrix $\overleftrightarrow{\bar{D}}^{(s)}(\vec{k}_\parallel)$ are given by

$$\bar{D}^{(s)}(\vec{k}_\parallel|aa') = \sum_{\kappa\alpha}\sum_{\kappa'\beta} V_\alpha(\kappa|\vec{k}_\parallel;sa\lambda)^* D_{\alpha\beta}(\kappa\kappa'|\vec{k}_\parallel)$$

$$\cdot\ V_\beta(\kappa'|\vec{k}_\parallel;sa'\lambda) \tag{195}$$

and are independent of λ. The matrix $\overleftrightarrow{D}^{(s)}(\vec{k}_{\parallel})$ appears f_s times along the diagonal of the matrix $\overleftrightarrow{D}(\vec{k}_{\parallel})$.

Because the eigenvalues of a matrix are unaffected by a similarity (unitary) transformation on the matrix, we see that the eigenvalues of $\overleftrightarrow{D}^{(s)}(\vec{k}_{\parallel})$ are just the squares of the frequencies of the normal modes which belong to the sth irreducible representation of $G_o(\vec{k}_{\parallel})$, and the degeneracy of these modes is equal to the dimensionality of the irreducible representation s.

Additional Degeneracies due to Time-Reversal Symmetry

Under special circumstances, time-reversal symmetry can give rise to extra degeneracies among the normal vibration modes of a crystal slab in addition to those arising from the multidimensionality of the irreducible representations of the space group $G_{\vec{k}_{\parallel}}$. In this section we present the conditions under which these extra degeneracies can occur.

In the previous section, we considered spatial symmetry alone in constructing a set of matrices which commute with the dynamical matrix $\overleftrightarrow{D}(\vec{k}_{\parallel})$. However, from Eq. (41), we see that $D(\vec{k}_{\parallel})$ has the property

$$\overleftrightarrow{D}(-\vec{k}_{\parallel}) = D^{*}(\vec{k}_{\parallel}) \tag{196}$$

which means that we can expand the number of matrices which commute with $\overleftrightarrow{D}(\vec{k}_{\parallel})$ if the point group of the crystal slab contains a rotational element $\overleftrightarrow{S}_-$ such that $\overleftrightarrow{S}_-\vec{k}_{\parallel} = -\vec{k}_{\parallel}$, that is, if $-\vec{k}_{\parallel}$ is in the star of $\vec{k}_{\parallel}$, where the star of $\vec{k}_{\parallel}$ is defined as the set of inequivalent wave vectors generated by applying the operations $\{\overleftrightarrow{S}\}$ of the point group of the crystal to the wave vector $\vec{k}_{\parallel}$.

If for the crystal slab in question, $-\vec{k}_{\parallel}$ is in the star of $\vec{k}_{\parallel}$ only for special values of $\vec{k}_{\parallel}$, the following considerations are applicable to these values of $\vec{k}_{\parallel}$ only.

To aid in the following analysis, we introduce the antiunitary operator $\overleftrightarrow{K}_0$ to represent the complex conjugate operation, and define it by its effect on an arbitrary vector $\vec{\Psi}$ in the 3r-dimensional space spanned by the eigenvectors of the dynamical matrix

$$\overleftrightarrow{K}_0\vec{\Psi} = \vec{\Psi}^* \tag{197}$$

Applying $\overleftrightarrow{K}_0$ once more gives $\overleftrightarrow{K}_0^2\vec{\Psi} = \vec{\Psi}$ and thus $\overleftrightarrow{K}_0^{-1} = \overleftrightarrow{K}_0$. Using $\overleftrightarrow{K}_0$ to perform a similarity transformation on $\overleftrightarrow{D}(\vec{k}_{\parallel})$, we have

$$\overleftrightarrow{K}_0\overleftrightarrow{D}(\vec{k}_{\parallel})\overleftrightarrow{K}_0^{-1} = \overleftrightarrow{D}^*(\vec{k}_{\parallel}) \tag{198}$$

Thus if $-\vec{k}_{\parallel}$ is in the star of $\vec{k}_{\parallel}$, Eqs. (193) and (166), together with (196), yield

$$\overleftrightarrow{K}_0\overleftrightarrow{\Gamma}(\vec{k}_{\parallel};\{\overleftrightarrow{S}\_|\vec{v}(S\_)\})\overleftrightarrow{D}(\vec{k}_{\parallel})\overleftrightarrow{\Gamma}^{-1}(\vec{k}_{\parallel};\{\overleftrightarrow{S}\_|\vec{v}(S\_)\})\overleftrightarrow{K}_0^{-1}$$

$$= \overleftrightarrow{K}_0\overleftrightarrow{D}(-\vec{k}_{\parallel})\overleftrightarrow{K}_0^{-1} = \overleftrightarrow{D}^*(-\vec{k}_{\parallel}) = \overleftrightarrow{D}(\vec{k}_{\parallel}) \tag{199}$$

i.e., the antiunitary matrix operator $\overleftrightarrow{K}_0\overleftrightarrow{\Gamma}(\vec{k}_{\parallel};\{\overleftrightarrow{S}\_|\vec{v}(S\_)\})$ commutes with the dynamical matrix $\overleftrightarrow{D}(\vec{k}_{\parallel})$.

In order to incorporate antiunitary operators such as $\overleftrightarrow{K}_0\overleftrightarrow{\Gamma}(\vec{k}_{\parallel};\{\overleftrightarrow{S}\_|\vec{v}(S\_)\})$ into the group theoretical discussion of the symmetry properties of $\overleftrightarrow{D}(\vec{k}_{\parallel})$ and its eigenvectors and eigenvalues, we must have at hand expressions for the products of $\overleftrightarrow{K}_0\overleftrightarrow{\Gamma}(\vec{k}_{\parallel};\{\overleftrightarrow{S}\_|\vec{v}(S\_)\})$ with the matrices

$\overleftrightarrow{\Gamma}(\vec{k}_{\parallel};\{\overleftrightarrow{R}|\vec{v}(R) + \vec{x}(m)\})$, and with other antiunitary matrix operators. It is not difficult to show (Ref. (72) that the following relations hold

$$\overleftrightarrow{K}_0\overleftrightarrow{\Gamma}\left(\vec{k}_{\parallel};\left\{\overleftrightarrow{S}_{-}|\vec{v}(S_{-}) + \vec{x}(m_1)\right\}\right) \cdot \overleftrightarrow{\Gamma}\left(\vec{k}_{\parallel};\left\{\overleftrightarrow{R}|\vec{v}(R) + \vec{x}(m_2)\right\}\right)$$

$$= \overleftrightarrow{K}_0\overleftrightarrow{\Gamma}\left(\vec{k}_{\parallel};\left\{S_{-}|\vec{v}(S_{-}) + \vec{x}(m_1)\right\}\right.$$

$$\left.\cdot \left\{\overleftrightarrow{R}|\vec{v}(R) + \vec{x}(m_2)\right\}\right) \qquad (200)$$

$$\overleftrightarrow{\Gamma}\left(\vec{k}_{\parallel};\left\{\overleftrightarrow{R}|\vec{v}(R) + \vec{x}(m_2)\right\}\right) \cdot \overleftrightarrow{K}_0\overleftrightarrow{\Gamma}\left(\vec{k}_{\parallel};\left\{\overleftrightarrow{S}|\vec{v}(S_{-}) + \vec{x}(m_1)\right\}\right)$$

$$= \overleftrightarrow{K}_0\overleftrightarrow{\Gamma}\left(\vec{k}_{\parallel};\left\{\overleftrightarrow{R}|\vec{v}(R) + \vec{x}(m_2)\right\}\right.$$

$$\left.\cdot \left\{\overleftrightarrow{S}_{-}|\vec{v}(S_{-}) + \vec{x}(m_1)\right\}\right) \qquad (201)$$

Also,

$$\overleftrightarrow{K}_0\overleftrightarrow{\Gamma}\left(\vec{k}_{\parallel};\left\{\overleftrightarrow{S}_{-}\overleftrightarrow{R}_1|\vec{v}(S_{-}R_1) + \vec{x}(m_1)\right\}\right)$$

$$\cdot \overleftrightarrow{K}_0\overleftrightarrow{\Gamma}\left(\vec{k}_{\parallel};\left\{S_{-}R_2|\vec{v}(S_{-}R_2) + \vec{x}(m_2)\right\}\right)$$

$$= \overleftrightarrow{\Gamma}\left(\vec{k}_{\parallel};\left\{\overleftrightarrow{S}_{-}\overleftrightarrow{R}_1|\vec{v}(S_{-}R_1) + \vec{x}(m_1)\right\}\right.$$

$$\left.\cdot \left\{\overleftrightarrow{S}_{-}\overleftrightarrow{R}_2|\vec{v}(S_{-}R_2) + \vec{x}(m_2)\right\}\right) \qquad (202)$$

i.e., the product of two antiunitary matrix operators is

a unitary matrix corresponding to an element of $G_{\vec{k}_\parallel}$ since $\overleftrightarrow{S}\_\overleftrightarrow{R}_1\overleftrightarrow{S}\_\overleftrightarrow{R}_2$ is an element of $G_0(\vec{k}_\parallel)$. Thus, the matrix operators $\overleftrightarrow{\Gamma}(\vec{k}_\parallel;\{\overleftrightarrow{R}|\vec{v}(R) + \vec{x}(m)\})$ and $\overleftrightarrow{K}_0\overleftrightarrow{\Gamma}(\vec{k}_\parallel;\{\overleftrightarrow{S}\_\overleftrightarrow{R}|\vec{v}(S\_R) + \vec{x}(m)\})$ form a symmetry group with a one-to-one relationship to the elements of the space group containing the crystal symmetry operations $\{\overleftrightarrow{R}|\vec{v}(R) + \vec{x}(m)\}$ and $\{\overleftrightarrow{S}\_\overleftrightarrow{R}|\vec{v}(S\_R) + \vec{x}(m)\}$. This space group, which is a sum of the space group $G_{\vec{k}_\parallel}$ plus the coset $\{\overleftrightarrow{S}\_|\vec{v}(S\_)\}G_{\vec{k}_\parallel}$, will be designated by the symbol $G_{\vec{k}_\parallel;-\vec{k}_\parallel}$. Note that $G_{\vec{k}_\parallel}$ is an invariant subgroup of $G_{\vec{k}_\parallel;-\vec{k}_\parallel}$.

The purely rotational elements $\{\overleftrightarrow{R}\}$ and $\{\overleftrightarrow{S}\_\overleftrightarrow{R}\}$ in the space group $G_{\vec{k}_\parallel;-\vec{k}_\parallel}$ taken by themselves form a point group $G_0(\vec{k}_\parallel;-\vec{k}_\parallel)$ in which $G_0(\vec{k}_\parallel)$ is an invariant subgroup. Just as it is convenient to introduce the unitary matrices $\{\overleftrightarrow{T}(\vec{k}_\parallel;\overleftrightarrow{R})\}$ by Eq. (169) in considering the consequences of spatial symmetry on the form of $\overleftrightarrow{D}(\vec{k}_\parallel)$, instead of working with the matrices $\{\overleftrightarrow{\Gamma}(\vec{k}_\parallel;\{\overleftrightarrow{R}|\vec{v}(R) + \vec{x}(m)\})\}$, it is also convenient to associate with each element $\overleftrightarrow{S}\_\overleftrightarrow{R}$ of the coset $\overleftrightarrow{S}\_G_0(\vec{k}_\parallel)$ an antiunitary matrix operator $\overleftrightarrow{T}(\vec{k}_\parallel;\overleftrightarrow{S}\_\overleftrightarrow{R})$, which is defined in terms of the matrix operator $\{\overleftrightarrow{K}_0\overleftrightarrow{\Gamma}(\vec{k}_\parallel;\{\overleftrightarrow{S}\_\overleftrightarrow{R}|\vec{v}(S\_R) + \vec{x}(m)\})\}$ by

$$
\begin{aligned}
\overleftrightarrow{T}(\vec{k}_\parallel;\overleftrightarrow{S}\_\overleftrightarrow{R}) &= \overleftrightarrow{K}_0 \exp(-i\vec{k}_\parallel \cdot [\vec{v}(S\_R) + \vec{x}(m)]) \\
&\quad \cdot \overleftrightarrow{\Gamma}(\vec{k}_\parallel;\{\overleftrightarrow{S}\_\overleftrightarrow{R}|\vec{v}(S\_R) + \vec{x}(m)\}) \\
&= \exp(i\vec{k}_\parallel \cdot [\vec{v}(S\_R) + \vec{x}(m)]) \\
&\quad \cdot \overleftrightarrow{K}_0\overleftrightarrow{\Gamma}(\vec{k}_\parallel;\{\overleftrightarrow{S}\_\overleftrightarrow{R}|\vec{v}(S\_R) + \vec{x}(m)\})
\end{aligned}
\tag{203}
$$

Explicitly, we have that

$$T_{\alpha\beta}(\kappa\kappa'|\vec{k}_{\parallel};\overleftrightarrow{S}\_\overleftrightarrow{R}) = (\overleftrightarrow{S}\_\overleftrightarrow{R})_{\alpha\beta}\delta\left(\kappa;F_0(\kappa';S\_R)\right) \cdot \exp\left(i\vec{k}_{\parallel}\cdot[\vec{x}(\kappa) - \overleftrightarrow{S}\_\overleftrightarrow{R}\vec{x}(\kappa')]\right)\overleftrightarrow{K}_0 \tag{204}$$

This definition is unique despite the fact that each rotational operation $\overleftrightarrow{S}\_\overleftrightarrow{R}_i$ is associated with an infinity of operations in the space group $G_{\vec{k}_{\parallel};-\vec{k}_{\parallel}}$. In order to keep in mind the antiunitary aspect of the matrix operator $\overleftrightarrow{T}(\vec{k}_{\parallel};\overleftrightarrow{S}\_\overleftrightarrow{R})$ in what follows, a rotational element in the coset $\overleftrightarrow{S}\_G_0(\vec{k}_{\parallel})$ will be denoted by $\overleftrightarrow{A}$ and the corresponding matrix operator by $\overleftrightarrow{T}(\vec{k}_{\parallel};\overleftrightarrow{A})$.

The products of antiunitary and unitary matrix operators $\overleftrightarrow{T}(\vec{k}_{\parallel};\ldots)$ can be obtained with the help of Eqs. (170), (171), (201), and (202) and their definitions (169) and (203), together with the relation

$$\overleftrightarrow{T}^*\left(\vec{k}_{\parallel};\{\overleftrightarrow{S}|\vec{v}(S) + \vec{x}(m)\}\right) = \overleftrightarrow{T}\left(-\vec{k}_{\parallel};\{\overleftrightarrow{S}|\vec{v}(S) + \vec{x}(m)\}\right) = \overleftrightarrow{T}\left(\overleftrightarrow{S}\_\vec{k}_{\parallel};\{\overleftrightarrow{S}|\vec{v}(S) + \vec{x}(m)\}\right) \tag{205}$$

If we introduce $\overleftrightarrow{\bar{R}}$ to denote an element in the point group $G_0(\vec{k}_{\parallel};-\vec{k}_{\parallel})$, we obtain the result that

$$\overleftrightarrow{T}(\vec{k}_{\parallel};\overleftrightarrow{\bar{R}}_i)\overleftrightarrow{T}(\vec{k}_{\parallel};\overleftrightarrow{\bar{R}}_j) = \phi(\vec{k}_{\parallel};\overleftrightarrow{\bar{R}}_i,\overleftrightarrow{\bar{R}}_j)\overleftrightarrow{T}(\vec{k}_{\parallel};\overleftrightarrow{\bar{R}}_i\overleftrightarrow{\bar{R}}_j) \tag{206}$$

where for $\overleftrightarrow{\bar{R}}_i = \overleftrightarrow{R}_i$

$$\phi(\vec{k}_{\parallel};\overleftrightarrow{R}_i,\overleftrightarrow{\bar{R}}_j) = \exp\left(i[\vec{k}_{\parallel}-\overleftrightarrow{R}_i^{-1}\vec{k}_{\parallel}]\cdot\vec{v}(\bar{R}_j)\right) \tag{207}$$

and for $\overleftrightarrow{\bar{R}}_i = \overleftrightarrow{A}_i$

$$\phi\left(\vec{k}_{\parallel};\overleftrightarrow{A}_i,\overleftrightarrow{\bar{R}}_j\right) = \exp\left(-i[\vec{k}_{\parallel} + \overleftrightarrow{A}_i^{-1}\vec{k}_{\parallel}] \cdot \vec{v}(\bar{R}_j)\right) \tag{208}$$

If $\vec{k}_{\parallel}$ either lies entirely within the Brillouin zone, or the space group $G_{\vec{k}_{\parallel};-\vec{k}_{\parallel}}$ of the crystal slab is symmorphic so that $\vec{v}(\bar{R})$ is zero, then the multiplier $\phi(\vec{k}_{\parallel};\overleftrightarrow{\bar{R}}_i,\overleftrightarrow{\bar{R}}_j)$ is unity.

From Eqs. (199) and (203), it is clear that the antiunitary matrix operators $\overleftrightarrow{T}(\vec{k}_{\parallel};\overleftrightarrow{S}\_\overleftrightarrow{R})$ commute with the partially Fourier-transformed dynamical matrix $\overleftrightarrow{D}(\vec{k}_{\parallel})$. In terms of the $\overleftrightarrow{\bar{R}}$ notation, this result and Eq. (172) may be combined into a single equation

$$\overleftrightarrow{D}(\vec{k}_{\parallel}) = \overleftrightarrow{T}^{-1}(\vec{k}_{\parallel};\overleftrightarrow{\bar{R}})\overleftrightarrow{D}(\vec{k}_{\parallel})\overleftrightarrow{T}(\vec{k}_{\parallel};\overleftrightarrow{\bar{R}}) \tag{209}$$

The matrices which transform the eigenvectors of an eigenvalue equation that is invariant under the operations of a group that contains antiunitary operators do not form a representation (ordinary or multiplier) of the group in the usual sense. We shall see that it will be necessary to introduce multiplier co-representation matrices to describe the transformation of the eigenvectors of $\overleftrightarrow{D}(\vec{k}_{\parallel})$ under the group of matrix operations $\overleftrightarrow{T}(\vec{k}_{\parallel};\overleftrightarrow{\bar{R}})$, where $\overleftrightarrow{\bar{R}}$ is a rotational element of $G_0(\vec{k}_{\parallel};-\vec{k}_{\parallel})$.

The extension of the analysis that leads from Eqs. (175) to (183) to include antiunitary matrix operators is straightforward. Operating on both sides of Eq. (182) from the left with the matrix operator $\overleftrightarrow{T}(\vec{k}_{\parallel};\overleftrightarrow{A})$, and using the fact that $\overleftrightarrow{T}(\vec{k}_{\parallel};\overleftrightarrow{A})$ commutes with $\overleftrightarrow{D}(\vec{k}_{\parallel})$, we obtain

$$\overleftrightarrow{D}(\vec{k}_{\parallel})\{\overleftrightarrow{T}(\vec{k}_{\parallel};\overleftrightarrow{A})\vec{v}(\vec{k}_{\parallel}sa\lambda)\} = \omega^2_{sa}(\vec{k}_{\parallel})\{\overleftrightarrow{T}(\vec{k}_{\parallel};\overleftrightarrow{A})\vec{v}(\vec{k}_{\parallel}sa\lambda)\} \tag{210}$$

Thus, if $\vec{v}(\vec{k}_{\parallel}sa\lambda)$ is an eigenvector of $\overleftrightarrow{D}(\vec{k}_{\parallel})$ with eigenvalue $\omega^2_{sa}(\vec{k}_{\parallel})$, then so is $\overleftrightarrow{T}(\vec{k}_{\parallel};\overleftrightarrow{A})\vec{v}(\vec{k}_{\parallel}sa\lambda)$, for every operation $\overleftrightarrow{A}$ of the coset $\overleftrightarrow{S}\_G_0(\vec{k}_{\parallel})$ of the point group $G_0(\vec{k}_{\parallel};-\vec{k}_{\parallel})$. Consequently, $\overleftrightarrow{T}(\vec{k}_{\parallel};\overleftrightarrow{A})\vec{v}(\vec{k}_{\parallel}sa\lambda)$ is a linear combination of the eigenvectors of $\overleftrightarrow{D}(\vec{k}_{\parallel})$ whose eigenvalues are equal to $\omega^2_{sa}(\vec{k}_{\parallel})$.

Let us return to the "j" index of Eq. (164) for labeling the branches of $\omega^2(\vec{k}_{\parallel})$. Since for any element $\overleftrightarrow{R}$ of the point group $G_0(\vec{k}_{\parallel};-\vec{k}_{\parallel})$ the matrix operator $\overleftrightarrow{T}(\vec{k}_{\parallel};\overleftrightarrow{R})$ commutes with the dynamical matrix $\overleftrightarrow{D}(\vec{k}_{\parallel})$, the effect of applying this operator to the eigenvector $\vec{v}(\vec{k}_{\parallel}j)$ can be expressed as

$$\overleftrightarrow{T}(\vec{k}_{\parallel};\overleftrightarrow{R})\vec{v}(\vec{k}_{\parallel}j) = \sum_{j'} \overleftrightarrow{H}_{j'j}(\vec{k}_{\parallel};\overleftrightarrow{R})\vec{v}(\vec{k}_{\parallel}j') \tag{211}$$

The sum over j' extends over all branches of the phonon spectrum at the point $\vec{k}_{\parallel}$ for which $\omega^2_{j'}(\vec{k}_{\parallel}) = \omega^2_j(\vec{k}_{\parallel})$.

In order to obtain the conditions for the existence of additional degeneracies due to time reversal symmetry, we investigate the linear dependence of the eigenvectors $\vec{v}(\vec{k}_{\parallel}sa\lambda)$ and

$$\bar{\vec{v}}(\vec{k}_{\parallel}sa\lambda) \equiv \overleftrightarrow{T}(\vec{k}_{\parallel};\overleftrightarrow{A}_0)\vec{v}(\vec{k}_{\parallel}sa\lambda) \tag{212}$$

where $\overleftrightarrow{A}_0$ is an arbitrary element of the co-set $\overleftrightarrow{S}\_G_0(\vec{k}_{\parallel})$. Clearly, if the two sets of eigenvectors are linearly independent, then there is an additional degeneracy due to time-reversal symmetry. The question of the linear dependence of the eigenvectors $\vec{v}(\vec{k}_\parallel s a\lambda)$ can be resolved by considering the relationship between the irreducible multiplier representations $\{\overleftrightarrow{\tau}^{(s)}(\vec{k}_\parallel;\overleftrightarrow{R})\}$ and $\{\overline{\overleftrightarrow{\tau}}^{(s)}(\vec{k}_\parallel;\overleftrightarrow{R})\}$ of the point group $G_0(\vec{k}_\parallel)$ which define the transformation properties of the eigenvectors $\vec{v}(\vec{k}_\parallel s a\lambda)$ and $\overline{\vec{v}}(\vec{k}_\parallel s a\lambda)$, respectively, under the operations of the unitary matrices $\overleftrightarrow{T}(\vec{k}_\parallel;\overleftrightarrow{R})$. The matrices $\{\overline{\overleftrightarrow{\tau}}^{(s)}(\vec{k}_\parallel;\overleftrightarrow{R})\}$ are expressible in terms of the matrices $\{\overleftrightarrow{\tau}^{(s)}(\vec{k}_\parallel;\overleftrightarrow{R})\}$ as (see Ref. 72)

$$\overline{\overleftrightarrow{\tau}}^{(s)}(\vec{k}_\parallel;\overleftrightarrow{R}) = \phi^*\left(\vec{k}_\parallel;\overleftrightarrow{A}_0,\overleftrightarrow{A}_0^{-1}R\right)\phi^*\left(\vec{k}_\parallel;\overleftrightarrow{A}_0^{-1}\overleftrightarrow{R},\overleftrightarrow{A}_0\right)\overleftrightarrow{\tau}^{(s)}\left(\vec{k}_\parallel;\overleftrightarrow{A}_0^{-1}\overleftrightarrow{R}\overleftrightarrow{A}_0\right)^* \tag{213}$$

$$= \phi^*\left(\vec{k}_\parallel;\overleftrightarrow{A}_0,\overleftrightarrow{A}_0^{-1}\overleftrightarrow{R}\overleftrightarrow{A}_0\right)\phi\left(\vec{k}_\parallel;\overleftrightarrow{R},\overleftrightarrow{A}_0\right)\overline{\overleftrightarrow{\tau}}^{(s)}\left(\vec{k}_\parallel;\overleftrightarrow{A}_0^{-1}\overleftrightarrow{R}\overleftrightarrow{A}_0\right) \tag{214}$$

Noting that the irreducible multiplier representations $\overleftrightarrow{\tau}^{(s)}$ and $\overline{\overleftrightarrow{\tau}}^{(s)}$ belong to the same factor system (72), they can be either equivalent or inequivalent.

If the irreducible multiplier representations $\overleftrightarrow{\tau}^{(s)}$ and $\overline{\overleftrightarrow{\tau}}^{(s)}$ are not equivalent (such representations will be referred to as being of the third type), the eigenvectors $\vec{v}(\vec{k}_\parallel s a\lambda)$ and $\overline{\vec{v}}(\vec{k}_\parallel s a\lambda)$ are orthogonal since they belong to different irreducible multiplier representations of the point group $G_0(\vec{k}_\parallel)$. Since $\vec{v}(\vec{k}_\parallel s a\lambda)$ and $\overline{\vec{v}}(\vec{k}_\parallel s a\lambda)$ are eigenvectors of the dynamical matrix $\overleftrightarrow{D}(\vec{k}_\parallel)$ with equal eigenvalues $\omega_{sa}^2(\vec{k}_\parallel)$, the f_s-fold degeneracy is doubled to $2f_s$ by time-reversal symmetry for this case.

Alternatively, $\overleftrightarrow{\tau}^{(s)}$ and $\overline{\overleftrightarrow{\tau}}^{(s)}$ are equivalent. Since they belong to the same factor system, they are related by a similarity transformation

$$\overleftrightarrow{\tau}^{(s)}(\vec{k}_{\parallel};\overleftrightarrow{R}) = \phi^{*}(\vec{k}_{\parallel};\overleftrightarrow{A}_0,\overleftrightarrow{A}_0^{-1}\overleftrightarrow{R})\phi^{*}(\vec{k}_{\parallel};\overleftrightarrow{A}_0^{-1}\overleftrightarrow{R},\overleftrightarrow{A}_0)\tau^{(s)}(\vec{k}_{\parallel};\overleftrightarrow{A}_0^{-1}\overleftrightarrow{R}\overleftrightarrow{A}_0)^{*}$$

$$= \overleftrightarrow{\beta}^{-1}\overleftrightarrow{\tau}^{(s)}(\vec{k}_{\parallel};\overleftrightarrow{R})\overleftrightarrow{\beta} \qquad (215)$$

where $\overleftrightarrow{\beta}$ is a unitary matrix. It can be shown (72) that the matrix $\overleftrightarrow{\beta}$ is unique up to a phase factor and that it may be expressed in terms of the irreducible multiplier representation matrices $\{\overleftrightarrow{\tau}^{(s)}(\vec{k}_{\parallel};\overleftrightarrow{R})\}$. It can also be shown that the property of the irreducible multiplier representation that causes the eigenvectors $\vec{v}(\vec{k}_{\parallel}sa\lambda)$ and $\overleftrightarrow{v}(\vec{k}_{\parallel}sa\lambda)$ to be linearly independent is reflected in the structure of the matrix $\overleftrightarrow{\beta}$ and is expressible as a condition on the matrix $\overleftrightarrow{\beta}\overleftrightarrow{\beta}^{*}$, i.e., $\overleftrightarrow{\beta}\overleftrightarrow{\beta}^{*}$ can have only two forms, namely

$$\overleftrightarrow{\beta}\overleftrightarrow{\beta}^{*} = \phi(\vec{k}_{\parallel};\overleftrightarrow{A}_0,\overleftrightarrow{A}_0)\overleftrightarrow{\tau}^{(s)}(\vec{k}_{\parallel};\overleftrightarrow{A}_0^{2}) \qquad (216)$$

or

$$\overleftrightarrow{\beta}\overleftrightarrow{\beta}^{*} = -\phi(\vec{k}_{\parallel};\overleftrightarrow{A}_0,\overleftrightarrow{A}_0)\overleftrightarrow{\tau}^{(s)}(\vec{k}_{\parallel};\overleftrightarrow{A}_0^{2}) \qquad (217)$$

These two cases will be referred to as of the first and second types, respectively. It is often the case that the element $\overleftrightarrow{A}_0$ is such that $\phi(\vec{k}_{\parallel};\overleftrightarrow{A}_0,\overleftrightarrow{A}_0)\overleftrightarrow{\tau}^{(s)}(\vec{k}_{\parallel};\overleftrightarrow{A}_0^{2})$ is equal to the unit matrix, i.e., $\overleftrightarrow{\beta}\overleftrightarrow{\beta}^{*} = \pm 1$. Then the matrix $\overleftrightarrow{\beta}$ is symmetric or antisymmetric, respectively. This is an example of what is meant by the structure of the matrix $\overleftrightarrow{\beta}$. It is straightforward (72) to show that the eigenvectors $\vec{v}(\vec{k}_{\parallel}sa\lambda)$ and $\overleftrightarrow{v}(\vec{k}_{\parallel}sa\lambda)$ are linearly independent when the matrix $\overleftrightarrow{\beta}$ satisfies Eq. (217). Thus time-reversal symmetry produces an additional degeneracy in this case. If Eq. (216) applies, time-reversal symmetry does not affect the degeneracy. In summary, if an irreducible multiplier representation $\{\overleftrightarrow{\tau}^{(s)}(\vec{k}_{\parallel};\overleftrightarrow{R})\}$ is of the first type, time-reversal symmetry does not produce any additional degeneracy. If it is of the second or third type, time-reversal symmetry leads to a doubling of the degeneracy obtained from the dimensionality of the matrices $\{\overleftrightarrow{\tau}^{(s)}(\vec{k}_{\parallel};\overleftrightarrow{R})\}$.

The criteria for determining the type to which a given irreducible multiplier representation $\{\overleftrightarrow{\tau}^{(s)}(\vec{k}_{\parallel};\overleftrightarrow{R})\}$ belongs may be cast in a convenient form by essentially using the orthogonality of the matrices $\{\overleftrightarrow{\tau}^{(s)}(\vec{k}_{\parallel};\overleftrightarrow{R})\}$ and $\{\overleftrightarrow{\tau}^{(s)}(\vec{k}_{\parallel};\overleftrightarrow{R})\}$ to arrive at the result

$$\sum_{\overleftrightarrow{A}} \phi(\vec{k}_{\parallel};\overleftrightarrow{A},\overleftrightarrow{A})\chi^{(s)}(\vec{k}_{\parallel};\overleftrightarrow{A}^2) = \begin{cases} h; \text{ first type} & (218a) \\ -h; \text{ second type} & (218b) \\ 0; \text{ third type} & (218c) \end{cases}$$

The sum is taken over all rotational elements $\overleftrightarrow{A}$ of the coset $\overleftrightarrow{S}\_G_0(\vec{k}_{\parallel})$.

The above criteria provide a complete solution to the problem of determining the additional degeneracies due to time-reversal symmetry in terms of characters $\{\chi^{(s)}(\vec{k}_{\parallel};\overleftrightarrow{R})\}$ and the multiplier $\phi(\vec{k}_{\parallel};\overleftrightarrow{A},\overleftrightarrow{A})$ without recourse to either the representation $\{\overleftrightarrow{\tau}^{(s)}(\vec{k}_{\parallel};\overleftrightarrow{R})\}$ or the matrix $\overleftrightarrow{\beta}$. The dimensions of the subspaces of eigenvectors, which are invariant under the symmetry operations $\{\overleftrightarrow{T}(\vec{k}_{\parallel};\overleftrightarrow{R})\}$ of the point group $G_0(\vec{k}_{\parallel};-\vec{k}_{\parallel})$, are f_s or $2f_s$ depending upon which irreducible multiplier representation they belong to for unitary symmetry operations. Thus, the dimensions of the irreducible multiplier co-representation matrices $\{\overleftrightarrow{Ⓗ}(\vec{k}_{\parallel};\overleftrightarrow{R})\}$ introduced in Eq. (211) are either f_s or $2f_s$. It is straightforward to construct an irreducible multiplier co-representation for each type of irreducible multiplier representation.

The corresponding irreducible multiplier co-representations will be referred to as being of the first, second, or third type also, and will be denoted by $\{\overleftrightarrow{Ⓗ}^{(s)}(\vec{k}_{\parallel};\overleftrightarrow{\mathbb{R}})\}$. We now express the matrices $\{\overleftrightarrow{Ⓗ}^{(s)}(\vec{k}_{\parallel};\overleftrightarrow{\mathbb{R}})\}$ in terms of the matrices $\{\overleftrightarrow{\tau}^{(s)}\}$ and $\overleftrightarrow{\beta}$

<u>First type</u>

$$\overleftrightarrow{Ⓗ}^{(s)}(\vec{k}_{\parallel};\overleftrightarrow{\mathbb{R}}) = \overleftrightarrow{\tau}^{(s)}(\vec{k}_{\parallel};\overleftrightarrow{\mathbb{R}}) \tag{219}$$

$$\overleftrightarrow{Ⓗ}^{(s)}(\vec{k}_{\parallel};\overleftrightarrow{\mathbb{A}}_0) = \overleftrightarrow{\beta} \tag{220}$$

$$\overleftrightarrow{Ⓗ}^{(s)}(\vec{k}_{\parallel};\overleftrightarrow{\mathbb{A}}) = \phi^*(\vec{k}_{\parallel};\overleftrightarrow{\mathbb{A}}\overleftrightarrow{\mathbb{A}}_0^{-1},\overleftrightarrow{\mathbb{A}}_0)\overleftrightarrow{\tau}^{(s)}(\vec{k}_{\parallel};\overleftrightarrow{\mathbb{A}}\overleftrightarrow{\mathbb{A}}_0^{-1})\overleftrightarrow{\beta} \tag{221}$$

<u>Second type</u>

$$\overleftrightarrow{Ⓗ}^{(s)}(\vec{k}_{\parallel};\overleftrightarrow{\mathbb{R}}) = \begin{pmatrix} \overleftrightarrow{\tau}^{(s)}(\vec{k}_{\parallel};\overleftrightarrow{\mathbb{R}}) & \overleftrightarrow{0} \\ \overleftrightarrow{0} & \overleftrightarrow{\tau}^{(s)}(\vec{k}_{\parallel};\overleftrightarrow{\mathbb{R}}) \end{pmatrix} \tag{222}$$

$$\overleftrightarrow{Ⓗ}^{(s)}(\vec{k}_{\parallel};\overleftrightarrow{\mathbb{A}}_0) = \begin{pmatrix} \overleftrightarrow{0} & \overleftrightarrow{\beta} \\ \overleftrightarrow{\beta} & \overleftrightarrow{0} \end{pmatrix} \tag{223}$$

$$\overleftrightarrow{Ⓗ}^{(s)}(\vec{k}_{\parallel};\overleftrightarrow{\mathbb{A}}) = \phi^*(\vec{k}_{\parallel};\overleftrightarrow{\mathbb{A}}\overleftrightarrow{\mathbb{A}}_0^{-1},\overleftrightarrow{\mathbb{A}}_0)\begin{pmatrix} \overleftrightarrow{0} & -\overleftrightarrow{\tau}^{(s)}(\vec{k}_{\parallel};\overleftrightarrow{\mathbb{A}}\overleftrightarrow{\mathbb{A}}_0^{-1})\overleftrightarrow{\beta} \\ \overleftrightarrow{\tau}^{(s)}(\vec{k}_{\parallel};\overleftrightarrow{\mathbb{A}}\overleftrightarrow{\mathbb{A}}_0^{-1})\overleftrightarrow{\beta} & \overleftrightarrow{0} \end{pmatrix} \tag{224}$$

Third type

$$\overleftrightarrow{\textcircled{H}}^{(s)}(\vec{k}_{\parallel};\overleftrightarrow{R}) = \begin{pmatrix} \overleftrightarrow{\tau}^{(s)}(\vec{k}_{\parallel};\overleftrightarrow{R}) & \overleftrightarrow{0} \\ \overleftrightarrow{0} & \overleftrightarrow{\bar{\tau}}^{(s)}(\vec{k}_{\parallel};\overleftrightarrow{R}) \end{pmatrix} \qquad (225)$$

$$\overleftrightarrow{\textcircled{H}}^{(s)}(\vec{k}_{\parallel};\overleftrightarrow{A}_{o}) = \begin{pmatrix} \overleftrightarrow{0} & \phi(\vec{k}_{\parallel};A_{o},A_{o})\overleftrightarrow{\tau}^{(s)}(\vec{k}_{\parallel};A_{o}^{2}) \\ \overleftrightarrow{1} & \overleftrightarrow{0} \end{pmatrix} \qquad (226)$$

$$\overleftrightarrow{\textcircled{H}}^{(s)}(\vec{k}_{\parallel};\overleftrightarrow{A}) = \begin{pmatrix} \overleftrightarrow{0} & \phi(\vec{k}_{\parallel};\overleftrightarrow{A},\overleftrightarrow{A}_{o})\overleftrightarrow{\tau}^{(s)}(\vec{k}_{\parallel};\overleftrightarrow{A}\overleftrightarrow{A}_{o}) \\ \phi^{*}(\vec{k}_{\parallel};\overleftrightarrow{A}_{o},\overleftrightarrow{A}_{o}^{-1}\overleftrightarrow{A})\overleftrightarrow{\tau}^{(s)}(\vec{k}_{\parallel};\overleftrightarrow{A}_{o}^{-1}A)^{*} & \overleftrightarrow{0} \end{pmatrix} \qquad (227)$$

SURFACE GREEN'S FUNCTION

The knowledge of the Green's function for a semi-infinite crystal or a crystal slab is central to the study of dynamical properties of atoms in crystal surfaces. Having this Green's function we can determine properties of an individual atom, such as its mean square amplitude or velocity, or of the surface region as a whole, such as its contribution to the specific heat of a crystal. With it we can also study the effects of its proximity to a free surface on the vibrational properties of an impurity atom, and the effects of an adsorbed layer of impurities on thermodynamic properties of a crystal.

The present section is therefore devoted to the determination of the Green's function for a crystal slab, for a surface treated as a defect, and for a semi-infinite continuum. Applications of these Green's functions to the calculation of correlation functions is also discussed.

Green's Function for a Crystal Slab

The time-independent equations of motion for a crystal slab in which the atomic positions are specified according to definition (A) can be written in matrix form according to

$$\sum_{\ell'\kappa'\beta} L_{\alpha\beta}(\ell\kappa;\ell'\kappa';\omega^2)u_\beta(\ell'\kappa') = 0 \tag{228}$$

where

$$L_{\alpha\beta}(\ell\kappa;\ell'\kappa';\omega^2) = M_\kappa\omega^2\delta_{\ell\ell'}\delta_{\kappa\kappa'}\delta_{\alpha\beta} - \Phi_{\alpha\beta}(\ell\kappa;\ell'\kappa') \tag{229}$$

The matrix $U_{\alpha\beta}(\ell\kappa;\ell'\kappa';\omega^2)$ inverse to the matrix $L_{\alpha\beta}(\ell\kappa;\ell'\kappa';\omega^2)$ is called the vibrational Green's function for the crystal slab

$$\sum_{\ell'\kappa'\beta} L_{\alpha\beta}(\ell\kappa;\ell'\kappa';\omega^2)\cdot U_{\beta\gamma}(\ell'\kappa';\ell''\kappa'';\omega^2)$$

$$= \sum_{\ell'\kappa'\beta} U_{\alpha\beta}(\ell\kappa;\ell'\kappa';\omega^2)L_{\beta\gamma}(\ell'\kappa';\ell''\kappa'';\omega^2) = \delta_{\ell\ell''}\delta_{\kappa\kappa''}\delta_{\alpha\gamma} \tag{230}$$

This Green's function can be expanded in terms of the eigenvectors $\{v_\alpha(\kappa|\vec{k}_\parallel j)\}$ of the dynamical matrix of the crystal slab $D_{\alpha\beta}(\kappa\kappa'|\vec{k}_\parallel)$, defined by Eq. (26). The result is

$$U_{\alpha\beta}(\ell\kappa;\ell'\kappa';\omega^2)$$

$$= \frac{1}{L^2(M_\kappa M_{\kappa'})^{\frac{1}{2}}} \sum_{\vec{k}_\parallel j} \frac{v_\alpha(\kappa|\vec{k}_\parallel j)v_\beta(\kappa'|\vec{k}_\parallel j)^*}{\omega^2 - \omega_j^2(\vec{k}_\parallel)} e^{i\vec{k}_\parallel \cdot (\vec{x}(\ell) - \vec{x}(\ell'))} \tag{231}$$

The calculation of $U_{\alpha\beta}(\ell\kappa;\ell'\kappa';\omega^2)$ from Eq. (231) requires the use of a high-speed computer for the generation of the eigenvectors $\{v_\alpha(\kappa|\vec{k}_\parallel j)\}$ and eigenfrequencies $\{\omega_j^2(\vec{k}_\parallel)\}$, and for the subsequent summations over $\vec{k}_\parallel$ and j. The number of layers in the slab being studied is limited by the memory and storage capacity of the computer. However, this method does provide a means for incorporating easily into calculations of $U_{\alpha\beta}(\ell\kappa;\ell'\kappa';\omega^2)$ effects of changes in force constants and of atomic masses in the surface layers of crystals such as occur in the presence of adsorbed layers, and of surface reconstruction and relaxation in interplanar spacings at crystal surfaces.

Let us note that the slab eigenfrequencies $\omega_j^2(\vec{k}_\parallel)$ in Eq. (27) are the poles of the Green's function $\overleftrightarrow{U}$, Eq. (231). So from the poles of the Green's function for a crystal slab, one can obtain all the excitations of a slab, the bulk modes which are also the poles of the bulk Green's function, and also the surface modes which appear as new poles in the Green's function $\overleftrightarrow{U}$ for a crystal slab.

Green's Function for a Semi-Infinite Crystal

The Surface as a Defect in a Crystal

An alternative to the preceding, purely numerical, approach to the determination of the Green's function for a semi-infinite crystal is provided by regarding the surface as a defect in an otherwise infinitely extended crystal.

For this purpose we use the results discussed on pp. 73-76 to write the time-independent equation for the Green's function in the form

$$\sum_{\ell'\beta}\left\{M\omega^2\delta_{\alpha\beta}\delta_{\ell\ell'} - \Phi^{(o)}_{\alpha\beta}(\ell\ell') + \Delta\Phi_{\alpha\beta}(\ell\ell')\right\} \times U_{\beta\gamma}(\ell'\ell'';\omega^2) = \delta_{\alpha\gamma}\delta_{\ell\ell''} \tag{232}$$

where, as before, we have assumed that we are dealing with a Bravais crystal, for simplicity. In terms of the Green's function $G_{\alpha\beta}(\ell\ell'';\omega^2)$ for the infinitely extended crystal, which is defined by Eq. (124), we can rewrite this equation in the form

$$U_{\alpha\beta}(\ell\ell';\omega^2) = G_{\alpha\beta}(\ell\ell';\omega^2) - \sum_{\ell''\gamma}\sum_{\ell'''\delta} G_{\alpha\gamma}(\ell\ell'';\omega^2)\Delta\Phi_{\gamma\delta}(\ell''\ell''') \times U_{\delta\beta}(\ell'''\ell';\omega^2) \tag{233}$$

We exploit the periodicity of the cut crystal in the directions parallel to the surface to analyze by Fourier series the Green's function $U_{\alpha\beta}(\ell\ell';\omega^2)$ in the coordinates parallel to the surface

$$U_{\alpha\beta}(\ell\ell';\omega^2) = \frac{1}{L^2}\sum_{\vec{k}_\parallel} U_{\alpha\beta}(\vec{k}_\parallel\omega\, \ell_3\ell_3')e^{i\vec{k}_\parallel \cdot (\vec{x}_\parallel(\ell) - \vec{x}_\parallel(\ell'))} \tag{234}$$

The Green's function $G_{\alpha\beta}(\ell\ell';\omega^2)$ can be expanded in a similar fashion, as we have already seen in Eq. (124)

$$G_{\alpha\beta}(\ell\ell';\omega^2) = \frac{1}{L^2}\sum_{\vec{k}_\parallel} G_{\alpha\beta}(\vec{k}_\parallel\omega|\ell_3\ell_3')e^{i\vec{k}_\parallel \cdot (\vec{x}_\parallel(\ell) - \vec{x}_\parallel(\ell'))} \tag{235}$$

When the expansions (234) and (235) are substituted into Eq. (233), the equation for the Fourier coefficients is

$$\begin{aligned} U_{\alpha\beta}(\vec{k}_\parallel\omega|\ell_3\ell_3') &= G_{\alpha\beta}(\vec{k}_\parallel\omega|\ell_3\ell_3') \\ &- \sum_{\ell_3''\gamma}\sum_{\ell_3'''\delta} G_{\alpha\gamma}(\vec{k}_\parallel\omega|\ell_3\ell_3'') \\ &\cdot \Delta\Phi_{\gamma\delta}(\vec{k}_\parallel|\ell_3''\ell_3''')U_{\delta\beta}(\vec{k}_\parallel\omega|\ell_3'''\ell_3') \end{aligned} \tag{240}$$

where the Fourier Coefficient $\Delta\Phi_{\alpha\beta}(\vec{k}_\parallel|\ell_3\ell_3')$ has been defined in Eq. (123b).

The solution of Eq. (240) can be carried out in at least three different ways, depending on the particular problem being considered. We sketch these methods of solution in the remainder of this section.

If the Green's function $U_{\alpha\beta}(\vec{k}_{\parallel}\omega|\ell_3\ell_3')$ is required only in the space of the surface, as previously defined, we can use the localized nature of the matrix $\Delta\Phi_{\alpha\beta}(\vec{k}_{\parallel}|\ell_3\ell_3')$ to advantage. We partition the matrix $U_{\alpha\beta}(\vec{k}_{\parallel}\omega|\ell_3\ell_3')$ in the same way as we partitioned the matrices $G_{\alpha\beta}(\vec{k}_{\parallel}\omega|\ell_3\ell_3')$ and $\Delta\Phi_{\alpha\beta}(\vec{k}_{\parallel}|\ell_3\ell_3')$

$$\overleftrightarrow{U}(\vec{k}_{\parallel}\omega) = \left(\begin{array}{c|c} \overleftrightarrow{u}(\vec{k}_{\parallel}\omega) & \overleftrightarrow{U}_{12}(\vec{k}_{\parallel}\omega) \\ \hline \overleftrightarrow{U}_{21}(\vec{k}_{\parallel}\omega) & \overleftrightarrow{U}_{22}(\vec{k}_{\parallel}\omega) \end{array}\right) \tag{241}$$

where all the elements of the matrix $\overleftrightarrow{u}(\vec{k}_{\parallel}\omega)$ are in the space of the surface. When this partitioned form of $\overleftrightarrow{U}(\vec{k}_{\parallel}\omega)$ is substituted into Eq. (240), together with Eqs. (128) and (129), we obtain the following set of four equations

$$\overleftrightarrow{u}(\vec{k}_{\parallel}\omega) = \overleftrightarrow{g}(\vec{k}_{\parallel}\omega) - \overleftrightarrow{g}(\vec{k}_{\parallel}\omega)\Delta\overleftrightarrow{\phi}(\vec{k}_{\parallel})\overleftrightarrow{u}(\vec{k}_{\parallel}\omega) \tag{242a}$$

$$\overleftrightarrow{U}_{12}(\vec{k}_{\parallel}\omega) = \overleftrightarrow{G}_{12}(\vec{k}_{\parallel}\omega) - \overleftrightarrow{g}(\vec{k}_{\parallel}\omega)\Delta\overleftrightarrow{\phi}(\vec{k}_{\parallel})\overleftrightarrow{U}_{12}(\vec{k}_{\parallel}\omega) \tag{242b}$$

$$\overleftrightarrow{U}_{21}(\vec{k}_{\parallel}\omega) = \overleftrightarrow{G}_{21}(\vec{k}_{\parallel}\omega) - \overleftrightarrow{G}_{21}(\vec{k}_{\parallel}\omega)\Delta\overleftrightarrow{\phi}(\vec{k}_{\parallel})\overleftrightarrow{u}(\vec{k}_{\parallel}\omega) \tag{242c}$$

$$\overleftrightarrow{U}_{22}(\vec{k}_{\parallel}\omega) = \overleftrightarrow{G}_{22}(\vec{k}_{\parallel}\omega) - \overleftrightarrow{G}_{21}(\vec{k}_{\parallel}\omega)\Delta\overleftrightarrow{\phi}(\vec{k}_{\parallel})\overleftrightarrow{U}_{12}(\vec{k}_{\parallel}\omega) \tag{242d}$$

Thus $\overleftrightarrow{u}(\vec{k}_{\parallel}\omega)$ can be obtained by inverting a matrix defined only in the space of the surface

$$\overleftrightarrow{u}(\vec{k}_{\parallel}\omega) = [\overleftrightarrow{I} + \overleftrightarrow{g}(\vec{k}_{\parallel}\omega)\overleftrightarrow{\Delta\phi}(\vec{k}_{\parallel})]^{-1}\overleftrightarrow{g}(\vec{k}_{\parallel}\omega) \qquad (243)$$

If $\vec{k}_{\parallel}$ is a point of high symmetry in the two-dimensional first Brillouin zone, the inversion can be simplied, because the matrix $\overleftrightarrow{g}(\vec{k}_{\parallel}\omega)\overleftrightarrow{\Delta\phi}(\vec{k}_{\parallel})$ can be block diagonalized in general by group theory methods.

We next note that the solution of Eq. (240) can be written formally as

$$U_{\alpha\beta}(\vec{k}_{\parallel}\omega|\ell_3\ell_3') = G_{\alpha\beta}(\vec{k}_{\parallel}\omega|\ell_3\ell_3') - \sum_{\ell_3''\gamma}\sum_{\ell_3'''\delta} G_{\alpha\gamma}(\vec{k}_{\parallel}\omega|\ell_3\ell_3'') \cdot T_{\gamma\delta}(\vec{k}_{\parallel}\omega|\ell_3''\ell_3''')G_{\delta\beta}(\vec{k}_{\parallel}\omega|\ell_3'''\ell_3') \qquad (244)$$

where the matrix $T_{\alpha\beta}(\vec{k}_{\parallel}\omega|\ell_3\ell_3')$ called the <u>scattering matrix</u>, is the solution of the equation

$$T_{\alpha\beta}(\vec{k}_{\parallel}\omega|\ell_3\ell_3') = \Delta\Phi_{\alpha\beta}(\vec{k}_{\parallel}|\ell_3\ell_3') - \sum_{\ell_3''\gamma}\sum_{\ell_3'''\beta} \Delta\Phi_{\alpha\gamma}(\vec{k}_{\parallel}|\ell_3\ell_3'') \cdot G_{\gamma\delta}(\vec{k}_{\parallel}\omega|\ell_3''\ell_3''')T_{\delta\beta}(\vec{k}_{\parallel}\omega|\ell_3'''\ell_3') \qquad (245)$$

The scattering matrix is of interest in its own right,

as we shall see in a later section, and in what follows we will focus our attention on it. This is not restrictive in any way, since knowledge of it is tantamount to knowledge of the Green's function $U_{\alpha\beta}(\vec{k}_{\parallel}\omega|\ell_3\ell_3')$, according to Eq. (244).

The matrix $T_{\alpha\beta}(\vec{k}_{\parallel}\omega|\ell_3\ell_3')$ is localized in the space of the surface, i.e., it can be partitioned in the form

$$\overleftrightarrow{T}(\vec{k}_{\parallel}\omega) = \left(\begin{array}{c|c} \overleftrightarrow{t}(\vec{k}_{\parallel}\omega) & \overleftrightarrow{0} \\ \hline \overleftrightarrow{0} & \overleftrightarrow{0} \end{array}\right) \tag{246}$$

It can therefore be written formally as

$$\begin{aligned} \overleftrightarrow{t}(\vec{k}_{\parallel}\omega) &= [\overleftrightarrow{I} + \Delta\overleftrightarrow{\phi}(\vec{k}_{\parallel})\overleftrightarrow{g}(\vec{k}_{\parallel}\omega)]^{-1}\Delta\overleftrightarrow{\phi}(\vec{k}_{\parallel}) \\ &= \Delta\overleftrightarrow{\phi}(\vec{k}_{\parallel})[\overleftrightarrow{I} + \overleftrightarrow{g}(\vec{k}_{\parallel}\omega)\Delta\overleftrightarrow{\phi}(\vec{k}_{\parallel})]^{-1} \end{aligned} \tag{247}$$

This result, substituted into Eq. (244), yields $U_{\alpha\beta}(\vec{k}_{\parallel}\omega|\ell_3\ell_3')$ for all $(\ell_3\alpha)$, $(\ell_3'\beta)$.

For some simple crystal models the matrix $\Delta\Phi_{\alpha\beta}(\vec{k}_{\parallel}|\ell_3\ell_3')$ can be written in separable form

$$\Delta\Phi_{\alpha\beta}(\vec{k}_{\parallel}|\ell_3\ell_3') = \sum_n f_\alpha^{(n)}(\vec{k}_{\parallel}|\ell_3)\, f_\beta^{(n)}(\vec{k}_{\parallel}|\ell_3')^* \tag{248}$$

where the hermiticity of $\Delta\Phi_{\alpha\beta}(\vec{k}_{\parallel}|\ell_3\ell_3')$

$$\Delta\Phi_{\alpha\beta}(\vec{k}_{\parallel}|\ell_3\ell_3') = \Delta\Phi_{\beta\alpha}(\vec{k}_{\parallel}|\ell_3'\ell_3)^* \tag{249}$$

has been taken into account explicitly. In such a case Eq. (245) can be solved readily to yield

$$T_{\alpha\beta}(\vec{k}_{\parallel}\omega|\ell_3\ell_3') = \sum_{mn} f_{\alpha}^{(m)}(\vec{k}_{\parallel}|\ell_3)[\overleftrightarrow{I} + \overleftrightarrow{G}(\vec{k}_{\parallel}\omega)]^{-1}_{mn} \cdot f_{\beta}^{(n)}(\vec{k}_{\parallel}|\ell_3')^{*} \tag{250}$$

where the elements of the matrix $\overleftrightarrow{G}(\vec{k}_{\parallel}\omega)$ are given by

$$G_{mn}(\vec{k}_{\parallel}\omega) = \sum_{\ell_3\alpha}\sum_{\ell_3'\beta} f_{\alpha}^{(m)}(\vec{k}_{\parallel}|\ell_3)^{*} G_{\alpha\beta}(\vec{k}_{\parallel}\omega|\ell_3\ell_3') \cdot f_{\beta}^{(n)}(\vec{k}_{\parallel}|\ell_3') \tag{251}$$

If the sum over n in Eq. (250) consists of only a few terms, this is a very effective way of obtaining $T_{\alpha\beta}(\vec{k}_{\parallel}\omega|\ell_3\ell_3')$.

From the preceding analysis we see that the poles of the Green's function $U_{\alpha\beta}(\vec{k}_{\parallel}\omega|\ell_3\ell_3')$ that do not coincide with those of $G_{\alpha\beta}(\vec{k}_{\parallel}\omega|\ell_3\ell_3')$, and hence give the frequencies of the normal modes perturbed by the introduction of the pair of surfaces through the perturbation $\Delta\Phi_{\alpha\beta}(\vec{k}_{\parallel}|\ell_3\ell_3')$, are given by the poles of the matrix $T_{\alpha\beta}(\vec{k}_{\parallel}\omega|\ell_3\ell_3')$. These are seen to be given by the zeros of the determinantal equations

$$|\overleftrightarrow{I} + \overleftrightarrow{g}(\vec{k}_{\parallel}\omega)\Delta\overleftrightarrow{\phi}(\vec{k}_{\parallel})| = 0 \tag{252a}$$

or

$$|\overleftrightarrow{I} + \overleftrightarrow{G}(\vec{k}_{\parallel}\omega)| = 0 \tag{252b}$$

depending on whether the form for $T_{\alpha\beta}(\vec{k}_{\parallel}\omega|\ell_3\ell_3')$ given by Eq. (247) or (250) is used.

The final method of obtaining the Green's function $U_{\alpha\beta}(\ell\ell';\omega^2)$ proceeds directly from Eq. (244), which can be rewritten in the form

$$U_{\alpha\beta}(\ell\ell';\omega^2) = G_{\alpha\beta}(\ell\ell';\omega^2) - \sum_{\ell''\gamma}\sum_{\ell'''\delta} G_{\alpha\gamma}(\ell\ell'';\omega^2)T_{\gamma\delta}(\ell''\ell''';\omega^2) \cdot G_{\delta\beta}(\ell'''\ell';\omega^2) \tag{253}$$

where

$$T_{\alpha\beta}(\ell\ell';\omega^2) = \Delta\Phi_{\alpha\beta}(\ell\ell') - \sum_{\ell''\gamma}\sum_{\ell'''\delta} \Delta\Phi_{\alpha\gamma}(\ell\ell'')G_{\gamma\delta}(\ell''\ell''';\omega^2) \cdot T_{\delta\beta}(\ell'''\ell';\omega^2) \tag{254}$$

To solve Eq. (253) we utilize expansions in the eigenvectors of the infinitely extended crystal. These are defined as follows. The equations of motion of an infinite Bravais crystal are

$$M\ddot{u}_{\alpha}(\ell) = -\sum_{\ell'\beta}\Phi^{(o)}_{\alpha\beta}(\ell\ell')u_{\beta}(\ell') \tag{255}$$

where $\ell = (\ell_1, \ell_2, \ell_3)$ are the three integers specifying an atomic position according to

$$\vec{x}(\ell) = \ell_1\vec{a}_1 + \ell_2\vec{a}_2 + \ell_3\vec{a}_3 \tag{256}$$

where the primitive translation vectors $\vec{a}_1$ and $\vec{a}_2$ lie in the x_1x_2-plane. The solutions of Eqs. (255) can be written in the form

$$u_\alpha(\ell;t) = \frac{e_\alpha(\vec{k}j)}{M^{\frac{1}{2}}} e^{i\vec{k}\cdot\vec{x}(\ell) - i\omega_j(\vec{k})t} \tag{257}$$

where $\vec{k}$ is a three-dimensional wave vector, whose $N = L^3$ allowed values are uniformly and densely distributed throughout the three-dimensional first Brillouin zone of the infinite crystal. The actual values of $\vec{k}$ are obtained on the assumption that the atomic displacements are periodic with the periodicity of a macrocrystal whose edges are defined by $L\vec{a}_1$, $L\vec{a}_2$, and $L\vec{a}_3$. The functions $\omega_j^2(k)$ and $e_\alpha(\vec{k}j)$ are the eigenvalues and corresponding unit eigenvectors of the dynamical matrix of the crystal

$$\sum_\beta D_{\alpha\beta}(\vec{k})e_\beta(\vec{k}j) = \omega_j^2(\vec{k})e_\alpha(\vec{k}j) \tag{258}$$

where $j = 1,2,3$ and

$$D_{\alpha\beta}(\vec{k}) = \frac{1}{M}\sum_{\ell'} \Phi_{\alpha\beta}^{(o)}(\ell\ell')e^{-i\vec{k}\cdot(\vec{x}(\ell) - \vec{x}(\ell'))} = D_{\beta\alpha}(\vec{k})^* \tag{259}$$

The eigenvectors $\{e_\alpha(\vec{k}j)\}$ are orthonormal and complete according to

$$\sum_{\alpha} e_{\alpha}(\vec{k}j) e_{\alpha}(\vec{k}j')^{*} = \delta_{jj'} \tag{260a}$$

$$\sum_{j} e_{\alpha}(\vec{k}j) e_{\beta}(\vec{k}j)^{*} = \delta_{\alpha\beta} \tag{260b}$$

The Green's function $G_{\alpha\beta}(\ell\ell';\omega^2)$ can be expanded in terms of the eigenfunctions $\{e_{\alpha}(\vec{k}j)\}$ according to

$$G_{\alpha\beta}(\ell\ell';\omega^2) = \frac{1}{NM} \sum_{\vec{k}j} \frac{e_{\alpha}(\vec{k}j) e_{\beta}(\vec{k}j)^{*}}{\omega^2 - \omega_j^2(\vec{k})} e^{i\vec{k} \cdot [\vec{x}(\ell) - \vec{x}(\ell')]} \tag{261}$$

We seek expansions of $U_{\alpha\beta}(\ell\ell';\omega^2)$ and $T_{\alpha\beta}(\ell\ell';\omega^2)$ in the form of double series

$$U_{\alpha\beta}(\ell\ell';\omega^2) = \frac{1}{NM} \sum_{\vec{k}j} \sum_{\vec{k}'j'} e_{\alpha}(\vec{k}j) u(\vec{k}j;\vec{k}'j';\omega^2) e_{\beta}(\vec{k}'j') \cdot e^{i\vec{k} \cdot \vec{x}(\ell) + i\vec{k}' \cdot \vec{x}(\ell')} \tag{262}$$

$$T_{\alpha\beta}(\ell\ell';\omega^2) = \frac{M}{N} \sum_{\vec{k}j} \sum_{\vec{k}'j'} e_{\alpha}(\vec{k}j) t(\vec{k}j;\vec{k}'j';\omega^2) e_{\beta}(\vec{k}'j') \cdot e^{i\vec{k} \cdot \vec{x}(\ell) + i\vec{k}' \cdot \vec{x}(\ell')} \tag{263}$$

since the former cannot depend on $\vec{x}(\ell)$ and $\vec{x}(\ell')$ only through their difference, in the presence of a surface. Substitution of Eqs. (262) and (263) into Eq. (253) yields the following relation among the Fourier coefficients

$$u(\vec{k}j;\vec{k}'j';\omega^2) = \frac{\Delta(\vec{k} + \vec{k}')\delta_{jj'}}{\omega^2 - \omega_j^2(\vec{k})} - \frac{1}{\omega^2 - \omega_j^2(\vec{k})}$$

$$\cdot\ t(\vec{k}j;\vec{k}'j';\omega^2)\frac{1}{\omega^2 - \omega_{j'}^2(\vec{k}')} \qquad (264)$$

The equation satisfied by the Fourier coefficient $t(\vec{k}j;\vec{k}'j';\omega^2)$ is

$$t(\vec{k}j;\vec{k}'j';\omega^2) = V(\vec{k}j;\vec{k}'j')$$

$$- \sum_{\vec{k}_1 j_1} \frac{V(\vec{k}j;-\vec{k}_1 j_1)}{\omega^2 - \omega_{j_1}^2(\vec{k}_1)}\, t(\vec{k}_1 j_1;\vec{k}'j';\omega^2) \qquad (265a)$$

where

$$V(\vec{k}j;\vec{k}'j') = \frac{1}{NM}\sum_{\ell\alpha}\sum_{\ell'\beta} e_\alpha(\vec{k}j)^*\Delta\Phi_{\alpha\beta}(\ell\ell')e_\beta(\vec{k}'j')^*$$

$$\cdot\ e^{-i\vec{k}\cdot\vec{x}(\ell) - i\vec{k}'\cdot\vec{x}(\ell')} \qquad (265b)$$

Because $\Delta\Phi_{\alpha\beta}(\ell\ell')$ depends on $\ell_1, \ell_2, \ell_1', \ell_2'$ only through the differences $\ell_1 - \ell_1'$ and $\ell_2 - \ell_2'$, we can rewrite the expansion for $V(\vec{k}j;\vec{k}'j')$ in the form

$$V(\vec{k}j;\vec{k}'j') = \frac{\Delta(\vec{k}_\parallel + \vec{k}_\parallel')}{LM}\sum_{\ell_1\ell_2}\sum_{\ell_3\alpha}\sum_{\ell_3'\beta} e_\alpha(\vec{k}j)^* \qquad (266a)$$

$$
\left.
\begin{aligned}
&\cdot\ \Delta\Phi_{\alpha\beta}(\ell\ell')e_{\beta}(\vec{k}'j')^{*} \\
&\cdot\ e^{-i\vec{k}_{\parallel}\cdot\left(\vec{x}_{\parallel}(\ell)-\vec{x}_{\parallel}(\ell')\right)} \\
&\cdot\ e^{-\ell_3\vec{k}\cdot\vec{a}_3 - i\ell_3'\vec{k}'\cdot\vec{a}_3}
\end{aligned}
\right\} \qquad (266a)
$$

$$
\begin{aligned}
&= \frac{\Delta(\vec{k}_{\parallel}+\vec{k}_{\parallel}')}{LM}\sum_{\ell_3\alpha}\sum_{\ell_3'\beta} e_{\alpha}(\vec{k}j)^{*} \\
&\cdot\ \Delta\Phi_{\alpha\beta}\left(\vec{k}_{\parallel}|\ell_3\ell_3'\right)e_{\beta}(\vec{k}'j')^{*} \\
&\cdot\ e^{-i\ell_3\vec{k}\cdot\vec{a}_3 - i\ell_3'\vec{k}'\cdot\vec{a}_3}
\end{aligned}
\qquad (266b)
$$

where the Fourier coefficient $\Delta\Phi_{\alpha\beta}\left(\vec{k}_{\parallel}|\ell_3\ell_3'\right)$ has been defined in Eq. (123b). The function $\Delta(\vec{k}_{\parallel}+\vec{k}_{\parallel}')$ equals unity if $\vec{k}_{\parallel}+\vec{k}_{\parallel}'$ equals a two-dimensional reciprocal lattice vector, but vanishes otherwise. Its presence is a reflection of the periodicity of the cut crystal in directions parallel to the surfaces.

We notice that the right-hand side of Eq. (266) has the form

$$
V(\vec{k}j;\vec{k}'j') = \frac{\Delta(\vec{k}_{\parallel}+\vec{k}_{\parallel}')}{L}\sum_{nn'} C_{nn'}(\vec{k}_{\parallel})V_n(\vec{k}j)V_{n'}(\vec{k}'j') \qquad (267)
$$

where n stands for the pair of indices (ℓ_3,α), and $C_{nn'}(\vec{k}_{\parallel})$ is a real, symmetric matrix. We now introduce a symmetric matrix $d_{nn'} = (\overleftrightarrow{C}^{\frac{1}{2}})_{nn'}$, so that

$$C_{nn'} = \sum_{n_1} d_{nn_1} d_{n_1 n'} \tag{268}$$

and the function

$$u_n(\vec{k}j) = \sum_{n_1} d_{nn_1} V_{n_1}(\vec{k}j) \tag{269}$$

The function $V(\vec{k}j;\vec{k}'j')$ can now be written as

$$V(\vec{k}j;\vec{k}'j') = \frac{\Delta(\vec{k}_\parallel + \vec{k}_\parallel')}{L} \sum_n u_n(\vec{k}j) u_n(\vec{k}'j') \tag{270}$$

The solution of Eq. (265) is now readily found, and is given by

$$t(\vec{k}j;\vec{k}'j';\omega^2) = \frac{\Delta(\vec{k}_\parallel + \vec{k}_\parallel')}{L} \sum_{mn} u_m(\vec{k}j) \cdot [\overleftrightarrow{I} + \overleftrightarrow{M}(\vec{k}_\parallel\omega)]^{-1}_{mn} u_n(\vec{k}'j') \tag{271}$$

where the elements of the matrix $\overleftrightarrow{M}(\vec{k}_\parallel\omega)$ are given by

$$M_{mn}(\vec{k}_\parallel\omega) = \frac{1}{L} \sum_{\vec{k}_1 j_1} \Delta\left(\vec{k}_\parallel - \vec{k}_\parallel^{(1)}\right) \frac{u_m\left(-\vec{k}_1 j_1\right) u_n\left(\vec{k}_1 j_1\right)}{\omega^2 - \omega^2_{j_1}\left(\vec{k}_1\right)} \tag{272}$$

The procedure just outlined for obtaining $t(\vec{k}j;\vec{k}'j';\omega^2)$ is practicable only for rather simple models of monatomic crystals. We will see an application of it later in this chapter.

Applications of Green's Functions

The usefulness of the Green's function $U_{\alpha\beta}(\ell\kappa;\ell'\kappa';\omega^2)$ stems from the fact that many dynamical properties of a crystal slab can be expressed in terms of it.

In particular, for the interpretation of a large number of experiments which probe the dynamical properties of crystals containing defects, it turns out to be very useful to evaluate time-dependent correlation functions of the type $\langle u_\alpha(\ell\kappa;t)u_\beta(\ell'\kappa';0)\rangle$, $\langle u_\alpha(\ell\kappa;t)p_\beta(\ell'\kappa';0)\rangle$, and $\langle p_\alpha(\ell\kappa;t)p_\beta(\ell'\kappa';0)\rangle$, Eq. (61). The physical interpretation of these correlation functions is simple only in the classical limit when they become real. For example, the function $\langle u_\alpha(\ell\kappa;t)u_\beta(\ell'\kappa';0)\rangle$ is a measure of the degree to which the α component of the displacement of the atom $(\ell\kappa)$ at time t depends on the fact that the atom $(\ell'\kappa')$ undergoes a displacement in the β direction at time $t = 0$. If there is no correlation between these displacements, the average of the product of these displacements reduces to the product of their averages. In the harmonic approximation, the average $\langle u_\alpha(\ell\kappa;t)\rangle$ vanishes, and the absence of correlation between the displacements $u_\alpha(\ell\kappa;t)$ and $u_\beta(\ell'\kappa';0)$ leads to the vanishing of the correlation function $\langle u_\alpha(\ell\kappa;t)u_\beta(\ell'\kappa';0)\rangle$.

In the quantum limit these correlation functions are complex. Their imaginary parts reflect the noncommutativity of the operators appearing in them.

The expressions for these correlation functions [Eq. (61)] are not very useful for the solution of physical problems, because their evaluation requires a knowledge of the exact eigenvectors and eigenvalues of the crystal perturbed by an impurity.

It turns out to be more useful to find first the Fourier transform of these correlation functions with respect to time [Eq. (67)], and to use the inverse Fourier transformation to find the correlation functions themselves.

Using now the expression (231) for the Green's function of a crystal slab, the expressions (67) are found to be

$$\int_{-\infty}^{+\infty} dt\ e^{i\omega t} \langle u_{\alpha}(\ell\kappa;t)u_{\beta}(\ell'\kappa';0)\rangle = 2\hbar\ \mathrm{sgn}\ \omega[n(\omega) + 1] \cdot \mathrm{Im}U_{\alpha\beta}(\ell\kappa;\ell'\kappa';\omega^2 - i\varepsilon) \tag{273a}$$

$$\int_{-\infty}^{+\infty} dt\ e^{i\omega t} \langle p_{\alpha}(\ell\kappa;t)p_{\beta}(\ell'\kappa';0)\rangle = 2\hbar\ M_{\kappa}M_{\kappa'}\ \omega^2 \cdot \mathrm{sgn}\ \omega[m(\omega) + 1]\mathrm{Im}\ U_{\alpha\beta}(\ell\kappa;\ell'\kappa';\omega^2 - i\varepsilon) \tag{273b}$$

$$\int_{-\infty}^{+\infty} dt\ e^{i\omega t} \langle u_{\alpha}(\ell\kappa;t)p_{\beta}(\ell'\kappa';0)\rangle = 2i\hbar\ M_{\kappa'}\omega\mathrm{sgn}\ \omega[n(\omega) + 1]\mathrm{Im}\ U_{\alpha\beta}(\ell\kappa;\ell'\kappa';\omega^2 - i\varepsilon) \tag{273c}$$

Carrying out the Fourier inversion, we obtain

$$\langle u_\alpha(\ell\kappa;t)u_\beta(\ell'\kappa';0)\rangle = \frac{\hbar}{\pi}\int_{-\infty}^{+\infty} d\omega\, e^{-i\omega t}\,\frac{\operatorname{sgn}\omega}{1-e^{-\beta\hbar\omega}}$$

$$\cdot\ \operatorname{Im} U_{\alpha\beta}(\ell\kappa;\ell'\kappa';\omega^2 - i\varepsilon) \quad (274a)$$

$$\langle p_\alpha(\ell\kappa;t)p_\beta(\ell'\kappa';0)\rangle = \frac{\hbar}{\pi}M_\kappa M_{\kappa'}\int_{-\infty}^{+\infty} d\omega\, e^{-i\omega t}\,\frac{\omega^2\operatorname{sgn}\omega}{1-e^{-\beta\hbar\omega}}$$

$$\cdot\ \operatorname{Im} U_{\alpha\beta}(\ell\kappa;\ell'\kappa';\omega^2 - i\varepsilon) \quad (274b)$$

$$\langle u_\alpha(\ell\kappa;t)p_\beta(\ell'\kappa';0)\rangle = \frac{i\hbar}{\pi}M_{\kappa'}\int_{-\infty}^{+\infty} d\omega\, e^{-i\omega t}\,\frac{\omega\operatorname{sgn}\omega}{1-e^{-\beta\hbar\omega}}$$

$$\cdot\ \operatorname{Im} U_{\alpha\beta}(\ell\kappa;\ell'\kappa';\omega^2 - i\varepsilon) \quad (274c)$$

Setting t = 0 in these expressions, we obtain the equal time-correlation functions

$$\langle u_\alpha(\ell\kappa)u_\beta(\ell'\kappa')\rangle = \frac{\hbar}{\pi}\int_0^{\infty} d\omega\, \coth\tfrac{1}{2}\beta\hbar\omega$$

$$\cdot\ \operatorname{Im} U_{\alpha\beta}(\ell\kappa;\ell'\kappa';\omega^2 - i\varepsilon) \quad (275a)$$

$$\langle p_\alpha(\ell\kappa)p_\beta(\ell'\kappa')\rangle = \frac{\hbar}{\pi}M_\kappa M_{\kappa'}\int_0^{\infty} d\omega\, \omega^2\coth\tfrac{1}{2}\beta\hbar\omega$$

$$\cdot\ \operatorname{Im} U_{\alpha\beta}(\ell\kappa;\ell'\kappa';\omega^2 - i\varepsilon) \quad (275b)$$

$$\langle u_\alpha(\ell\kappa)p_\beta(\ell'\kappa')\rangle$$

$$= \frac{i\hbar}{\pi} M_{\kappa'} \int_0^\infty d\omega\ \omega\ \mathrm{Im}\ U_{\alpha\beta}(\ell\kappa;\ell'\kappa';\omega^2 - i\varepsilon)$$

$$= \frac{i\hbar}{2} \delta_{\ell\ell'}\delta_{\kappa\kappa'}\delta_{\alpha\beta} \tag{275c}$$

In writing Eq. (275c) we have used the explicit Eq. (231) for $U_{\alpha\beta}(\ell\kappa;\ell'\kappa';\omega^2 - i\varepsilon)$.

Alternate expressions for the equal time-correlation functions can be obtained from Eqs. (64) with the use of Eq. (231) for the Green's function $\overleftrightarrow{U}$

$$\langle u_\alpha(\ell\kappa)u_\beta(\ell'\kappa')\rangle = -k_BT \sum_{n=-\infty}^{\infty} U_{\alpha\beta}\left(\ell\kappa;\ell'\kappa';-\Omega_n^2\right) \tag{276a}$$

$$\langle p_\alpha(\ell\kappa)p_\beta(\ell'\kappa')\rangle = M_\kappa M_{\kappa'} k_BT \sum_{n=-\infty}^{\infty}$$

$$\cdot \left\{\frac{\delta_{\ell\ell'}\delta_{\kappa\kappa'}\delta_{\alpha\beta}}{M_\kappa} + \Omega_n^2 U_{\alpha\beta}\left(\ell\kappa;\ell'\kappa';-\Omega_n^2\right)\right\} \tag{276b}$$

where $\Omega_n = 2\pi n\ k_BT/\hbar$.

The convenience of these expressions for the evaluation of the equal time-correlation functions lies in the fact that they bypass the necessity of performing the analytic continuation of $U_{\alpha\beta}(\ell\kappa;\ell'\kappa';\omega^2)$ to complex ω^2

required for the evaluation of Eqs. (275). This analytic continuation is not always easily carried out.

In this fashion the evaluation of the given correlation functions in all cases reduces to the problem of determining elements of the Green's function, $U_{\alpha\beta}(\ell\kappa;\ell'\kappa';\omega^2)$, which can be done by any of the methods described in this section.

Continuum Green's Functions

The Green's functions for a crystal slab or for a semi-infinite crystal discussed in the preceding subsections have all been obtained on the basis of lattice theory. However, as we will see explicitly later in this chapter, certain kinds of problems in the dynamical theory of semi-infinite solids can be treated effectively by the use of dynamical Green's functions obtained on the basis of elasticity theory. In this subsection we introduce these Green's functions and show how they can be obtained explicitly and analytically in some simple cases.

We assume that the semi-infinite elastic medium occupies the region $x_3 \geq 0$ and is bounded by a planar, stress-free surface at the plane $x_3 = 0$.

The equations of motion of the medium are

$$\rho\ddot{u}_\alpha(\vec{x},t) = \sum_\beta \frac{\partial}{\partial x_\beta} T_{\alpha\beta}(\vec{x},t), \quad \alpha = 1,2,3 \tag{277}$$

where ρ is the mass density of the medium and $T_{\alpha\beta}(\vec{x},t)$ is the stress tensor

$$T_{\alpha\beta}(\vec{x},t) = \sum_{\mu\nu} C_{\alpha\beta\mu\nu} \frac{\partial}{\partial x_\nu} u_\mu(\vec{x},t) \tag{278}$$

The $\{C_{\alpha\beta\mu\nu}\}$ are the elastic moduli of the medium, written as the components of a fourth-rank tensor, and $u_\alpha(\vec{x},t)$ is the α Cartesian component of the displacement field at position $\vec{x}$ and time t.

Equations (277) and (278) must be supplemented by boundary conditions. The condition that the surface $x_3 = 0$ be stress-free is expressed by the equations

$$\sum_{\mu\nu} C_{\alpha 3\mu\nu} \frac{\partial}{\partial x_\nu} u_\mu(\vec{x},t)\Big|_{x_3 = 0} = 0, \quad \alpha = 1,2,3 \tag{279}$$

In addition, we require that $u_\alpha(\vec{x},t)$ satisfy outgoing or exponentially decaying wave conditions at $x_3 = +\infty$.

We now make the substitution $u_\alpha(\vec{x},t) = \exp(-i\omega t) \cdot v_\alpha(\vec{x})/\rho^{\frac{1}{2}}$, and find that the time-independent amplitudes $\{v_\alpha(\vec{x})\}$ obey the set of differential equations

$$\omega^2 v_\alpha(\vec{x}) + \frac{1}{\rho} \sum_{\beta\mu\nu} C_{\alpha\beta\mu\nu} \frac{\partial^2}{\partial x_\beta \partial x_\nu} v_\mu(\vec{x}) = 0, \qquad x_3 > 0, \ \alpha = 1,2,3 \tag{280}$$

subject to the boundary conditions

$$\sum_{\mu\nu} C_{\alpha 3\mu\nu} \frac{\partial}{\partial x_\nu} v_\mu(\vec{x})\Big|_{x_3 = 0} = 0, \quad \alpha = 1,2,3 \tag{281}$$

at the plane $x_3 = 0$, and outgoing or exponentially decaying wave conditions at $x_3 = +\infty$.

There is, in general, an infinity of solutions to the set of differential equations (280) and boundary conditions (281) and we label them by the index s = 1,2,3,... These equations and boundary conditions can therefore be rewritten so as to display this fact explicitly

$$-\frac{1}{\rho}\sum_{\beta\mu\nu} C_{\alpha\beta\mu\nu}\frac{\partial^2}{\partial x_\beta \partial x_\nu} v_\mu^{(s)}(\vec{x}) = \omega_s^2 v_\alpha^{(s)}(\vec{x}),$$

$$x_3 > 0, \quad \alpha = 1,2,3 \qquad (282)$$

$$\sum_{\mu\nu} C_{\alpha 3\mu\nu}\frac{\partial}{\partial x_\nu} v_\mu^{(s)}(\vec{x})\Big|_{x_3 = 0} = 0, \quad \alpha = 1,2,3 \qquad (283)$$

The partial differential operator appearing on the left-hand side on Eq. (282), supplemented by the boundary conditions at the plane $x_3 = 0$ and at infinity can be shown to be Hermitian. For if $f_\alpha^{(s)}(\vec{x})$ and $g_\beta^{(s')}(\vec{x})$ are two different functions satisfying the boundary conditions (283) we have that

$$-\frac{1}{\rho}\sum_{\alpha\beta\mu\nu} C_{\alpha\beta\mu\nu}\int d^3x\ f_\alpha^{(s)}(\vec{x})^* \frac{\partial^2}{\partial x_\beta \partial x_\nu} g_\mu^{(s')}(\vec{x})$$

$$= -\frac{1}{\rho}\sum_{\alpha\beta\mu\nu} C_{\alpha\beta\mu\nu}\int d^3x\ \frac{\partial}{\partial x_\beta}\left\{f_\alpha^{(s)}(\vec{x})^* \frac{\partial}{\partial x_\nu} g_\mu^{(s')}(\vec{x})\right\}$$

$$+\frac{1}{\rho}\sum_{\alpha\beta\mu\nu} C_{\alpha\beta\mu\nu}\int d^3x\ \frac{\partial f_\alpha^{(s)}(\vec{x})^*}{\partial x_\beta}\frac{\partial g_\mu^{(s')}(\vec{x})}{\partial x_\nu} \qquad (284)$$

where the integration extends over the volume of the semi-infinite solid. The first integral can be transformed into a surface integral which vanishes because of the boundary condition (283) at the surface $x_3 = 0$, because of periodic boundary conditions in the transverse directions, and because of the vanishing boundary conditions at $x_3 = +\infty$. We thus have that

$$-\frac{1}{\rho}\sum_{\alpha\beta\mu\nu} C_{\alpha\beta\mu\nu}\int d^3x\; f_\alpha^{(s)}(\vec{x})^*\,\frac{\partial^2}{\partial x_\beta \partial x_\nu}\,g_\mu^{(s')}(\vec{x})$$

$$=\frac{1}{\rho}\sum_{\alpha\beta\mu\nu} C_{\alpha\beta\mu\nu}\int d^3x\;\frac{\partial}{\partial x_\nu}\left\{\frac{\partial f_\alpha^{(s)}(\vec{x})^*}{\partial x_\beta}\,g_\mu^{(s')}(\vec{x})\right\}$$

$$-\frac{1}{\rho}\sum_{\alpha\beta\mu\nu} C_{\alpha\beta\mu\nu}\int d^3x\; g_\mu^{(s')}(\vec{x})\,\frac{\partial^2 f_\alpha^{(s)}(\vec{x})^*}{\partial x_\beta \partial x_\nu} \tag{285}$$

The first integral can again be transformed into a surface integral, which vanishes for the same reasons as before. Here we must use the symmetry property of the elastic moduli $C_{\alpha\beta\mu\nu} = C_{\mu\nu\alpha\beta}$. Using this property once again, together with the changes of dummy summation variables $\alpha \leftrightarrow \mu$ and $\beta \leftrightarrow \nu$, we obtain finally

$$-\frac{1}{\rho}\sum_{\alpha\beta\mu\nu}\int d^3x\; f_\alpha^{(s)}(\vec{x})^*\,C_{\alpha\beta\mu\nu}\,\frac{\partial^2}{\partial x_\beta \partial x_\nu}\,g_\mu^{(s')}(\vec{x})$$

$$=-\frac{1}{\rho}\sum_{\alpha\beta\mu\nu}\int d^3x\; g_\alpha^{(s')}(\vec{x})\,C_{\alpha\beta\mu\nu}\,\frac{\partial^2}{\partial x_\beta \partial x_\nu}\,f_\mu^{(s)}(\vec{x})^* \tag{286}$$

It follows from this result that the eigenfunctions $\{v_\alpha^{(s)}(\vec{x})\}$ can be chosen to be orthonormal and complete

$$\left.\begin{aligned}&\sum_\alpha\int d^2x_\parallel\int_0^\infty dx_3\; v_\alpha^{(s)}(\vec{x})^* v_\alpha^{(s')}(\vec{x}) = \delta_{ss'}\\[2ex]&\sum_s v_\alpha^{(s)}(\vec{x})^* v_\beta^{(s)}(\vec{x}\,') = \delta_{\alpha\beta}\,\delta(\vec{x}-\vec{x}\,')\end{aligned}\right\} \tag{287}$$

They can also be used as the basis for a normal coordinate transformation to phonon creation and destruction operators. For if we set

$$u_\alpha(\vec{x}) = \sum_s \left(\frac{\hbar}{2\rho\omega_s}\right)^{\frac{1}{2}} \left[v_\alpha^{(s)}(\vec{x})b_s + v_\alpha^{(s)}(\vec{x})^* b_s^\dagger\right] \qquad (288)$$

$$p_\alpha(\vec{x}) = \rho\dot{u}_\alpha(\vec{x}) = \rho\, \frac{1}{i}\sum_s \left(\frac{\hbar\omega_s}{2\rho}\right)^{\frac{1}{2}} \left[v_\alpha^{(s)}(\vec{x})b_s - v_\alpha^{(s)}(\vec{x})^* b_s^\dagger\right] \qquad (289)$$

where $b_s^\dagger$ and b_s are creation and destruction operators for phonons in the mode s, satisfying the commutation relations

$$\left[b_s, b_{s'}\right] = \left[b_s^\dagger, b_{s'}^\dagger\right] = 0, \quad \left[b_s, b_{s'}^\dagger\right] = \delta_{ss'} \qquad (290)$$

then the vibrational Hamiltonian for the semi-infinite medium

$$H = \sum_\alpha \int d^3x\; \tfrac{1}{2}\rho\dot{u}_\alpha^2(\vec{x}) + \sum_{\alpha\beta\mu\nu} \int d^3x\; \tfrac{1}{2}\, \frac{\partial u_\alpha(\vec{x})}{\partial x_\beta}\, C_{\alpha\beta\mu\nu}\, \frac{\partial u_\mu(\vec{x})}{\partial x_\nu} \qquad (291)$$

assumes the simple form

$$H = \sum_s \hbar\omega_s \left(b_s^\dagger b_s + \tfrac{1}{2}\right) \qquad (292)$$

From the equations of motion

$$i\hbar\dot{b}_s = [b_s, H] = \hbar\omega_s b_s \tag{293}$$

$$i\hbar\dot{b}_s^\dagger = [b_s^\dagger, H] = -\hbar\omega_s b_s^\dagger \tag{294}$$

we find that the Heisenberg representation operators $b_s(t)$ and $b_s^\dagger(t)$ are given by

$$b_s(t) = \exp\left(i\frac{t}{\hbar}H\right)b_s \exp\left(-i\frac{t}{\hbar}H\right) = b_s e^{-i\omega_s t} \tag{295}$$

$$b_s^\dagger(t) = \exp\left(i\frac{t}{\hbar}H\right)b_s^\dagger \exp\left(-i\frac{t}{\hbar}H\right) = b_s^\dagger e^{i\omega_s t} \tag{296}$$

In addition, from the form of the Hamiltonian (292) and the commutation relations obeyed by the operators b_s and $b_s^\dagger$, which imply that $b_s^\dagger b_s$ and $b_{s'}^\dagger b_{s'}$ commute for all s and s', so that the partial Hamiltonians $H_s = \hbar\omega_s\left(b_s^\dagger b_s + \frac{1}{2}\right)$ corresponding to the individual normal modes can be simultaneously diagonalized, it follows by standard methods that the statistical averages of the products of creation and destruction operators are given by

$$\langle b_s^\dagger b_{s'}\rangle = \delta_{ss'} n_s \tag{297a}$$

$$\langle b_s b_{s'}^\dagger\rangle = \delta_{ss'}\left(n_s + 1\right) \tag{297b}$$

$$\langle b_s b_{s'}\rangle = \langle b_s^\dagger b_{s'}^\dagger\rangle = 0 \tag{297c}$$

where

$$n_s = \left\{\exp \beta\hbar\omega_s - 1\right\}^{-1}, \quad \beta = 1/k_B T \tag{298}$$

With these results in hand we can turn now to the introduction of the dynamical Green's function tensor $D_{\alpha\beta}(\vec{x},\vec{x}'|t - t')$ which is defined as the solution of the equations

$$\sum_{\mu}\left\{-\delta_{\alpha\mu}\frac{\partial^2}{\partial t^2} + \frac{1}{\rho}\sum_{\beta\nu} C_{\alpha\beta\mu\nu}\frac{\partial^2}{\partial x_\beta \partial x_\nu}\right\} D_{\mu\gamma}(\vec{x},\vec{x}'|t - t')$$

$$= \delta_{\alpha\gamma}\delta(\vec{x} - \vec{x}')\delta(t - t')$$

$$x_3 > 0,\ x_3' > 0,\ \alpha,\gamma = 1,2,3 \qquad (299a)$$

subject to the following boundary conditions at the surface $x_3 = 0$

$$\frac{1}{\rho}\sum_{\mu\nu} C_{\alpha 3\mu\nu}\frac{\partial}{\partial x_\nu} D_{\mu\gamma}(\vec{x},\vec{x}'|t - t')\Big|_{x_3 = 0} = 0$$

$$x_3' > 0,\ \alpha,\gamma = 1,2,3 \qquad (299b)$$

In addition, $D_{\alpha\beta}(\vec{x},\vec{x}'|t - t')$ obeys outgoing exponentially decaying wave conditions at $x_3 = +\infty$.

The time translation invariance of our system allows us to Fourier transform $D_{\alpha\beta}(\vec{x},\vec{x}'|t - t')$ with respect to time according to

$$D_{\alpha\beta}(\vec{x},\vec{x}'|t - t') = \int\frac{d\omega}{2\pi} D_{\alpha\beta}(\vec{x},\vec{x}'|\omega)e^{-i\omega(t - t')} \qquad (300)$$

where the Fourier coefficient $D_{\alpha\beta}(\vec{x},\vec{x}'|\omega)$ is the solution of the equation

$$\sum_{\mu}\left\{\omega^2\delta_{\alpha\mu} + \frac{1}{\rho}\sum_{\beta\nu} C_{\alpha\beta\mu\nu}\frac{\partial^2}{\partial x_\beta \partial x_\nu}\right\}D_{\mu\gamma}(\vec{x},\vec{x}'|\omega) = \delta_{\alpha\gamma}\delta(\vec{x} - \vec{x}'),$$

$$x_3 > 0,\ x_3' > 0,\ \alpha,\gamma = 1,2,3 \qquad (301a)$$

that satisfies the boundary conditions

$$\frac{1}{\rho}\sum_{\mu\nu} C_{\alpha 3\mu\nu}\frac{\partial}{\partial x_\nu} D_{\mu\gamma}(\vec{x},\vec{x}'|\omega)\Big|_{x_3 = 0} = 0$$

$$x_3' > 0,\ \alpha,\gamma = 1,2,3 \qquad (301b)$$

at the plane $x_3 = 0$, and outgoing or exponentially decaying wave conditions at $x_3 = +\infty$.

It follows from Eqs. (280), (281), and the defining equations (301) that $D_{\alpha\beta}(\vec{x},\vec{x}'|\omega)$ can be represented in the form

$$D_{\alpha\beta}(\vec{x},\vec{x}'|\omega) = \sum_{s}\frac{v_\alpha^{(s)}(\vec{x})v_\beta^{(s)}(x')^*}{\omega^2 - \omega_s^2} = D_{\beta\alpha}(\vec{x}',\vec{x}|\omega)^* \qquad (302)$$

The dynamical Green's function for an infinitely extended medium is defined as the solution of Eqs. (302), or of Eqs. (299a) and (301a), which satisfies outgoing or exponentially decaying wave condition at $x_3 = \pm\infty$.

If it were necessary to obtain $D_{\alpha\beta}(\vec{x},\vec{x}'|\omega)$ by solving the eigenvalue problem (280) for the eigenvalues $\{\omega_s^2\}$ and the corresponding eigenfunctions $\{v_\alpha^{(s)}(\vec{x})\}$, substituting them into Eq. (302), and evaluating the sum on s, this would be a formidable calculation indeed.

Fortunately, however, it is possible to obtain $D_{\alpha\beta}(x,x'|\omega)$ directly from the differential equation and associated boundary conditions (301), and this bypasses the need for obtaining the $\{\omega_s^2\}$ and $\{v_\alpha^{(s)}(\vec{x})\}$ and for evaluating the sum on s in Eq. (302). This can be done as follows.

Because our systems of interest possess infinitesimal translational invariance in directions parallel to the surface we can Fourier analyze $D_{\alpha\beta}(\vec{x},\vec{x}'|\omega)$ according to

$$D_{\alpha\beta}(\vec{x},\vec{x}'|\omega) = \int \frac{d^2k_\parallel}{(2\pi)^2} D_{\alpha\beta}(\vec{k}_\parallel\omega|x_3x_3')e^{i\vec{k}_\parallel \cdot (\vec{x}_\parallel - \vec{x}_\parallel')} \tag{303}$$

where $\vec{k}_\parallel = \hat{i}_1k_1 + \hat{i}_2k_2$ and $\vec{x}_\parallel = \hat{i}_1x_1 + \hat{i}_2x_2$, and $\hat{i}_1$ and $\hat{i}_2$ are unit vectors in the 1- and 2-directions, respectively. Combining Eqs. (303) and (301) we find that the Fourier coefficient $D_{\alpha\beta}(\vec{k}_\parallel\omega|x_3x_3')$ is the solution of the following system of ordinary differential equations

$$\sum_\mu \left\{\delta_{\alpha\mu}\omega^2 + \frac{1}{\rho}\sum_{\beta\nu} C_{\alpha\beta\mu\nu}\left[(1 - \delta_{\beta3})ik_\beta + \delta_{\beta3}\frac{d}{dx_3}\right]\right.$$

$$\left.\times\left[(1 - \delta_{\nu3})ik_\nu + \delta_{\nu3}\frac{d}{dx_3}\right]\right\} \cdot D_{\mu\gamma}(\vec{k}_\parallel\omega|x_3x_3'),$$

$$= \delta_{\alpha\gamma}\delta(\vec{x} - \vec{x}'),$$

$$x_3 > 0,\ x_3' > 0,\ \alpha,\gamma = 1,2,3 \tag{304a}$$

subject to the boundary conditions

$$\frac{1}{\rho}\sum_{\mu\nu} C_{\alpha 3\mu\nu} \cdot \left[(1 - \delta_{\nu 3}) ik_{\nu} + \delta_{\nu 3} \frac{d}{dx_3}\right]$$

$$\cdot\ D_{\mu\gamma}(\vec{k}_{\parallel}\omega|x_3x_3')\Big|_{x_3 = 0} = 0,$$

$$x_3' > 0,\ \alpha,\gamma = 1,2,3 \quad (304b)$$

at the plane $x_3 = 0$, and outgoing or exponentially decaying wave conditions at $x_3 = +\infty$.

The Fourier coefficient $D_{\alpha\beta}^{(\infty)}(\vec{k}_{\parallel}\omega|x_3x_3')$ of the Green's function for an infinitely extended medium is the solution of Eq. (304a) that satisfies outgoing or exponentially decaying wave conditions at $x_3 = \pm\infty$. It is convenient to regard $D_{\alpha\beta}^{(\infty)}(\vec{k}_{\parallel}\omega|x_3x_3')$ as the particular integral of Eq. (304a), and to determine the complementary function as the solution of the homogeneous equation corresponding to Eq. (304a), which, when added to $D_{\alpha\beta}^{(\infty)}(\vec{k}_{\parallel}\omega|x_3x_3')$ ensures the satisfaction of the boundary conditions (304b).

Calculations of the Fourier coefficients $D_{\alpha\beta}(\vec{k}_{\parallel}\omega|x_3x_3')$ have recently been carried out for an elastically isotropic medium, bounded by a planar, stress-free surface (Ref. 77), and for an hexagonal medium, bounded by a planar, stress-free surface parallel to the basal plane (i.e., perpendicular to the axis of six-fold symmetry) (Ref. 78). The reader is referred to these articles for the details of the calculations. Because of their utility in a variety of calculations dealing with the dynamical properties of solid surfaces, as will be demonstrated explicitly later, we record here the explicit expressions for these coefficients for the case of an elastically isotropic medium (77). In this case (and in the case of the hexagonal medium mentioned above) the isotropy of the medium in the plane of the surface makes it possible to express $D_{\alpha\beta}(\vec{k}_\parallel\omega|x_3x_3')$ in terms of a smaller set of simpler Green's functions $d_{\alpha\beta}(k_\parallel\omega|x_3x_3')$, which depend on the two-dimensional wave vector $\vec{k}_\parallel$ only through its magnitude. The relation between the $D_{\alpha\beta}(\vec{k}_\parallel\omega|x_3x_3')$ and the $d_{\alpha\beta}(k_\parallel\omega|x_3x_3')$ is

$$D_{\alpha\beta}(\vec{k}_\parallel\omega|x_3x_3') = \sum_{\mu\nu} S^{-1}_{\alpha\mu}(\hat{k}_\parallel)d_{\mu\nu}(k_\parallel\omega|x_3x_3')S_{\nu\beta}(\hat{k}_\parallel) \quad (305)$$

where the matrix $\overleftrightarrow{S}(\hat{k}_\parallel)$ appearing in this equation is given by

$$\left.\begin{aligned} \overleftrightarrow{S}(\hat{k}_\parallel) &= \begin{bmatrix} \hat{k}_1 & \hat{k}_2 & 0 \\ -\hat{k}_2 & \hat{k}_1 & 0 \\ 0 & 0 & 1 \end{bmatrix} \\ \overleftrightarrow{S}^{-1}(\hat{k}_\parallel) &= \begin{bmatrix} \hat{k}_1 & -\hat{k}_1 & 0 \\ \hat{k}_2 & \hat{k}_1 & 0 \\ 0 & 0 & 1 \end{bmatrix} \end{aligned}\right\} \quad (306)$$

where $\hat{k}_\alpha = (k_\alpha/k_\parallel)$ $(\alpha = 1,2)$. The nonzero elements of the tensor $d_{\mu\nu}(k_\parallel\omega|x_3x_3')$ are given by

$$d_{11}(k_\parallel \omega | x_3 x_3') = -\frac{k_\parallel^2}{2\alpha_\ell \omega^2}\left[e^{-\alpha_\ell |x_3 - x_3'|} - \varepsilon e^{-\alpha_t |x_3 - x_3'|}\right]$$

$$- \frac{k_\parallel^2}{2\alpha_\ell \omega^2} \frac{1}{r_+}\left[r\_e^{-\alpha_\ell (x_3 + x_3')}\right.$$

$$+ e^{-\alpha_\ell x_3 - \alpha_t x_3'} + e^{-\alpha_t x_3 - \alpha_\ell x_3'}$$

$$\left. + \varepsilon r\_e^{-\alpha_t (x_3 + x_3')}\right] \qquad (307a)$$

$$d_{13}(k_\parallel \omega | x_3 x_3') =$$

$$\frac{ik_\parallel}{2\omega^2} \operatorname{sgn}(x_3 - x_3')\left[e^{-\alpha_t |x_3 - x_3'|} - e^{-\alpha_\ell |x_3 - x_3'|}\right]$$

$$+ \frac{ik_\parallel}{2\omega^2} \frac{1}{r_+}\left[r\_e^{-\alpha_\ell (x_3 + x_3')} + \frac{1}{\varepsilon} e^{-\alpha_\ell x_3 - \alpha_t x_3'} + e^{-\alpha_t x_3 - \alpha_\ell x_3'}\right.$$

$$\left. + r\_e^{-\alpha_t (x_3 + x_3')}\right] \qquad (307b)$$

$$d_{22}(k_\parallel \omega | x_3 x_3') = \frac{-1}{2\alpha_t c_t^2} e^{-\alpha_t |x_3 - x_3'|}$$

$$- \frac{1}{2\alpha_t c_t^2} e^{-\alpha_t (x_3 + x_3')} \qquad (307c)$$

$$d_{31}(k_\parallel \omega | x_3 x_3') =$$

$$\frac{ik_\parallel}{2\omega^2}\,\mathrm{sgn}(x_3 - x_3')\left[e^{-\alpha_t|x_3 - x_3'|} - e^{-\alpha_\ell|x_3 - x_3'|}\right]$$

$$-\frac{ik_\parallel}{2\omega^2}\frac{1}{r_+}\Bigg[r_- e^{-\alpha_\ell(x_3 + x_3')} + e^{-\alpha_\ell x_3 - \alpha_t x_3'} + \frac{1}{\varepsilon}e^{-\alpha_t x_3 - \alpha_\ell x_3'}$$

$$+ r_- e^{-\alpha_t(x_3 + x_3')}\Bigg] \qquad (307d)$$

$$d_{33}(k_\parallel \omega | x_3 x_3') = \frac{-k_\parallel^2}{2\alpha_t\omega^2}\left[-\varepsilon e^{-\alpha_\ell|x_3 - x_3'|} + e^{-\alpha_t|x_3 - x_3'|}\right]$$

$$-\frac{k_\parallel^2}{2\alpha_t\omega^2}\frac{1}{r_+}\Bigg[\varepsilon r_- e^{-\alpha_\ell(x_3 + x_3')}$$

$$+ e^{-\alpha_\ell x_3 - \alpha_t x_3'} + e^{-\alpha_t x_3 - \alpha_\ell x_3'}$$

$$+ r_- e^{-\alpha_t(x_3 + x_3')}\Bigg] \qquad (307e)$$

where

$$\alpha_{\ell,t} = \left(k_\parallel^2 - \frac{\omega^2}{c_{\ell,t}}\right)^{\frac{1}{2}}, \quad k_\parallel > \frac{\omega}{c_{\ell,t}}$$

$$= -i\left(\frac{\omega^2}{c_{\ell,t}^2} - k_\parallel^2\right)^{\frac{1}{2}}, \quad k_\parallel < \frac{\omega}{c_{\ell,t}} \qquad (308)$$

$$\varepsilon = \frac{\alpha_t \alpha_\ell}{k_\parallel^2}, \quad r_\pm = \frac{4\alpha_t \alpha_\ell k_\parallel^2 \pm (\alpha_t^2 + k_\parallel^2)^2}{4\alpha_t \alpha_\ell (\alpha_t^2 + k_\parallel^2)} \tag{309}$$

and c_t and c_ℓ are the speeds of transverse and longitudinal waves in the medium. In each of the above expressions for the $d_{\alpha\beta}(k_\parallel \omega | x_3 x_3')$ the terms on the first line give the expression for $d_{\alpha\beta}^{(\infty)}(k_\parallel \omega | x_3 x_3')$, the corresponding Fourier coefficient for an infinitely extended medium. We also note that the two expressions for $\alpha_{\ell,t}$ given by Eq. (308) can be combined into one, if we write

$$\alpha_{\ell,t} = \left(k_\parallel^2 - \frac{(\omega + i\eta)^2}{c_{\ell,t}^2} \right)^{\frac{1}{2}} \tag{310}$$

where η is a positive infinitesimal, and the branch cut for the square root is taken along the negative real axis.

We conclude this section by pointing out that inasmuch as Eq. (302) shows us that the Green's function $D_{\alpha\beta}(\vec{x}, \vec{x}' | \omega)$ considered as a function of ω has simple poles at the normal mode frequencies $\{\omega_s\}$ of the semi-infinite medium, the Fourier coefficient $D_{\alpha\beta}(\vec{k}_\parallel \omega | x_3 x_3')$, or equivalently $d_{\alpha\beta}(k_\parallel \omega | x_3 x_3')$, has simple poles at the normal mode frequencies of the semi-infinite medium characterized by a given value of the two-dimensional wave vector $\vec{k}_\parallel$. In particular, $D_{\alpha\beta}(\vec{k}_\parallel \omega | x_3 x_3')$ should have a simple pole at the dispersion relation for Rayleigh surface waves. That this is indeed the case can be seen explicitly from the results given by Eq. (307). Each of the coefficients d_{11}, d_{13}, d_{31}, and d_{33}, and hence all of the coefficients $D_{\alpha\beta}(\vec{k}_\parallel \omega | x_3 x_3')$, has a simple pole at the zeros of r_+. From Eqs. (307) and (308) we see that this condition can be expressed in the form

$$\left(2k_\parallel^2 - \frac{\omega^2}{c_t^2}\right)^2 = 4\left(k_\parallel^2 - \frac{\omega^2}{c_t^2}\right)^{\frac{1}{2}}\left(k_\parallel^2 - \frac{\omega^2}{c_\ell^2}\right)^{\frac{1}{2}} k_\parallel^2 \qquad (311)$$

The Ansatz $\omega = c_R k_\parallel$ in Eq. (311) yields immediately Eq. (84) which we have seen earlier is the equation for the determination of the speed of Rayleigh waves.

SURFACE SPECIFIC HEAT, ENTROPY, AND DENSITY OF STATES

If we compare the dynamical matrix for an infinitely extended three-dimensional crystal, in which the atomic displacements satisfy periodic boundary conditions in a macrocrystal containing N primitive unit cells, with the dynamical matrix for a finite crystal containing the same number of atoms and bounded by free surfaces, we find that they differ in the rows and columns corresponding to atoms which are closer to the crystal surface than the range of the interatomic forces. For interatomic potentials of finite range the number of such rows or columns is proportional to the surface area of the crystal. According to Ledermann's theorem (2), the number of eigenvalues of the dynamical matrix which enter or leave a given frequency interval when we replace periodic boundary conditions on the atomic displacements by those appropriate to a finite crystal is of the order of the number of such rows or columns. Consequently, the frequency distribution functions for cyclic and finite crystals should differ by an amount proportional to the ratio of the surface area of the finite crystal to its volume. Since the frequency distribution function of a cyclic crystal has no contribution proportional to this ratio, this result has the consequence that every extensive vibrational property of a finite crystal, including the thermodynamic functions, should have a contribution proportional to the surface area of the crystal, in addition to the contribution which is proportional to the volume of the crystal. This conclusion is independent of whether or not there are surface waves associated with the free surfaces of the crystal.

Surface Specific Heat

The surface contribution to the specific heat of a finite crystal has been studied extensively (79-80), both theoretically and experimentally. In a later section we present a calculation of this contribution for a semi-infinite isotropic elastic medium (81), bounded by a single, stress-free plane boundary and also for a hexagonal medium with its stress-free surface parallel to the basal plane (78, 82). Since elasticity theory is the long-wavelength limit of lattice theory for acoustic vibration modes, we can expect that the latter calculation will yield correctly only that part of the surface specific heat contributed by the long-wavelength acoustic modes, namely its low-temperature limit. A calculation of the surface contribution to the specific heat of a simple model of a crystal, which is valid for all temperatures is presented in the chapter on Examples.

The surface specific heat of a crystal can be calculated numerically, as opposed to analytically, in the following way. The normal mode frequencies $\{\omega_j(\vec{q}_\parallel)\}$ can be obtained for a slab-shaped crystal through the numerical solution of Eqs. (26) and (27) or, in the case of ionic crystals, Eqs. (103) and (102). From a knowledge of the normal mode frequencies all the thermodynamic functions of the crystal slab can be obtained in the harmonic approximation by a direct summation over the normal modes. The specific heat of the slab is given by

$$C_v(T) = k_B \sum_{\vec{q}_{\parallel}} \sum_{j} \frac{\left(\frac{1}{2}\beta\hbar\omega_j(\vec{q}_{\parallel})\right)}{\sinh^2 \frac{1}{2}\beta\hbar\omega_j(\vec{q}_{\parallel})} \tag{312}$$

Similarly, the thermodynamic functions for an infinite, three-dimensional crystal can be calculated by direct summation over all the normal modes. For example, the specific heat of such a crystal is also given by Eq. (312), provided $\vec{q}_{\parallel}$ is replaced by a three-dimensional wave vector, whose values allowed by periodic boundary conditions are restricted to the first Brillouin zone for the crystal, and the branch index j assumes the values 1,2,...,3r, where r is now the number of atoms in a primitive unit cell. Once the thermodynamic functions for crystals with and without surfaces, and possessing the same number of degrees of freedom, have been calculated, the surface contribution to any thermodynamic function can be calculated in the following way. Let u be the value per atom of some thermodynamic function U for a crystal with surfaces, i.e., u = U/N, where N is the number of particles in the crystal, and U stands for the internal energy E, the entropy S, the Helmholtz free energy F, or the specific heat at constant volume C_v. Let u^b and U^b be the same quantities in an infinite crystal without surfaces. Then if U^s, the surface contribution to U, is written in the form

$$U^s = Su^s \tag{313}$$

where S is the surface area of the crystal with surfaces (twice the area of one surface for a slab-shaped crystal), we have

$$u^s = \frac{N}{S}[u - u^b] \tag{314}$$

The surface specific heat vanishes at 0°K because both C_v and C_v^b vanish in the limit as $T \rightarrow 0$. It also vanishes in the limit of high temperatures because in this limit each normal vibration mode contributes the classical amount k_B to the specific heat whether or not a surface is present. Thus both C_v and C_v^b approach $3Nk_B$ in the high-temperature limit. Consequently, the surface specific heat as a function of temperature possesses a maximum.

Allen and deWette (83) have used the method just described to obtain the surface contributions to the thermodynamic functions for the (100), (110), and (111) surfaces of the noble gas solids neon, argon, krypton, and xenon. The same method was used (84) to calculate the surface specific heat for a (100) surface of NaCl. The results are in rather good agreement with the experimental results (84) for NaCl powder.

Surface Entropy

The vibrational contribution is also important for the evaluation of the surface entropies at high temperatures. It can be estimated by an expansion around the Einstein approximation (85,86) (see also Chapter 4, p. 345). The precision of this calculation is increased if one takes due account of the translation symmetry parallel to the surface (87). This improvement is obtained if one does an Einstein-like approximation on the dynamical matrix $D_{\alpha\beta}(\kappa\kappa'|\vec{k}_{\parallel})$, given by Eq. (26).

Specific calculations were done for simple models (85,86) for models including central potentials between first and second nearest neighbors, and also angular interactions, for (001), (110), and (111) surfaces of eight fcc and eleven bcc crystals (87) and also for a realistic model of graphite (88). The results are in agreement with the order of magnitude of the experimental data (89).

With this expansion around the Einstein approximation it is easy also to obtain the contribution to the entropy from surface defects, like steps and kinks (90).

The vibrational contribution to the surface entropies was calculated at all temperatures by a Green's function method (86) (see Chapter 4 and Fig. 4.2) as well as by numerical methods (83,84). By the Green's function method it was also possible to obtain, as a function of temperature, the entropy of a surface step (91).

Surface Free Energy and Density of States

In Refs. 83, 84, and 86, the phonon contribution to the surface free energy was also calculated as a function of temperature. However, the electronic contribution to this thermodynamic quantity is, in general, several orders of magnitude larger (86).

Let us note finally that the surface contribution to the phonon density of states is observed for small particle powders by neutron scattering (92). One obtains, as a qualitative behavior for the variation of the density of states for a given bulk band: increase for the lower frequency modes and decrease for the higher ones (83,84,93).

SURFACE MEAN SQUARE DISPLACEMENTS

The thermal vibrations of surface atoms may be investigated experimentally using low-energy electron diffraction (LEED) or the Mössbauer effect. The former method, in particular, has been used quite extensively during the last decade.

Within the framework of single-scattering or kinematic theory, the peak intensity of a LEED diffraction spot is specified as a function of temperature by (94)

$$I = I_o \sigma_o f_o^2 \left| \sum_{\ell} \alpha(\ell) e^{i\vec{Q} \cdot \vec{x}(\ell)} e^{-M(\ell)} \right|^2 \qquad (315)$$

where I_o is the incident intensity, σ_o is the Thomson factor, f_o is the atomic scattering factor, $\alpha(\ell)$ is a transmission factor, $\vec{Q}$ is the difference $\vec{K}' - \vec{K}$ of the scattered wave vector $\vec{K}'$ and the incident wave vector $\vec{K}$ of the electron, and $\vec{x}(\ell)$ is the equilibrium position vector of the atom in unit cell ℓ. The quantity $M(\ell)$ is given by

$$M(\ell) = \tfrac{1}{2}\langle [\vec{Q} \cdot \vec{u}(\ell)]^2 \rangle \qquad (316)$$

where $\vec{u}(\ell)$ is the displacement of atom ℓ from its equilibrium position and the angular brackets denote an average over a canonical ensemble. For simplicity, we have restricted our attention to a monatomic crystal with one atom per unit cell. The factor $\exp[-2M(\ell)]$ is known as the Debye-Waller factor.

The essential ingredient of the Debye-Waller factor is the mean square displacement $\langle u^2 \rangle$ which can be related to an effective Debye temperature Θ by

$$\langle u^2 \rangle = \left(3\hbar^2 T / M k_B \Theta^2\right) \Phi(\Theta / T) \qquad (317)$$

where M is the atomic mass and

$$\Phi(x) = \tfrac{1}{4}x + \frac{1}{x} \int_0^x \frac{t\,dt}{e^t - 1} \qquad (318)$$

In the high-temperature limit, $\Phi(\Theta/T) \rightarrow 1$, and $\langle u^2 \rangle$ is simply proportional to T.

For sufficiently low energy electrons (< 50 eV) the transmission factors $\alpha(\ell)$ decrease very rapidly with increasing distance from the surface. Consequently, such low-energy electrons are scattered primarily from the surface layer of atoms, and the scattered intensity is then dominated by the Debye-Waller factor for surface atoms. However, the strong scattering by the surface atoms brings into question the single-scattering approximation contained in Eq. (315). Duke and Laramore (95) have incorporated lattice vibrational effects into a multiple-scattering formalism of LEED and have shown that the temperature dependence of the diffracted-beam peak intensity is not as simple as that predicted by Eqs. (315) and (316). Jepsen, Marcus, and Jona (96) have carried out a specific calculation for the (111) surface of silver to assess the importance of multiple-scattering effects. They found that the effective surface Debye temperature is rather well specified by the use of kinematic theory for electron energies in the 40-50 eV range, but deviations of 10-15% from the proper value appear at lower energies. It should be mentioned that it is frequently possible to extract the single-scattering part from LEED intensity profiles by averaging several profiles at constant momentum transfer as suggested by Lagally, Ngoc, and Webb (97). In this way reliable values of surface Debye temperatures can be obtained.

We now develop the theory of surface atom mean square displacements in the harmonic approximation. The equations of motion can be written as

$$M_\kappa \ddot{u}_\alpha(\ell\kappa) = - \sum_{\ell\kappa'\beta} \Phi_{\alpha\beta}(\ell\kappa;\ell'\kappa')u_\beta(\ell'\kappa') \qquad (319)$$

where $u_\alpha(\ell\kappa)$ is the αth Cartesian component of displacement of the κth atom in the ℓth unit cell, M_κ is the mass of the κth kind of atom, and the quantities $\Phi_{\alpha\beta}(\ell\kappa;\ell'\kappa')$ are the harmonic coupling coefficients. If we make the transformation

$$v_\alpha(\ell\kappa) = M_\kappa^{\frac{1}{2}} u_\alpha(\ell\kappa) \qquad (320)$$

the equations of motion can be rewritten as

$$\ddot{v}_\alpha(\ell\kappa) = - \sum_{\ell'\kappa'\beta} D_{\alpha\beta}(\ell\kappa;\ell'\kappa')v_\beta(\ell'\kappa') \qquad (321)$$

where the quantities $D_{\alpha\beta}(\ell\kappa;\ell'\kappa')$ are the elements of the dynamical matrix and are given by

$$D_{\alpha\beta}(\ell\kappa;\ell'\kappa') = (M_\kappa M_{\kappa'})^{-\frac{1}{2}} \Phi_{\alpha\beta}(\ell\kappa;\ell'\kappa') \qquad (322)$$

The presence of a free surface affects the values of appropriate elements of both the force-constant matrix and the dynamical matrix. In the first place, certain elements are zero because interactions are eliminated between atoms on opposite sides of the boundary plane defining the surface. Second, the coupling constant associated with two atoms near the surface and on the same side of the boundary plane may have values different from the bulk values.

We write the normal coordinate transformation which diagonalizes the dynamical matrix in the form

$$v_{\alpha}(\ell\kappa) = \sum_{p} e_{p\alpha}(\ell\kappa)Q_{p} \qquad (323)$$

where Q_p is the normal coordinate for mode p. The eigenvector components $e_{p\alpha}(\ell\kappa)$ satisfy the equation

$$\sum_{\vec{\ell}'\kappa'\beta} \left\{D_{\alpha\beta}(\ell\kappa;\ell'\kappa') - \omega_p^2\delta_{\ell\ell'}\delta_{\kappa\kappa'}\delta_{\alpha\beta}\right\}e_{p\beta}(\ell'\kappa') = 0 \qquad (324)$$

where ω_p is the normal mode frequency for mode p. The mean square displacement component can be expressed as

$$\langle u_{\alpha}^2(\ell\kappa)\rangle = (1/M_{\kappa})\sum_{p} |e_{p\alpha}(\ell\kappa)|^2\langle|Q_p|^2\rangle \qquad (325)$$

where use has been made of the result

$$\langle Q_p^{*}Q_{p'}\rangle = \langle|Q_p|^2\rangle\delta_{pp'} \qquad (326)$$

At this point we utilize quantum statistical mechanics and write for a harmonic oscillator

$$\langle|Q_p|^2\rangle = \bar{\varepsilon}(\omega_p)/\omega_p^2 \qquad (327)$$

where $\bar{\varepsilon}(\omega_p)$ is the mean energy of the mode and is given by

$$\bar{\varepsilon}(\omega_p) = \tfrac{1}{2}\hbar\omega_p\coth(\hbar\omega_p/2k_BT) \qquad (328)$$

These results can now be employed to rewrite Eq. (325) in the form

$$\langle u_\alpha^2(\ell\kappa)\rangle = (1/M_\kappa)\sum_p |e_{p\alpha}(\ell\kappa)|^2\ \bar{\varepsilon}(\omega_p)/\omega_p^2 \tag{329}$$

Equation (329) is a useful expression for calculating mean square displacements. However, another useful relation can be obtained by using the following well-known theorem of matrices (94)

$$\sum_p f(\omega_p^2)\, e_{p\alpha}(\ell\kappa) e_{p\beta}(\ell'\kappa') = [f(\overleftrightarrow{D})]_{\ell\kappa\alpha,\ell'\kappa'\beta} \tag{330}$$

The mean square displacement then becomes

$$\langle u_\alpha^2(\ell\kappa)\rangle = (\hbar/2M_\kappa)\,[\overleftrightarrow{D}^{-\frac{1}{2}}\coth(\hbar D^{\frac{1}{2}}/2k_BT)\,]_{\ell\kappa\alpha,\ell\kappa\alpha} \tag{331}$$

In the limits of high and low temperatures, we obtain the results

$$\langle u_\alpha^2(\ell\kappa)\rangle \approx (k_BT/M_\kappa)\,[\overleftrightarrow{D}^{-1}]_{\ell\kappa\alpha,\ell\kappa\alpha} \qquad T > \Theta \tag{332a}$$

$$\langle u_\alpha^2(\ell\kappa)\rangle \approx (\hbar/2M_\kappa)\,[\overleftrightarrow{D}^{-\frac{1}{2}}]_{\ell\kappa\alpha,\ell\kappa\alpha} \qquad T = 0°K \tag{332b}$$

Qualitatively, we see from Eqs. (332) that the mean square displacement increases as the coupling constants decrease. Since the coupling constants for surface atoms tend to be smaller than for bulk atoms, the mean square displacements of surface atoms tend to be larger than those of bulk atoms. Furthermore, for monatomic crystals, we see that the mean square displacement is independent of the atomic mass and proportional to T at high temperatures, and varies as $M_\kappa^{-\frac{1}{2}}$ in the extreme low-temperature limit.

It is convenient in making actual calculations to introduce periodic boundary conditions in the two directions (say, the 1 and 2 directions) parallel to the surface. In terms of a two-dimensional wave vector $\vec{q}_\parallel = (q_1, q_2)$ the eigenvector components can be written as

$$e_{p\alpha}(\ell\kappa) = (1/N_s)^{\frac{1}{2}} e_{j\alpha}(\ell_3\kappa; q_1 q_2) \exp[i(q_1\ell_1 + q_2\ell_2)] \quad (333)$$

where N_s is the number of unit cells in a surface layer and j identifies the normal modes for given $\vec{q}_\parallel$. Using Eq. (333) we can write Eq. (329) in the alternative form

$$\langle u_\alpha^2(\ell\kappa)\rangle = (1/M_\kappa N_s) \sum_{\vec{q}_\parallel j} |e_{j\alpha}(\ell_3\kappa; \vec{q}_\parallel)|^2 \bar{\varepsilon}[\omega(\vec{q}_\parallel j)]/\omega^2(\vec{a}_\parallel j) \quad (334)$$

If we define the elements of a reduced dynamical matrix by

$$[\overleftrightarrow{D}(q_1 q_2)]_{\ell_3\kappa\alpha, \ell'_3\kappa'\beta} = \sum_{\ell_1, \ell_2} D_{\alpha\beta}(\ell\kappa; \ell'\kappa') \cdot e^{i[q_1(\ell_1' - \ell_1) + q_2(\ell_2' - \ell_2)]} \quad (335)$$

then $\langle u_\alpha^2(\ell\kappa)\rangle$ can also be written as

$$\langle u_\alpha^2(\ell\kappa)\rangle = \left(\hbar/2M_\kappa N_S\right)\sum_{\vec{q}_\parallel}\left[\overleftrightarrow{D}^{-\frac{1}{2}}(\vec{q}_\parallel)\right.$$

$$\left.\cdot \coth\{\hbar\overleftrightarrow{D}^{\frac{1}{2}}(\vec{q}_\parallel)/2k_BT\}\right]_{\ell_3\kappa\alpha,\ell_3\kappa\alpha} \qquad (336)$$

For a crystal L atomic layers thick, the reduced dynamical matrix is 3rL x 3rL in size, where r is the number of atoms per unit cell. On the other hand, the original dynamical matrix is $3rLN_s \times 3rLN_s$. For a crystal 20 atomic layers on each edge, we see that the reduced dynamical matrix is smaller by a factor of 400 than the original dynamical matrix. This means that considerable computer time can be saved by using the reduced dynamical matrix.

A number of other approaches have been presented for the calculation of mean square displacements. Maradudin and Melngailis (98) have used a Green's function method to treat the mean square displacement components at the (100) surface of a simple cubic lattice with nearest and next nearest neighbor interactions. They found an anisotropy in the mean square displacement components at the surface in contrast to the bulk. Masri and Dobrzynski (99) have written the dynamical matrix $\overleftrightarrow{D}$ as the sum of a diagonal part $\overleftrightarrow{d}$ and an off-diagonal part $\overleftrightarrow{R}$

$$\overleftrightarrow{D} = \overleftrightarrow{d} + \overleftrightarrow{R} \qquad (337)$$

Substitution of Eq. (337) into Eq. (331) and expansion in a matrix power series in $\overleftrightarrow{R}$ yields a series whose leading term and first few correction terms can easily be evaluated. Applications of this method are given in Chapter 4, p. 349. A continuum approach to the problem of surface mean square displacements has been presented by Dennis and Huber, Kalashnikov, Wallis, Maradudin and

Dobrzynski, and Lajzerowicz and Dobrzynski (100). The enhancement of the surface mean square displacement over the bulk and the anisotropy at the surface were confirmed.

We now present specific results for the mean square displacements at the (100) surface of nickel, a face-centered cubic crystal. The calculations were carried out (101) using Eq. (336) and a model with nearest-neighbor central forces. This model gives a rather good fit to the bulk phonon dispersion curves. The results are shown in Fig. 2.4. One sees clearly the increase in both

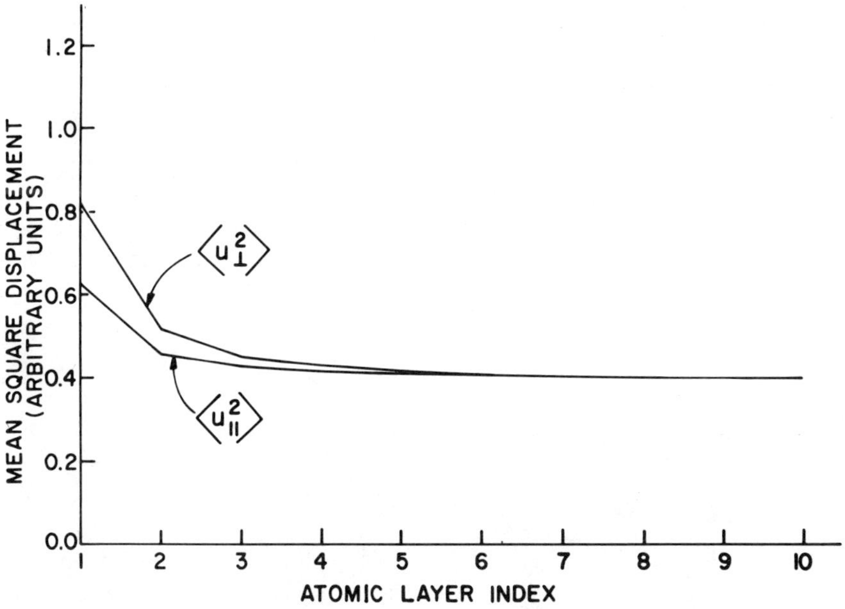

Fig. 2.4.--Theoretical mean square displacement components plotted against atomic layer index for the (100) surface of nickel. After Clark et al. (101).

the perpendicular and parallel mean square displacement components as one approaches the surface. The anisotropy at the surface is also evident. Results of a similar

A rather large number of experimental investigations have been directed toward the measurement of surface mean square displacements. These include the work of MacRae (103), of Jones, Mc Kinney, and Webb (104), and of Somorjai and coworkers (105) on face-centered cubic metals, and the work of Kaplan and Somorjai (106) and of Tabor and coworkers (107) on body-centered cubic metals. These studies confirm in a qualitative manner the theoretically predicted enhancement of the surface mean square displacements and the surface anisotropy. However, quantitative agreement between theory and experiment requires that one take into account the changes (108) in the surface force constants from the bulk values and multiple-scattering effects. A particularly noteworthy case is that of crystalline xenon. The (111) surface has been studied experimentally by Ignatiev and Rhodin (109) and analyzed by Tong, Rhodin, and Ignatiev (110). The scattering is highly kinematic, and the surface mean square displacements extracted from the data are in excellent agreement with theoretical predictions (101,102).

INTERACTION OF DEFECTS WITH CRYSTAL SURFACES

Because a free boundary surface can be regarded as a defect in an otherwise perfect cyclic crystal, we expect that there exists an energy of interaction between a point defect in a crystal and a free surface. This interaction energy can be defined as the difference between the vibrational self energy of a point defect in a crystal with a free surface and the vibrational self energy of the same defect when it is sufficiently far from the free surface that it becomes independent of its location in the crystal. The latter self energy is the same as the self energy of the defect in the perfect cyclic crystal. The interaction energy, consequently, is a function of the distance of the point defect from the crystal surface.

In addition, the existence of a free surface can affect dynamical properties of an impurity atom in the crystal, or adsorbed on its surface, for example the frequencies of any exceptional vibration modes associated with the impurity, or the mean square amplitude or mean square velocity of the impurity.

Finally, impurity atoms can affect dynamical properties of the crystal which are associated with the presence of a free surface. For example, a monatomic layer or impurities on a crystal surface can change the dispersion relation for surface vibration modes associated with the unperturbed surface, and can change the surface contribution to any of the thermodynamic functions of the crystal.

In this section we outline theories of each of these consequences of the interactions between defects of impurity atoms and crystal surface. Simple examples will be given in Chapter 4.

The Energy of Interaction of a Defect with a Surface

We confine this discussion to the absolute zero of temperature. The energy of interaction between the point defect and the crystal surface therefore is the zero point energy of interaction.

Let us denote by $\overleftrightarrow{U}(\omega^2)$ the Green's function for a crystal slab, and by $\overleftrightarrow{G}(\omega^2)$ the Green's function for the corresponding cyclic crystal containing the same number of atoms. It is a well-known result that the change in the zero-point energy of the crystal slab due to the introduction of the point defect into it is (111)
nature have been obtained for surfaces of rare gas solids by Allen and De Wette (102).

$$\Delta E_s = \frac{1}{2\pi i} \int_C \frac{\hbar z}{2} \, d\ln |\overleftrightarrow{I} - \overleftrightarrow{U}(z^2)\delta\overleftrightarrow{L}(z^2)| \qquad (338)$$

where C is any closed counterclockwise contour which encloses the zeros of $|\overleftrightarrow{I} - \overleftrightarrow{U}\delta\overleftrightarrow{L}|$, and where $\delta\overleftrightarrow{L}(\omega^2)$ describes the perturbation of the equations of motion of the crystal by the presence of the point defect. The change in the zero-point energy of the cyclic crystal due to the introduction of the same defect into it is

$$\Delta E_c = \frac{1}{2\pi i} \int_C \frac{\hbar z}{2} \, d\ln |\overleftrightarrow{I} - \overleftrightarrow{G}(z^2)\delta\overleftrightarrow{L}(z^2)| \qquad (339)$$

The difference between these two energy changes is the zero-point energy of interaction between the point defect and the crystal surface. It is straightforward (111) to show that it is given by

$$\Delta E_0 = -\frac{\hbar}{2\pi} \int_0^\infty y \, d\ln D(-y^2) \qquad (340)$$

where

$$D(\omega^2) = \frac{|\overleftrightarrow{I} - \overleftrightarrow{U}(\omega^2)\delta\overleftrightarrow{L}(\omega^2)|}{|\overleftrightarrow{I} - \overleftrightarrow{G}(\omega^2)\delta\overleftrightarrow{L}(\omega^2)|} \qquad (341)$$

As a simple example we calculate on p. 378 the energy of interaction of an isotopic impurity at the ℓth lattice site of a monatomic simple cubic lattice with a free surface at $\ell_3 = 1$. In this case

$$\Delta E_0 = -\frac{3^{\frac{1}{2}}}{8\pi^2}\frac{\hbar\omega_L\varepsilon}{(2\ell_3 - 1)^4} \qquad (342)$$

where $\varepsilon = 1 - (M'/M)$ and M' is the mass of the impurity while M is the mass of the atom it replaces; ω_L is the maximum bulk frequency. Hence the interaction energy is inversely proportional to the fourth power of the distance of the defect from the boundary. The interaction is one of attraction if $0 < M' < M$, and it is repulsive if $M < M' < \infty$.

This is qualitatively expected. The interaction of an isotope with a free boundary is equivalent to the interaction between the isotope and a very light impurity at the surface of the crystal. If the isotope is heavier than a normal lattice atom, a repulsion exists; when the isotopic defect is lighter, it interacts with a light impurity boundary and an attraction ensues. Furthermore, the interactions are $o(\varepsilon)$ rather than $o(\varepsilon^2)$; hence the interactions of isotopes with boundaries are not of the "image" type which exist in electrostatics and hydrodynamics.

We therefore find not only that an ordering process should exist at absolute zero temperature, but that a coating or "frosting" of light isotopes should develop in a solid isotopic mixture, leaving the heavier atomic species inside. It would be interesting to leave a hydrogen-deuterium mixture in liquid helium bath at low temperatures, for a long period to observe whether the separation process would require days or years. Note that the energy of the boundary attraction diminishes as the inverse fourth power of the distance while the attraction energy between a pair of isotopic impurities in the bulk varies as the inverse seventh power (111).

If the state of perfect order is to exist at low temperatures, we would expect holes in a lattice to be attracted to a free boundary and hence expelled from a crystal. These qualitative expectations have been confirmed by calculations of Montroll and Potts (112), who showed that the interaction of a hole with a free boundary in a three-dimensional lattice is attractive and inversely proportional to the square of the distance of the hole from the boundary.

Conclusions similar to those described here for the interactions of defects with the boundary surfaces of crystals have also been obtained by Yamahuzi and Tanaka (113).

Effects of Crystal Surfaces on Dynamical Properties of Impurity Atoms

If the Green's function $\overleftrightarrow{U}(\omega^2)$ for an otherwise perfect crystal possessing free surfaces has been determined, it can be used to study the effects of surfaces on the vibrations of impurity atoms in the crystal.

If the dynamical properties of an impurity atom or defect are described by a matrix $\delta\overleftrightarrow{L}(\omega^2)$, we can introduce a function $\Delta(\omega^2)$ for a crystal slab (111)

$$\Delta(\omega^2) = |\overleftrightarrow{I} - \overleftrightarrow{U}(\omega^2)\delta\overleftrightarrow{L}(\omega^2)| \tag{343}$$

If we define two real functions $\Delta^{(1)}(\omega^2)$ and $\Delta^{(2)}(\omega^2)$ of the real variable ω^2 by

$$\Delta(\omega^2 \pm io) = \Delta^{(1)}(\omega^2) \pm i\Delta^{(2)}(\omega^2) \tag{344}$$

then the equation whose solutions are the frequencies of

any localized, gap, and resonance modes associated with the impurity atom or defect is

$$\Delta^{(1)}(\omega^2) = 0 \tag{345}$$

All of the group theoretic methods (111) for the factorization, and hence simplification, of Eq. (343) for the case of a defect in an otherwise perfect crystal, in which case $\overleftrightarrow{U}(\omega^2)$ is replaced by $\overleftrightarrow{G}(\omega^2)$, can be applied to the simplification of Eq. (343) itself. In applying the group theoretic and symmetry arguments to this end it must be kept in mind that the symmetry group of the defect now must reflect the presence of the free surfaces of the slab, and in general is a subgroup of the symmetry group of the same defect in an otherwise perfect crystal.

The qualitative nature of the dependence of the localized mode frequencies associated with a substitutional impurity atom on the distance of the impurity from a free surface can be predicted from general arguments, without any detailed calculations. Because a free surface can be obtained by softening (to zero) a large number of force constants in the crystal, the normal mode frequencies of a crystal with free surfaces, including the localized mode frequencies, will be depressed compared with the corresponding frequencies in a cyclic crystal. Moreover, because the point symmetry of the crystal about the defect site is lowered when the impurity is near a surface, if the localized modes are degenerate, the possibility exists that this degeneracy is split. Finally, the depression and splitting of the localized mode frequencies tend to zero with increasing distance of the impurity from the free surface.

These qualitative results have been confirmed by the results of detailed calculations by Ashkin (114) for an isotopic impurity close to an (001) surface of a simple cubic crystal with nearest and next nearest neighbor, central force interactions. When the impurity is in the surface of the crystal the triply degenerate localized mode to which this impurity gives rise when it is in the bulk of the crystal is split into a doubly degenerate and a nondegenerate mode, of which the latter is associated with the vibration of the impurity normal to the surface and has the higher frequency. However, the localized modes have already taken their bulk character by the time the impurity is only one atom plane into the crystal. This is probably due both to the short range of the interatomic forces in the crystal model studied by Ashkin, and to the fact that only very-high-frequency localized modes (and therefore highly localized modes) were considered by him. Both of these facts would make the impurity atom insensitive to the crystal surface, even close to it. The splitting of the degenerate localized modes for an impurity in the surface of a crystal could be observed in the one phonon cross-section for the resonant absorption or emission of γ-rays by nuclei bound in the surface of a crystal.

It should be remarked that the qualitative conclusions we have drawn for localized modes in the preceding paragraphs apply as well to the resonance modes associated with heavy impurities in the surface layers of a crystal.

The Green's function approach is not the only one which has been shown to be useful in the solution of problems connected with the interaction of impurity atoms with crystal surfaces. Hori and Asahi (115) have applied the transfer matrix method (116) to the study of the localized vibration modes which arise when a sufficiently light isotopic impurity is present at one of the ends of a linear chain or in a corner of a two-dimensional lattice.

For a discussion of this approach to such problems the reader is referred to the original papers of these authors.

Grimley (117) considered an adatom on a (100) surface of a simple model of a simple cubic lattice. He obtained an expression for the localized mode frequencies using a Green's function technique.

If we denote the Green's function for crystal with free surfaces into which a point defect has been introduced by $\overleftrightarrow{U}'(\omega^2)$ the equation determining this function is

$$\overleftrightarrow{U}' = \overleftrightarrow{U} + \overleftrightarrow{U}\, \delta\overleftrightarrow{L}\, \overleftrightarrow{U} \qquad (346)$$

For the calculation of the mean square displacement or the mean square velocity of the impurity atom, on the basis of Eqs. (276a) and (276b), respectively, we need to know $\overleftrightarrow{U}'(\omega^2)$ only in the space of the impurity atom (111). The solution of Eq. (346) is particularly simply obtained in the case that the defect is an isotopic impurity (118). More valuable is a solution of Eq. (346) for the calculation of the mean square displacement of an impurity in which force constant changes accompanying the introduction of the impurity are taken into account. Some results of such a calculation have been obtained in particular by an expansion about the Einstein approximation (see the section on Surface Mean Square Displacements) (119-121). They are of particular interest for the theory of the Mössbauer effect for a resonance nucleus which, as is commonly the case, is an impurity in the surface of a crystal (122), and also for the variation with temperature of the LEED and atom scattering intensities.

The first experimental work which bears on the interaction of impurity atoms with crystal surfaces seems to be that of Pliskin and Eischens (123). These authors have apparently observed infrared absorption by vibration modes associated with hydrogen atoms adsorbed on platinum. Two different absorption peaks were seen, and they were attributed to two different types of bonding of the hydrogen atoms to the platinum surface. When deuterium was substituted for hydrogen, the frequencies of these peaks were shifted downward by a factor of 1.39, i.e., by very nearly the square root of the ratio of the mass of a deuterium atom to the mass of a hydrogen atom. Since that time, a great deal of experimental work has been devoted to the vibration of adsorbed atoms and molecules on crystal surfaces, by inelastic low-energy electron diffraction, neutron diffraction, infrared, and Raman techniques (124).

Effects of Impurity Atoms on Dynamical Properties of Crystal Surfaces

The decay length for Rayleigh surface waves is of the order of their wavelength parallel to the crystal surface. Consequently, the dispersion relation for a Rayleigh wave should be largely insensitive to perturbations of a crystal whose scale is small compared with the wavelength of the Rayleigh wave. Because it is the long-wavelength vibrations which determine the low-temperature themodynamic functions of a crystal, the latter will also not be affected by the presence of defects whose spatial extent is small, of the order of a lattice parameter, e.g., point defects. Consequently, in the study of the effects of defects on the properties of surface vibration modes, or on crystalline properties in which the dynamical properties of crystal surfaces play the dominant role, the emphasis has been on the study of the effects of an adsorbed monolayer of impurity atoms on the physical properties of interest. There is a theoretical advantage in studying the effects of such defects on the dynamical properties of crystal surfaces as compared with point defects because the translational periodicity of the crystal parallel to the free surface is preserved by such defects. Therefore, the calculations of the dispersion curves for surface modes can be carried out by the same methods that are employed for the calculation of the Green's function for a crystal with an adsorbed monolayer of impurities.

The effects of point defects on the surface modes in a two-dimensional lattice have been the subject of several theoretical studies.

By the use of the transfer matrix method Hori and Asahi (125) have shown that if all the atoms on one edge of a two-dimensional crystal are replaced by light isotopes of the same mass, surface modes can arise in which the atomic displacements decay exponentially with increasing distance into the crystal from the free surface, and whose frequencies can be inside or outside the band of frequencies allowed the crystal without the edge of impurities.

The latter problem has also been studied from a somewhat different standpoint and by a different method by Kaplan (126). He was interested in the consequences for the Rayleigh surface (in the present context, edge) modes on a semi-infinite two-dimensional lattice of a row of isotopic impurities of the same mass along one of the free edges of the lattice. He found that the impurities do not affect the dispersion curve for surface modes in the long-wavelength limit. However, at shorter wavelengths the presence of the surface impurity layer can alter the frequency of the surface mode, or can even suppress the mode; new surface modes can appear whose frequencies are higher or lower than the frequencies of the bulk of the crystal for the same wave vector.

The effects on the normal mode frequencies due to an adsorbed layer of impurity atoms on the surface of a three-dimensional crystal were studied first by Dobrzynski and Mills (127). For more details see p. 353. Dobrzynski and Mills used the simple cubic crystal with nearest neighbor, central and noncentral forces in their investigations, and assumed that the atoms in the adsorbed layer, deposited on a (001) free surface, differ from those of the substrate only in their masses. They determined the Green's function $\overleftrightarrow{U}'(\omega^2)$ for the perturbed crystal, where $\delta\overleftrightarrow{L}(\omega^2)$ now describes the effects of changing the masses of all of the atoms in the surface layer of the crystal, and used this result to demonstrate the existence of impurity-induced surface vibrational modes with frequencies above the vibrational band of the bulk crystal as well as below it, and of resonance modes with frequencies in the vibrational band of the bulk crystal for certain ranges of values of the impurity mass. Inasmuch as the crystal model used in these calculations does not admit Rayleigh surface modes, the influence of the adsorbed layer on Rayleigh waves could not be determined. However, Dobrzynski and Mills also calculated the effect of the adsorbed layer on the low-temperature specific heat of the crystal and found that it does not alter the coefficient of the T^2 term from the value it has for the crystal slab in the absence of an adsorbed layer [see Eq. (523)]. The effects of the adsorbed layer first show up in the specific heat in the coefficient of the T^3 term. This result is consistent with the qualitative argument that the long-wavelength modes which determine the coefficient of the T^2 term do not feel a perturbation whose spatial extent normal to the surface is small compared with their wavelengths.

The result is also valid when the effect of varying the force constants within and near the adsorbed monolayer is taken into account (128). They also studied the change in entropy due to adsorption. Indeed, the vibrational contribution is essential for the evaluation of the adsorption entropies (129-131), which can be obtained by measuring the desorption rate of the adsorbed particles as a function of temperature (129) or by analyzing the adsorption isotherms determined by Auger spectroscopy (131).

The localized modes of vibration due to the adsorption of a monolayer were calculated for several geometries: (001) surfaces of bcc crystals with application to H on W in particular (130,132); (001) surface of a simple cubic crystal with central forces between pairs of nearest and second nearest neighbor atoms (133); a fcc crystal bounded by (111) (134) and (100) (135,136) faces with application to adsorption of a rare gas monolayer on another rare gas solid. In Ref. 135 the effect of a static relaxation of the distance between substrate and adsorbate was taken into account. The effect of a superstructure on the localized modes of vibration of an adsorbed monolayer was also studied first on a (2 x 1) monolayer superstructure on a (001) surface of a simple cubic crystal (137). The main physical effects one can expect are described in the section on Examples. Specific calculations were done for monolayers of H on (001) W (138), S and O on (001) Ni (64,139-140), Xenon on (0001) graphite (141), and for Argon and Nitrogen monolayers adsorbed on graphite (142).

The main qualitative effects displayed in all these studies are that a heavy monolayer diminishes the frequencies of the localized modes and new modes may appear below the bulk bands for a given value of $\vec{k}_{\parallel}$. A light monolayer gives the opposite effects, and in particular more optical surface modes appear above the bulk bands. An increase (or decrease) in the surface force constants has a similar effect to a light (or heavy) monolayer. In particular the variation of the frequency of a Rayleigh wave when a monolayer is adsorbed is small, because the Rayleigh wave penetration is proportional to length, which is much larger than the thickness of the adsorbed monolayer. However, this variation can be detected as very precise measures of the Rayleigh wave frequencies can be carried out (143).

The effect of anharmonicity and of temperature on the localized modes of an adsorbed monolayer was recently studied (144) with the approach described for a free surface, on p. 202. It was found in the cases of (001) (144) and (111) (145) monolayers of one rare gas adsorbed on another rare gas substrate that some acoustic modes may become soft at a given temperature, which could be the signature of a second-order phase transition to a new superstructure. The variation with temperature of the optical localized mode due to a monolayer of O on (001) Ni was also estimated (140) and found to be measurable with the present experimental precisions.

These high-frequency modes due to adsorption are measured through the inelastic scattering of low-energy electrons, atoms or neutrons, as well as by infrared and Raman techniques (124).

Finally, let us mention that the mean square displacement of an atom within a physisorbed monolayer diverges logarithmically (146) when the ratio between the force constant coupling the physisorbed atoms to the substrate and the force constant coupling the physisorbed atoms among themselves is diverging (see p. 373).

INDIRECT INTERACTIONS OF ADATOMS ON A CRYSTAL SURFACE

Chemisorption, and to a lesser degree physisorption, has been extensively studied in several laboratories in recent years. The well-defined structures within the surface adlayer are often different in structure from the surface layer of the substrate. Of fundamental importance in understanding these various patterns is an understanding of the interaction energy between two adatoms.

This interaction can be direct and it can also be indirect, through the substrate. The direct, dipole-dipole, interaction between two adatoms on a metal surface has been determined by Kohn and Lau (147). The indirect interaction between two adatoms on a metal substrate, mediated by the conduction electrons, has been discussed by Grimley (148,149) and by Einstein and Schrieffer (150). The indirect interaction of two adatoms through the phonon field of the substrate has been studied by Schick and Campbell (151) and by Cunningham et al. (152,153). Recently, Lau and Kohn (154) have investigated the interaction between two adatoms mediated by the elastic distortion of the substrate to which each gives rise, to which the interaction studied by Cunningham et al. is the leading quantum correction. Lau and Kohn have shown that the interaction caused by the elastic distortion of the substrate is three or four orders of magnitude larger than that arising from the phonon field. More recently the interaction between two adatoms mediated by the elastic distortion of the substrate has also been studied by Stoneham (155), who obtains results qualitatively similar to those of Lau and Kohn.

In this section we outline a theory of the indirect interaction of two adatoms on a crystal surface which incorporates in a unified way both the indirect interaction arising from the elastic distortion of the substrate by the adatoms and that arising from the phonon field.

Rather than maintaining complete generality in this discussion we specialize immediately to the case of a semi-infinite Bravais crystal occupying the lower half space $x_3 \le 0$ [definition (B.2), p. 20]. The lattice sites of this crystal will be given by the vectors $\vec{x}(\ell) = \ell_1\vec{a}_1 + \ell_2\vec{a}_2 + \ell_3\vec{a}_3$, where the primitive translation vectors $\vec{a}_1$ and $\vec{a}_2$ are parallel to the surface of the crystal, while $\vec{a}_3$ has a component normal to the surface, and $\ell_3 \le 0$. The rest positions of the two adatoms, assumed to be identical, are given by $\vec{x}_{\parallel}(1) + \hat{x}_3 r_0$ and $\vec{x}_{\parallel}(2) + \hat{x}_3 r_0$, where $\vec{x}_{\parallel}(i)$ $(i = 1,2)$ have only components parallel to the surface and do not necessarily coincide with any of the vectors $\ell_1\vec{a}_1 + \ell_2\vec{a}_2$. That is, an adatom is not assumed to be situated directly above one of the atoms of the substrate, but can be situated above an interstitial position. The distance r_0 is the distance above the surface of the crystal at which the interaction energy of the adatom and the crystal, for a given $\vec{x}_{\parallel}(i)$, is a minimum, when the atoms of the crystal are held fixed in the rest positions they have in the absence of the adatom. The fact that each of the adatoms is at the same distance above the surface implies that each occupies a position with respect to the atoms of the substrate which can be sent into the other by one of the symmetry operations of the semi-infinite crystal. In particular, we will assume that the two vectors $\vec{x}_{\parallel}(1)$ and $\vec{x}_{\parallel}(2)$ differ by one of the vectors $\ell_1\vec{a}_1 + \ell_2\vec{a}_2$.

The potential energy of the semi-infinite crystal and of the two adatoms which are interacting with the crystal can be written in the harmonic approximation as

$$\Phi = \tfrac{1}{2}\sum_{\ell\alpha}\sum_{\ell'\beta} \Phi_{\alpha\beta}(\ell\ell')\xi_\alpha(\ell)\xi_\beta(\ell') + \sum_{i}\sum_{\ell\alpha} F_\alpha(i|\ell)\xi_\alpha(\ell)$$
$$+ \tfrac{1}{2}\sum_{i}\sum_{\ell\alpha}\sum_{\ell'\beta} A_{\alpha\beta}(i|\ell\ell')\xi_\alpha(\ell)\xi_\beta(\ell')$$
$$+ \sum_{\ell\alpha}\sum_{i\beta} B_{\alpha\beta}(\ell i)\xi_\alpha(\ell)\eta_\beta(i) + \tfrac{1}{2}\sum_{i}\sum_{\alpha\beta} C_{\alpha\beta}\eta_\alpha(i)\eta_\beta(i) \tag{347}$$

In this expression $\xi_\alpha(\ell)$ is the α Cartesian component of the displacement of the ℓth atom of the semi-infinite crystal from the rest position it would occupy in the absence of the adatoms; $\eta_\alpha(i)$ is the α Cartesian component of the displacement of the ith adatom ($i = 1,2$) from its rest position. The terms linear in the $\{\xi_\alpha(\ell)\}$ arise from the forces exerted by the adatoms on the atoms of the crystal. Note that there are no terms linear in the $\{\eta_\alpha(i)\}$. This is because the rest position of the ith adatom, $\vec{x}_\parallel(i) + \hat{x}_3 r_0$, is determined from the condition that there be no net force acting on that adatom when the atoms of the crystal are in the rest positions they would have in the absence of the adatom $\{\xi_\alpha(\ell) \equiv 0\}$ and it is at its rest position. Note that we also have not included any terms corresponding to the direct interaction of two adatoms.

From Eq. (347) the direct effect of each adatom on the substrate is seen to be twofold: it perturbs the atomic force constants of the substrate in its immediate vicinity, thorugh the $\{A_{\alpha\beta}(i|\ell\ell')\}$; and it exerts forces on the atoms of the substrate through the $\{F_\alpha(i|\ell)\}$. We will see below that there is also an indirect perturbation of the atomic force constants of the substrate through the $\{B_{\alpha\beta}(\ell i)\}$.

The kinetic energy of the system is given by

$$T = \tfrac{1}{2}M \sum_{\ell\alpha} \dot{\xi}_\alpha^2(\ell) + \tfrac{1}{2}m \sum_{i\alpha} \dot{\eta}_\alpha^2(i) \tag{348}$$

where M is the mass of an atom of the crystal, and m is the mass of an adatom. The equations of motion of the crystal and of the adatoms are

$$M\ddot{\xi}_\alpha(\ell) = -\frac{\partial\Phi}{\partial\xi_\alpha(\ell)} = -\sum_{\ell'\beta} \Phi_{\alpha\beta}(\ell\ell')\xi_\beta(\ell') - \sum_i F_\alpha(i|\ell)$$
$$-\sum_i \sum_{\ell'\beta} A_{\alpha\beta}(i|\ell\ell')\xi_\beta(\ell') - \sum_{i\beta} B_{\alpha\beta}(\ell i)\eta_\beta(i) \tag{349a}$$

$$m\ddot{\eta}_\alpha(i) = -\frac{\partial\Phi}{\partial\eta_\alpha(i)} = -\sum_\beta C_{\alpha\beta}\eta_\beta(i) - \sum_{\ell'\beta} B_{\beta\alpha}(\ell' i)\xi_\beta(\ell') \tag{349b}$$

To solve these equations we write each of the displacements $\xi_\alpha(\ell)$ and $\eta_\alpha(i)$ as the sum of two contributions

$$\xi_\alpha(\ell) = \xi_\alpha^{(0)}(\ell) + u_\alpha(\ell;t) \tag{350a}$$

$$\eta_\alpha(i) = \eta_\alpha^{(0)}(i) + v_\alpha(i;t) \tag{350b}$$

In each case the first term gives the static relaxation in the rest positions of the atoms of the crystal and of the adatoms; the second describes the dynamical displacements of these atoms and adatoms from their new rest positions. When Eqs. (350) are substituted into Eq. (349), the latter can be rewritten in the forms

$$M\ddot{u}_\alpha(\ell) = -\sum_{\ell'\beta} \{\Phi_{\alpha\beta}(\ell\ell') + \sum_i A_{\alpha\beta}(i|\ell\ell')\}u_\beta(\ell') - \sum_{i\beta} B_{\alpha\beta}(\ell i)v_\beta(i) \tag{351a}$$

$$m\ddot{v}_\alpha(i) = -\sum_\beta C_{\alpha\beta}v_\beta(i) - \sum_{\ell'\beta} B^T_{\alpha\beta}(i\ell')u_\beta(\ell') \tag{351b}$$

provided that the two auxiliary equations

$$0 = -\sum_{\ell'\beta} \{\Phi_{\alpha\beta}(\ell\ell') + \sum_i A_{\alpha\beta}(i|\ell\ell')\}\xi^{(0)}_\beta(\ell') \sum_{i\beta} B_{\alpha\beta}(\ell i)\eta^{(0)}_\beta(i) - \sum_i F_\alpha(i|\ell) \tag{352a}$$

$$0 = -\sum_\beta C_{\alpha\beta}\eta^{(0)}_\beta(i) - \sum_{\ell'\beta} B^T_{\alpha\beta}(i\ell')\xi^{(0)}_\beta(\ell') \tag{352b}$$

are also satisfied.

We consider the latter pair of equations first. To simplify the analysis slightly we will assume that the adatom force constant $C_{\alpha\beta}$ is diagonal, and we write it as

$$C_{\alpha\beta} = \delta_{\alpha\beta} m\omega_\alpha^2 \tag{353}$$

We can then solve Eq. (352b) for $\eta^{(0)}_\alpha(i)$, with the result that

$$\eta^{(0)}_\alpha(i) = -\frac{1}{m\omega_\alpha^2}\sum_{\ell'\beta} B^T_{\alpha\beta}(i\ell')\xi^{(0)}_\beta(\ell') \tag{354}$$

When Eq. (354) is substituted into Eq. (352a) the equations for the static displacements of the atoms of the semi-infinite crystal take the form

$$\sum_{\ell'\beta}\left\{\Phi_{\alpha\beta}(\ell\ell') + \sum_{i} A_{\alpha\beta}(i|\ell\ell') - \sum_{i\lambda}\frac{B_{\alpha\lambda}(\ell i)B^{T}_{\lambda\beta}(i\ell')}{m\omega_{\lambda}^{2}}\right\}$$

$$\xi_{\beta}^{(0)}(\ell') = -\sum_{i} F_{\alpha}(i|\ell) \tag{355}$$

If we introduce the matrix $U_{\alpha\beta}(\ell\ell')$ as the solution of the equation

$$\sum_{\ell'\beta}\left\{\Phi_{\alpha\beta}(\ell\ell') + \sum_{i} A_{\alpha\beta}(i|\ell\ell') - \sum_{i\lambda}\frac{B_{\alpha\lambda}(\ell i)B^{T}_{\lambda\beta}(i\ell')}{m\omega_{\lambda}^{2}}\right\}$$

$$U_{\beta\gamma}(\ell'\ell'') = \delta_{\alpha\gamma}\delta_{\ell\ell''} \tag{356}$$

then the formal solution of Eq. (355) is

$$\xi_{\alpha}^{(0)}(\ell) = -\sum_{i}\sum_{\ell'\beta} U_{\alpha\beta}(\ell\ell')F_{\beta}(i|\ell') \tag{357}$$

These results enable us to obtain the energy of interaction of the two adatoms through the elastic distortion of the substrate. From Eqs. (347), (350), and (352) we see that the static potential energy of our system is

$$\Phi_{S} = \tfrac{1}{2}\sum_{\ell\alpha}\sum_{\ell'\beta}\left\{\Phi_{\alpha\beta}(\ell\ell') + \sum_{i} A_{\alpha\beta}(i|\ell\ell')\right\}\xi_{\alpha}^{(0)}(\ell)\xi_{\beta}^{(0)}(\ell')$$

$$+\sum_{i}\sum_{\ell\alpha} F_{\alpha}(i|\ell)\xi_{\alpha}^{(0)}(\ell) + \sum_{\ell\alpha}\sum_{i\beta} B_{\alpha\beta}(\ell i)\xi_{\alpha}^{(0)}(\ell)\eta_{\beta}^{(0)}(i)$$

$$+ \tfrac{1}{2}\sum_{i}\sum_{\alpha\beta} C_{\alpha\beta}\eta_{\alpha}^{(0)}(i)\eta_{\beta}^{(0)}(i) \tag{358}$$

We now multiply Eq. (352a) by $\frac{1}{2}\xi_{\alpha}^{(0)}(\ell)$ and sum over ℓ and α. We next multiply Eq. (352b) by $\frac{1}{2}\eta_{\alpha}^{(0)}(i)$ and sum over i and α. In this way we obtain the relations

$$\tfrac{1}{2}\sum_{\ell\alpha}\sum_{\ell'\beta}\left\{\Phi_{\alpha\beta}(\ell\ell') + \sum_{i} A_{\alpha\beta}(i|\ell\ell')\right\}\xi_{\alpha}^{(0)}(\ell)\xi_{\beta}^{(0)}(\ell')$$

$$+ \tfrac{1}{2}\sum_{\ell\alpha}\sum_{i\beta} B_{\alpha\beta}(\ell i)\xi_{\alpha}^{(0)}(\ell)\eta_{\beta}^{(0)}(i)$$

$$= -\tfrac{1}{2}\sum_{\ell\alpha}\sum_{i} F_{\alpha}(i|\ell)\xi_{\alpha}^{(0)}(\ell) \tag{359a}$$

$$\tfrac{1}{2}\sum_{i}\sum_{\alpha\beta} C_{\alpha\beta}\eta_{\alpha}^{(0)}(i)\eta_{\beta}^{(0)}(i) + \tfrac{1}{2}\sum_{\ell\alpha}\sum_{i\beta} B_{\alpha\beta}(\ell i)\xi_{\alpha}^{(0)}(\ell)\eta_{\beta}^{(0)}(i) = 0$$

$$\tag{359b}$$

Combining Eqs. (358) and (359) we find that the static potential energy of our system takes the simple form

$$\Phi_S = \tfrac{1}{2}\sum_{i}\sum_{\ell\alpha} F_{\alpha}(i|\ell)\xi_{\alpha}^{(0)}(\ell) \tag{360}$$

Equation (360) has a simple physical interpretation. It represents the work done on the crystal by the external forces $\{F_\alpha(i|\ell)\}$ acting through the displacements to which they give rise, with the forces increasing linearly from zero values to their full values.

If we substitute Eq. (357) into Eq. (360) we obtain for the static potential energy

$$\Phi_S = -\tfrac{1}{2}\sum_{ij}\sum_{\ell\alpha}\sum_{\ell'\beta} F_\alpha(i|\ell)U_{\alpha\beta}(\ell\ell')F_\beta(j|\ell') \qquad (361)$$

The result given by Eq. (361) is the total, static potential energy of our system in the presence of both adatoms above the substrate. The energy of interaction of the two adatoms is obtained by subtracting from this expression the static potential energy of our system associated with the presence of each adatom separately above the substrate. Without going into the details of the calculation, which parallels the derivation of Eq. (361), we present only the final result for each of the latter two energies. If only adatom 1 is present above the substrate, the static potential energy of our system is

$$\Phi_S^{(1)} = -\tfrac{1}{2}\sum_{\ell\alpha}\sum_{\ell'\beta} F_\alpha(1|\ell)U^{(1)}_{\alpha\beta}(\ell\ell')F_\beta(1|\ell') \qquad (362a)$$

where the matrix $U^{(1)}_{\alpha\beta}(\ell\ell')$ is the solution of the equation

$$\sum_{\ell'\beta}\left\{\Phi_{\alpha\beta}(\ell\ell') + A_{\alpha\beta}(1|\ell\ell') - \sum_\lambda \frac{B_{\alpha\lambda}(\ell 1)B^T_{\lambda\beta}(1\ell')}{m\omega_\lambda^2}\right\}$$

$$\cdot \; U^{(1)}_{\beta\gamma}(\ell'\ell'') = \delta_{\alpha\gamma}\delta_{\ell\ell''} \tag{362b}$$

If only adatom 2 is present above the substrate, the static potential energy of our system is given by the expression

$$\Phi^{(2)}_S = -\tfrac{1}{2}\sum_{\ell\alpha}\sum_{\ell'\beta} F_\alpha(2|\ell)U^{(2)}_{\alpha\beta}(\ell\ell')F_\beta(2|\ell') \tag{363a}$$

where the matrix $U^{(2)}_{\alpha\beta}(\ell\ell')$ is the solution of the equation

$$\sum_{\ell'\beta}\left\{\Phi_{\alpha\beta}(\ell\ell') + A_{\alpha\beta}(2|\ell\ell') - \sum_\lambda \frac{B_{\alpha\gamma}(\ell 2)B^T_{\lambda\beta}(2\ell')}{m\omega^2_\lambda}\right\}$$

$$\cdot \; U^{(2)}_{\beta\gamma}(\ell'\ell'') = \delta_{\alpha\gamma}\delta_{\ell\ell''} \tag{363b}$$

The static energy of interaction of the two adatoms is therefore given by

$$\Phi_{S;int.} = \Phi_S - \Phi^{(1)}_S - \Phi^{(2)}_S$$

$$= -\tfrac{1}{2}\sum_{\ell\alpha}\sum_{\ell'\beta'}\left\{F_\alpha(1|\ell)\left[U_{\alpha\beta}(\ell\ell') - U^{(1)}_{\alpha\beta}(\ell\ell')\right]F_\beta(1|\ell')\right.$$

$$+ F_\alpha(1|\ell)U_{\alpha\beta}(\ell\ell')F_\beta(2|\ell') + F_\alpha(2|\ell)U_{\alpha\beta}(\ell\ell')F_\beta(1|\ell')$$

$$\left. + F_\alpha(2|\ell)\left[U_{\alpha\beta}(\ell\ell') - U^{(2)}_{\alpha\beta}(\ell\ell')\right]F_\beta(2|\ell')\right\} \tag{364}$$

This result can be simplified somewhat. The difference $U_{\alpha\beta}(\ell\ell') - U^{(i)}_{\alpha\beta}(\ell\ell')$ (i = 1,2) can be written compactly as

$$U_{\alpha\beta}(\ell\ell') - U^{(i)}_{\alpha\beta}(\ell\ell') = -\sum_{\ell''\gamma}\sum_{\ell'''\delta} U^{(i)}_{\alpha\gamma}(\ell\ell'')M^{(\bar{i})}_{\gamma\delta}(\ell''\ell''') \cdot U_{\delta\beta}(\ell'''\ell') \tag{365}$$

where

$$M^{(i)}_{\alpha\beta}(\ell\ell') = A_{\alpha\beta}(i|\ell\ell') - \sum_{\lambda}\frac{B_{\alpha\lambda}(\ell i)B^{T}_{\lambda\beta}(i\ell')}{m\omega^2_{\lambda}} \tag{366}$$

We have used the notation $\bar{1} = 2$, $\bar{2} = 1$. In addition, the matrix $U_{\alpha\beta}(\ell\ell')$ can readily be shown to be symmetric: $U_{\alpha\beta}(\ell\ell') = U_{\beta\alpha}(\ell'\ell)$. It follows, therefore, that the static, indirect energy of interaction of the two adatoms is given by

$$\begin{aligned}\Phi_{s;int.} = &-\sum_{\ell\alpha}\sum_{\ell'\beta} F_{\alpha}(1|\ell)U_{\alpha\beta}(\ell\ell')F_{\beta}(2|\ell') \\ &+ \tfrac{1}{2}\sum_{\ell\alpha}\sum_{\ell'\beta}\sum_{\ell''\gamma}\sum_{\gamma'''\delta}\Big\{F_{\alpha}(1|\ell)U^{(1)}_{\alpha\gamma}(\ell\ell'')M^{(2)}_{\gamma\delta}(\ell''\ell''') \\ &\cdot U_{\delta\beta}(\ell'''\ell')F_{\beta}(1|\ell') + F_{\alpha}(2|\ell)U^{(2)}_{\alpha\gamma}(\ell\ell'')M^{(1)}_{\gamma\delta}(\ell''\ell''') \\ &\cdot U_{\delta\beta}(\ell'''\ell')F_{\beta}(2|\ell')\Big\}\end{aligned} \tag{367}$$

A calculation of the energy of interaction of two adatoms based on Eq. (367) has not been carried out yet. Lau and Kohn (154) have evaluated the interaction energy by neglecting the coefficients $\{A_{\alpha\beta}(i|\ell\ell')\}$ and $\{B_{\alpha\beta}(\ell i)\}$, in which case the matrices $\{M^{(i)}_{\alpha\beta}(\ell\ell')\}$ vanish identically, and the Green's function $U_{\alpha\beta}(\ell\ell')$ is replaced by the Green's function $U^{(0)}_{\alpha\beta}(\ell\ell')$, which is the static Green's function for a semi-infinite crystal with a free surface. The latter Green's function is the solution of the equation

$$\sum_{\ell'\beta} \Phi_{\alpha\beta}(\ell\ell')U^{(0)}_{\beta\gamma}(\ell'\ell'') = \delta_{\alpha\gamma}\delta_{\ell\ell''} \qquad (368)$$

In addition, Lau and Kohn (154), Stoneham (155), and Maradudin and Wallis (156) replaced the semi-infinite crystal by a semi-infinite, isotropic elastic continuum bounded by a planar, stress-free surface. They find that the interaction energy (367) varies as $|\vec{x}_{\parallel}(1) - \vec{x}_{\parallel}(2)|^{-3}$, for large $|\vec{x}_{\parallel}(1) - \vec{x}_{\parallel}(2)|$, and is of the order of 0.2 eV when the two adatoms are separated along the surface by a distance of the order of the lattice parameter a.

We turn now to the indirect interaction of the two adatoms mediated by the phonons of the substrate. We begin by assuming harmonic time dependences, proportional to $\exp(-i\omega t)$, for the dynamical displacements $u_\alpha(\ell;t)$ and $v_\alpha(i;t)$. Then with the aid of Eq. (353) we can solve Eq. (351b) to obtain $v_\alpha(i)$ as a function of $u_\alpha(\ell)$

$$v_\alpha(i) = \frac{1}{m(\omega^2 - \omega_\alpha^2)} \sum_{\ell'\beta} B^T_{\alpha\beta}(i\ell')u_\beta(\ell') \qquad (369)$$

When this result is substituted into Eq. (351a), we obtain the time independent equations of motion of the atoms of the substrate alone

$$M\omega^2 u_\alpha(\ell) = \sum_{\ell'\beta}\left\{\Phi_{\alpha\beta}(\ell\ell') + \sum_i A_{\alpha\beta}(i|\ell\ell')\right\}u_\beta(\ell')$$

$$+ \sum_{\ell'\beta}\sum_{i\lambda}\frac{B_{\alpha\lambda}(\ell i)B^T_{\lambda\beta}(i\ell')}{m\left(\omega^2 - \omega_\lambda^2\right)}\,u_\beta(\ell') \qquad (370)$$

Equation (370) can be rewritten formally as

$$\sum_{\ell'\beta}\left[L_{\alpha\beta}(\ell\ell';\omega^2) - \delta L_{\alpha\beta}(\ell\ell';\omega^2)\right]u_\beta(\ell') = 0 \qquad (371)$$

where

$$L_{\alpha\beta}(\ell\ell';\omega^2) = M\omega^2\delta_{\ell\ell'}\delta_{\alpha\beta} - \Phi_{\alpha\beta}(\ell\ell') \qquad (372)$$

and

$$\delta L_{\alpha\beta}(\ell\ell';\omega^2) = \sum_i\left[A_{\alpha\beta}(i|\ell\ell') + \sum_\lambda\frac{B_{\alpha\lambda}(\ell i)B^T_{\lambda\beta}(i\ell')}{m\left(\omega^2 - \omega_\lambda^2\right)}\right]$$

$$\equiv \sum_i \delta L^{(i)}_{\alpha\beta}(\ell\ell';\omega^2) \qquad (373)$$

Equation (371) can be rewritten in its turn in the form

$$u_\alpha(\ell) = \sum_{\ell'\beta}\sum_{\ell''\gamma} U_{\alpha\beta}(\ell\ell';\omega^2)\delta L_{\beta\gamma}(\ell'\ell'';\omega^2)u_\gamma(\ell'') \qquad (374)$$

where the matrix $U_{\alpha\beta}(\ell\ell';\omega^2)$ is the dynamical Green's function for a semi-infinite crystal. It is the solution of the equation

$$\sum_{\ell'\beta} L_{\alpha\beta}(\ell\ell';\omega^2)U_{\beta\gamma}(\ell'\ell'';\omega^2) = \delta_{\alpha\gamma}\delta_{\ell\ell''} \tag{375}$$

The solvability condition for Eq. (374) is the vanishing of the determinant of the coefficients

$$\Delta_{12}(\omega^2) = |\overleftrightarrow{I} - \overleftrightarrow{U}(\omega^2)\delta\overleftrightarrow{L}(\omega^2)| = 0 \tag{376}$$

This equation gives the frequencies of the normal modes of the slab which are perturbed by the presence of the two adatoms.

The expression for $\Delta_{12}(\omega^2)$ can be simplified considerably. On the assumption that each adatom interacts only with atoms of the substrate in its immediate vicinity, the elements of the matrix $\delta L^{(i)}_{\alpha\beta}(\ell\ell';\omega^2)$ are nonzero only for sites ℓ and ℓ' in the immediate vicinity of the vector $\vec{x}_{\parallel}(i)$. If we define the lattice sites of the substrate directly touched by the presence of the ith adatom as the defect space of that adatom, then by suitably labeling rows and columns of the matrix $\overleftrightarrow{\delta L}(\omega^2)$ we can partition it in the form

$$\overleftrightarrow{\delta L}(\omega^2) = \begin{bmatrix} \overleftrightarrow{\delta L}^{(1)}(\omega^2) & \overleftrightarrow{0} & \overleftrightarrow{0} \\ \overleftrightarrow{0} & \overleftrightarrow{\delta L}^{(2)}(\omega^2) & \overleftrightarrow{0} \\ \overleftrightarrow{0} & \overleftrightarrow{0} & \overleftrightarrow{0} \end{bmatrix} \tag{377}$$

The matrix $\overleftrightarrow{U}(\omega^2)$ can be partitioned in the same way

$$\overleftrightarrow{U}(\omega^2) = \begin{bmatrix} \overleftrightarrow{u}_{11}(\omega^2) & \overleftrightarrow{u}_{12}(\omega^2) & \overleftrightarrow{u}_{13}(\omega^2) \\ \overleftrightarrow{u}_{21}(\omega^2) & \overleftrightarrow{u}_{22}(\omega^2) & \overleftrightarrow{u}_{23}(\omega^2) \\ \overleftrightarrow{u}_{31}(\omega^2) & \overleftrightarrow{u}_{32}(\omega^2) & \overleftrightarrow{u}_{33}(\omega^2) \end{bmatrix} \tag{378}$$

It follows that the determinant $\Delta_{12}(\omega^2)$, defined by Eq. (376), simplifies to

$$\Delta_{12}(\omega^2) = \begin{vmatrix} \overleftrightarrow{I} - \overleftrightarrow{u}_{11}(\omega^2)\delta\overleftrightarrow{L}^{(1)}(\omega^2) & -\overleftrightarrow{u}_{12}(\omega^2)\delta\overleftrightarrow{L}^{(2)}(\omega^2) \\ -\overleftrightarrow{u}_{21}(\omega^2)\delta\overleftrightarrow{L}^{(1)}(\omega^2) & \overleftrightarrow{I} - \overleftrightarrow{u}_{22}(\omega^2)\delta\overleftrightarrow{L}^{(2)}(\omega^2) \end{vmatrix} \tag{379}$$

In the case that only adatom 1 is present above the substrate, the corresponding defect determinant, which we will denote by $\Delta_1(\omega^2)$, is given by

$$\Delta_1(\omega^2) = \left|\overleftrightarrow{I} - \overleftrightarrow{u}_{11}(\omega^2)\delta\overleftrightarrow{L}^{(1)}(\omega^2)\right| \tag{380}$$

When only adatom 2 is present above the substrate, the corresponding defect determinant, $\Delta_2(\omega^2)$, is given by

$$\Delta_2(\omega^2) = \left|\overleftrightarrow{I} - \overleftrightarrow{u}_{22}(\omega^2)\delta\overleftrightarrow{L}^{(2)}(\omega^2)\right| \tag{381}$$

If we are concerned with some additive function of the normal mode frequencies of our system, i.e.

$$S = \sum_s f(\omega_s) \tag{382}$$

then it is a well known result that the "interaction-S" between two defects in the system is (157)

$$S_{12} = \frac{1}{2\pi i}\int_C f(\zeta)\,d\ell n\,\Delta(\zeta^2) \tag{383}$$

where

$$\Delta(\omega^2) = \frac{\Delta_{12}(\omega^2)}{\Delta_1(\omega^2)\Delta_2(\omega^2)} \tag{384}$$

and where C is any closed, counterclockwise contour that encloses the zeros of $\Delta(\zeta^2)$, $\Delta_1(\zeta^2)$, $\Delta_2(\zeta^2)$, but none of the poles of $f(\zeta)$.

In the present case it is the interaction Helmholtz-free energy that is being sought. In the limit of low temperatures this function can be expanded in powers of the absolute temperature as (157)

$$\Delta F(T) = \Delta E_0 - \frac{k_BT}{\pi}\left[\frac{\pi^2}{6}\left(\frac{k_BT}{\hbar\omega_L}\right)\Omega(0) - \frac{\pi^4}{90}\left(\frac{k_BT}{\hbar\omega_L}\right)^3\Omega^{(2)}(0) + \frac{\pi^6}{945}\left(\frac{k_BT}{\hbar\omega_L}\right)^5\Omega^{(4)}(0) - \ldots\right] \tag{385}$$

where $\Omega^{(n)}(0)$ is the nth derivative at the origin of the function $\Omega(f)$ defined by

$$\Omega(f) = \frac{d}{df}\,\ell n\,\Delta\left(-\omega_L^2 f^2\right) \tag{386}$$

in which ω_L is the largest normal mode frequency of the substrate in the absence of the two adatoms. In Eq. (385) ΔE_0 is the zero point energy of interaction between the two adatoms, and is given by

$$\Delta E_0 = -\frac{\hbar\omega_L}{2\pi}\int_0^\infty f\Omega(f)df = \frac{\hbar\omega_L}{2\pi}\int_0^\infty \ell n\,\Delta\left(-\omega_L^2 f^2\right)df \tag{387}$$

The second form of Eq. (387) is obtained by integrating the first by parts and using Eq. (386).

In the high-temperature limit a convenient expression for the interaction free energy is given by

$$\Delta F(T) = \frac{k_B T}{2} \ln \left(\frac{|\overleftrightarrow{M}_0|}{|\overleftrightarrow{M}|} \Delta(0) \right) + k_B T \sum_{n=1}^{\infty} \ln \left(\frac{|\overleftrightarrow{M}_0|}{|\overleftrightarrow{M}|} \Delta\left(-\omega_n^2\right) \right) \tag{388}$$

where $\omega_n = n\left(2\pi k_B T/\hbar\right)$, and $\overleftrightarrow{M}$ and $\overleftrightarrow{M}_0$ are the (diagonal) matrices whose elements are the masses of the atoms in the perturbed and unperturbed substrates, respectively. In the present case, the perturbation of the vibrations of the substrate by the presence of the two adatoms is expressed by a (frequency-dependent) change in the atomic force constants of the substrate in the vicinity of each adatom. The masses of the atoms of the substrate are left unaltered by such a perturbation, so that in the present case $|\overleftrightarrow{M}| = |\overleftrightarrow{M}_0|$. In addition, in the present case, it can be shown that $\ln \Delta(0) = 0$, so that Eq. (388) takes the simpler form

$$\Delta F(T) = k_B T \sum_{n=1}^{\infty} \ln \Delta\left(-\omega_n^2\right) \tag{389}$$

The preceding results have been used by Cunningham et al. (152,153) for a simple lattice dynamical model to study the Helmholtz-free energy of interaction of two identical adatoms mediated by the phonons of the substrate. They find that the zero-point energy of interaction is negative, varies inversely with the seventh power of their separation along the surface, is proportional to the square of the ratio of the mass of the adatoms to the mass of an atom of the substrate, $(m/M)^2$, and is of the order of $10^{-4}\ \hbar\omega_L$ in magnitude when the two adatoms are separated by a distance of the order of an interatomic spacing. The leading temperature-dependent contribution to the interaction free energy at low temperatures is proportional to the sixth power of the absolute temperature and is also negative. In the high-temperature limit, the free energy is again negative and decreases to zero with increasing temperature as $T^{-(2p-1)}$; here $p = 2(m_1 + m_2 + 1)$ and m_1 and m_2 are defined by $\vec{x}_{\parallel}(1) - \vec{x}_{\parallel}(2) = (m_1 a_0, m_2 a_0)$, where a_0 is the lattice parameter. At all temperatures the free energy of interaction is found to be attractive.

The existing calculations of the indirect interaction of two adatoms on a solid surface, mediated both by the elastic distortion of the substrate and by the phonons of the substrate, can be refined along the lines indicated in this section. However, the general properties of both types of interaction are now beginning to be understood on the basis of the existing work.

SCATTERING OF PHONONS BY CRYSTAL SURFACES

At very low temperatures the phonons of principal interest in calculating the lattice thermal conductivity are acoustical phonons of very long wavelength. The most important scattering mechanism for such phonons is scattering by the crystal boundaries. Other possible phonon scattering mechanisms become ineffective in the limit of very long wavelengths, and the mean free path for phonons becomes comparable to the linear dimensions of the crystal (158). An important quantity, therefore, in calculations of lattice thermal conductivities at very low temperatures is the inverse relaxation time, $\tau_{\vec{k}j}^{-1}$ for the scattering of a phonon with wave vector $\vec{k}$ and branch index j from crystal surfaces. We may define this quantity as the inverse of the time required for the deviation $\delta n_{\vec{k}j}$ of the occupation number $n_{\vec{k}j}$ from its equilibrium value $\bar{n}_{\vec{k}j}$ at a given temperature to relax to 1/e of its initial value when all of the other modes have the thermal equilibrium values for their occupation numbers. The inverse relaxation time can be expressed in terms of the so-called Casimir length L by the relation $\tau_{\vec{k}j}^{-1} = c/L$ where c is an average speed of sound (159). The Casimir length is roughly a typical dimension of the crystal.

The first step in calculating $\tau_{\vec{k}j}^{-1}$ is to obtain the rate $W(\vec{k}'j' \to \vec{k}j)$ at which a lattice wave $\vec{u}^{(0)}(\vec{k}'j')$ is scattered by a crystal surface into the wave $\vec{u}^{(0)}(\vec{k}j)$. Using the Fermi golden rule, we can express this rate as (160)

$$W(\vec{k}'j' \to \vec{k}j) = \frac{2\pi}{\hbar}\left|\left(\vec{u}^{(0)}(\vec{k}'j'),\overleftrightarrow{T}\left(\ell\kappa,\ell'\kappa';\omega_j^2(\vec{k}) + i0\right)\vec{u}^{(0)}(\vec{k}j)\right)\right|^2 \cdot \delta\left(\hbar\omega_j(\vec{k}) - \hbar\omega_{j'}(\vec{k}')\right) \qquad (390)$$

where $\overleftrightarrow{T}(\ell\kappa,\ell'\kappa';\omega_j^2(\vec{k}))$ is the scattering matrix for boundary scattering and $\omega_j(\vec{k})$ is the frequency of mode $(\vec{k}j)$. The lattice wave displacement field component $u_\alpha^{(0)}(\vec{k}j)$ can be chosen to have the form

$$u_\alpha^{(0)}(\ell\kappa,\vec{k}j) = \left(\frac{\hbar}{2NM_\kappa\omega_j(\kappa)}\right)^{\frac{1}{2}} w_\alpha(\kappa|\vec{k}j)e^{i\vec{k}\cdot\vec{x}(\ell\kappa)} \qquad (391)$$

where $w_\alpha(\kappa|\vec{k}j)$ is an eigenvector component of the dynamical matrix of the periodic crystal. If the time-independent equations of motion for the crystal with boundaries are written in the form

$$(\overleftrightarrow{L} - \delta\overleftrightarrow{L})\vec{u} = 0 \tag{392}$$

where $\overleftrightarrow{L}$ refers to the periodic crystal and $-\delta\overleftrightarrow{L}$ is the change produced by the boundaries, the scattering matrix $\overleftrightarrow{T}(\ell\kappa,\ell'\kappa';\omega_j^2(\vec{k}))$ satisfies the equation

$$\overleftrightarrow{T} = \delta\overleftrightarrow{L} + \delta\overleftrightarrow{L}\overleftrightarrow{G}\overleftrightarrow{T} \tag{393}$$

where $\overleftrightarrow{G}$ is the Green's function for the periodic crystal. We shall evaluate $\overleftrightarrow{T}$ for specific cases later.

We now relate the inverse relaxation time to the scattering rate $W(\vec{k}'j' \to \vec{k}j)$. The time rate of change of the occupation number $n_{\vec{k}j}$ is given by

$$\frac{dn_{\vec{k}j}}{dt} = \sum_{\vec{k}'j'} \left\{ n_{\vec{k}'j'}W(\vec{k}'j' \to \vec{k}j) - n_{\vec{k}j}W(\vec{k}j \to \vec{k}'j')\right\} \tag{394}$$

For the cases of interest here, we have the symmetry relation

$$W(\vec{k}'j' \to \vec{k}j) = W(\vec{k}j \to \vec{k}'j') \tag{395}$$

so that Eq. (394) becomes

$$\frac{dn_{\vec{k}j}}{dt} = \sum_{\vec{k}'j'} W(\vec{k}'j' \to \vec{k}j)\left(n_{\vec{k}'j'} - n_{\vec{k}j}\right) \tag{396}$$

We note that the right-hand side of this equation vanishes when $(\vec{k}'j') = (\vec{k}j)$; consequently, we can explicitly exclude the contributions of these terms from the sums over $\vec{k}'$ and j'. Recalling the definition of the inverse relaxation time, we write

$$\delta n_{\vec{k}j} = n_{\vec{k}j} - \bar{n}_{\vec{k}j} \tag{397}$$

and set $n_{\vec{k}'j'} = \bar{n}_{\vec{k}'j'}$ for $(\vec{k}'j') \neq (\vec{k}j)$. Equation (396) then takes the form

$$\frac{d}{dt}\,\delta n_{\vec{k}j} = \sum_{\vec{k}'j'}{}' W(\vec{k}'j' \to \vec{k}j)\left(\bar{n}_{\vec{k}'j'} - \bar{n}_{\vec{k}j} - \delta n_{\vec{k}j}\right) \tag{398}$$

where the prime on the sum means that the term for $(\vec{k}'j') = (\vec{k}j)$ is excluded. From Eq. (390) we see that $W(\vec{k}'j' \to \vec{k}j)$ vanishes unless $\omega_{j'}(\vec{k}') = \omega_j(\vec{k})$. Since $\bar{n}_{\vec{k}j}$ depends on $\vec{k}$ and j only through its dependence on $\omega_j(\vec{k})$, we see that Eq. (398) reduces to the simple form

$$\frac{d}{dt}\,\delta n_{\vec{k}j} = -\delta n_{\vec{k}j} \sum_{\vec{k}'j'}{}' W(\vec{k}'j' \to \vec{k}j) \tag{399}$$

$$= -\frac{1}{\tau_{\vec{k}j}}\,\delta n_{\vec{k}j} \tag{400}$$

Thus, the inverse relaxation time is given by

$$\tau_{\vec{k}j}^{-1} = \sum_{\vec{k}'j'}{}' W(\vec{k}'j' \to \vec{k}j) \tag{401}$$

We now proceed to evaluate the scattering matrix $\overleftrightarrow{T}$ by first expressing it as a double Fourier transform

$$T_{\alpha\beta}(\ell\kappa,\ell'\kappa';\omega^2) = \frac{(M_\kappa M_{\kappa'})^{\frac{1}{2}}}{N} \sum_{\vec{k}_1 j_1} \sum_{\vec{k}_2 j_2} w_\alpha(\kappa|\vec{k}_1 j_1)$$

$$\cdot\ t(\vec{k}_1 j_1, \vec{k}_2 j_2; \omega^2) w_\beta(\kappa'|\vec{k}_2 j_2)$$

$$\cdot\ e^{i\vec{k}_1 \cdot \vec{x}(\ell\kappa) + i\vec{k}_2 \cdot \vec{x}(\ell'\kappa')} \tag{402}$$

Substituting Eq. (402) into Eq. (390) gives

$$W(\vec{k}'j' \to \vec{k}j) = \frac{\pi}{2} \frac{|t(\vec{k}'j', -\vec{k}j, \omega_j^2(\vec{k}) + i0)|^2}{\omega_j(\vec{k})\omega_{j'}(\vec{k}')}$$

$$\cdot\ \delta(\omega_j(\vec{k}) - \omega_{j'}(\vec{k}')) \tag{403}$$

Combining Eqs. (401) and (403), we obtain

$$\tau_{\vec{k}j}^{-1} = \frac{\pi}{2} \frac{1}{\omega_j^2(\vec{k})} \sum_{\vec{k}'j'}{}' |t(\vec{k}'j', -\vec{k}j; \omega_j^2(\vec{k}) + i0)|^2$$

$$\cdot\ \delta(\omega_j(\vec{k}) - \omega_{j'}(\vec{k}')) \tag{404}$$

To calculate the inverse relaxation time explicitly, we use the model of a monatomic simple cubic crystal with nearest neighbor central and noncentral interactions (161). When the central and noncentral force constants are equal and the crystal has the form of a slab with faces perpendicular to the z-axis, one finds that

$$|t(\vec{k}'j',\vec{k}j;\omega_j^2(\vec{k}) + i0)|^2 = (\omega_L^4/9L^2)$$
$$\cdot\ \delta_{jj'},\delta(k_1 - k_1')\delta(k_2 - k_2')$$
$$\cdot\ \cos^2(a_0k_3/2)\sin^2(a_0k_3'/2) \tag{405}$$

where ω_L is the largest normal mode frequency of the periodic crystal, L is the slab thickness, and a_0 is the lattice constant. If this result is now substituted into Eq. (404), we obtain the inverse relaxation time in the form

$$\tau_{\vec{k}j}^{-1} = \frac{c}{L}\frac{\omega_L}{\sqrt{3}}\frac{|\sin(a_0k_3/2)\cos(a_0k_3/2)|}{\omega(\vec{k})} \tag{406}$$

where $c = a_0\omega_L/2\sqrt{3}$ is the speed of sound for this crystal model and

$$\omega(\vec{k}) = (\omega_L/\sqrt{3})\Big[\sin^2(a_0k_1/2) + \sin^2(a_0k_2/2)$$
$$+ \sin^2(a_0k_3/2)\Big]^{\frac{1}{2}} \tag{407}$$

Differentiation of Eq. (407) yields the result

$$\frac{\partial\omega(\vec{k})}{\partial k_z} = \frac{a_0\omega_L^2\sin(a_0k_3/2)\cos(a_0k_3/2)}{6\omega(\vec{k})} \tag{408}$$

which can be used to re-express Eq. (406) as

$$\tau_{\vec{k}j}^{-1} = \frac{1}{L}\left|\frac{\partial\omega(\vec{k})}{\partial k_3}\right| \tag{409}$$

$$= \frac{c(\vec{k})}{L} \tag{410}$$

where $c(\vec{k}) = |\partial\omega(\vec{k})/\partial k_3|$ is the z-component of the phonon group velocity. Thus, for this model, the Casimir length is simply the slab thickness.

We see from Eqs. (408) and (409) that for $\vec{k}_{\parallel} \neq 0$ $\left[\vec{k}_{\parallel} = (k_1, k_2, 0)\right]$, $\tau_{\vec{k}j}^{-1}$ vanishes when $k_3 = 0$ and when $k_3 = \pi/a_0$. The former result is a direct consequence of the fact that phonons with $k_3 = 0$ and $\vec{k}_{\parallel} \neq 0$ propagate parallel to the free surfaces and are not scattered by them. In the latter case the z-component of the phonon group velocity vanishes for k_3 on the Brillouin zone boundary, irrespective of the values of k_1 and k_2. Phonons for which the component of group velocity normal to the free surfaces vanishes will not be scattered by these surfaces.

One can also evaluate the scattering matrix $t(\vec{k}j, \vec{k}'j';\omega^2)$ analytically for the simple cubic crystal with nearest and next-nearest neighbor central force interactions (161). The inverse relaxation time is found to be expressible in the form

$$\tau_{\vec{k}j}^{-1} = \frac{1}{L}\left|\frac{\partial\omega_j(\vec{k})}{\partial k_3}\right|, \quad \vec{k} = (0,0,k_3) \tag{411}$$

This result is analogous to that obtained for the simple cubic lattice with nearest neighbor central and noncentral forces. In the long-wavelength limit, Eq. (411) becomes

$$\tau_{\vec{k}j}^{-1} = \frac{c_j}{L} \tag{412}$$

where $c_1 = a_0\omega_L/2$ and $c_{2,3} = a_0\omega_L/2\sqrt{3}$ are the speeds of sound for longitudinal and transverse waves propagating in the z-direction.

The foregoing results suggest that in the long-wavelength limit it is a good approximation to represent the inverse relaxation time for boundary scattering by

$$\tau_{\vec{k}j}^{-1} = \frac{c_j(\theta,\phi)}{L}, \quad |\vec{k}| \to 0 \tag{413}$$

where $c_j(\theta,\phi)$ is the speed of sound for elastic waves of polarization j, and θ and ϕ are the polar and azimuthal angles of the wave vector $\vec{k}$ with respect to the direction normal to the free surfaces.

STATIC RELAXATION AND THERMAL EXPANSION AT A CRYSTAL SURFACE

The free surface of a semi-infinite crystal can be created by setting to zero all interactions which cross a plane passing through an infinite lattice parallel to the surface plane but not containing any particles. The particles in the surface layer and in the adjacent interior layers are then acted upon by unbalanced forces, and consequently suffer displacements to new equilibrium positions.

We now undertake the calculation of the displacements of the mean positions of atoms near the surface from the mean positions that these atoms would have in the bulk of the crystal. If these displacements are determined by minimizing the static energy, we shall call them the static displacements. If they are determined by minimizing the total free energy, including vibrational contributions, we shall call them the dynamic displacements.

Surface Atomic Displacements

The positions of the atoms are specified by definition (B.2) and Eqs. (9) and (10). However, for simplicity we shall treat here only the case of a monatomic crystal with one atom per unit cell. The generalization to several atoms per unit cell and to the case of a crystal slab is straightforward: one has only to add the index κ each time ℓ appears.

Let us begin by expanding the potential energy of a semi-infinite Bravais crystal in powers of the displacements of the atoms from the equilibrium positions they would have had if they formed part of an infinitely extended, or cyclic, crystal instead of a semi-infinite one

$$\Phi = \Phi_0 + \sum_{\substack{\ell \\ \alpha}} \Phi_\alpha(\ell)\xi_\alpha(\ell) + \frac{1}{2} \sum_{\substack{\ell\ell' \\ \alpha\beta}} \Phi_{\alpha\beta}(\ell\ell')\xi_\alpha(\ell)\xi_\beta(\ell')$$
$$+ \frac{1}{6} \sum_{\substack{\ell\ell'\ell'' \\ \alpha\beta\gamma}} \Phi_{\alpha\beta\gamma}(\ell\ell'\ell'')\xi_\alpha(\ell)\xi_\beta(\ell')\xi_\gamma(\ell'') + \ldots \quad (414)$$

In this expression Φ_0 is the value of the static potential energy and $\xi_\alpha(\ell)$ is the α Cartesian component of the displacement of the ℓth atom. The first-order atomic force constants $\{\Phi_\alpha(\ell)\}$ are nonzero only for sites ℓ near the surface, because a first-rank tensor invariant under the operations of the point group of the lattice site to which it refers must vanish identically for an infinitely extended Bravais crystal, which includes the inversion among the symmetry operations at each site. $\Phi_{\alpha\beta}(\ell\ell')$ and $\Phi_{\alpha\beta\gamma}(\ell\ell'\ell'')$ are the harmonic and cubic anharmonic force constants, respectively.

We suppose that $\xi_\alpha(\ell)$ is the resultant of three terms, the first of which describes a homogeneous deformation of the crystal, the second of which represents the dynamical displacements of the atoms due to the free surfaces, and the third of which describes arbitrary displacements of the atoms from their new positions in the deformed crystal

$$\xi_\alpha(\ell) = s_\alpha(\ell) + d_\alpha(\ell) + u_\alpha(\ell) \tag{415}$$

where

$$s_\alpha(\ell) = \sum_\lambda \varepsilon_{\alpha\lambda} x_\lambda(\ell) \tag{416}$$

The vector $\vec{x}(\ell)$ is the vector to the equilibrium position of the ℓth atom in the undeformed crystal. The parameters $\{\varepsilon_{\alpha\lambda}\}$ describe a homogeneous deformation of the crystal. Let us now substitute Eq. (415) into Eq. (414) and collect terms in powers of the displacement components $\{u_\alpha(\ell)\}$. For simplicity, we write

$$\eta_\alpha(\ell) = s_\alpha(\ell) + d_\alpha(\ell) \tag{417}$$

We then obtain

$$\Phi = \Phi_S + \Phi_D \tag{418}$$

where

$$\Phi_S = \Phi_0 + \sum_{\ell\alpha} \Phi_\alpha(\ell)\eta_\alpha(\ell) + \tfrac{1}{2} \sum_{\substack{\ell\ell' \\ \alpha\beta}} \Phi_{\alpha\beta}(\ell\ell')\eta_\alpha(\ell)\eta_\beta(\ell')$$

$$+ \frac{1}{6} \sum_{\substack{\ell\ell'\ell'' \\ \alpha\beta\gamma}} \Phi_{\alpha\beta\gamma}(\ell\ell'\ell'')\eta_\alpha(\ell)\eta_\beta(\ell')\eta_\gamma(\ell'') + \ldots \quad (419)$$

and

$$\Phi_D = \sum_{\ell\alpha} \hat{\Phi}_\alpha(\ell)u_\alpha(\ell) + \tfrac{1}{2} \sum_{\substack{\ell\ell' \\ \alpha\beta}} \hat{\Phi}_{\alpha\beta}(\ell\ell')u_\alpha(\ell)u_\beta(\ell')$$

$$+ \frac{1}{6} \sum_{\substack{\ell\ell'\ell'' \\ \alpha\beta\gamma}} \hat{\Phi}_{\alpha\beta\gamma}(\ell\ell'\ell'')u_\alpha(\ell)u_\beta(\ell')u_\gamma(\ell'') + \ldots \quad (420)$$

with

$$\hat{\Phi}_\alpha(\ell) = \Phi_\alpha(\ell) + \sum_{\ell'\beta} \Phi_{\alpha\beta}(\ell\ell')\eta_\beta(\ell')$$

$$+ \tfrac{1}{2} \sum_{\substack{\ell'\ell'' \\ \beta\gamma}} \Phi_{\alpha\beta\gamma}(\ell\ell'\ell'')\eta_\beta(\ell')\eta_\gamma(\ell'') + \ldots \quad (421)$$

$$\hat{\Phi}_{\alpha\beta}(\ell\ell') = \Phi_{\alpha\beta}(\ell\ell') + \sum_{\ell''\gamma} \Phi_{\alpha\beta\gamma}(\ell\ell'\ell'')\eta_\gamma(\ell'') + \ldots \quad (422)$$

$$\hat{\Phi}_{\alpha\beta\gamma}(\ell\ell'\ell'') = \Phi_{\alpha\beta\gamma}(\ell\ell'\ell'') + \ldots \quad (423)$$

To first order in the deformation parameters $\{\eta_\alpha(\ell)\}$, the vibrational contribution to the Helmholtz free energy obtained from the dynamical part of the crystal potential energy, Eq. (420), can be written (162) in the form

$$F(T) = F^{(0)}(T) + \tfrac{1}{2} \sum_{\substack{\ell\ell'\ell'' \\ \alpha\beta\gamma}} \Phi_{\alpha\beta\gamma}(\ell\ell'\ell'') \cdot \langle u_\alpha(\ell)u_\beta(\ell')\rangle \eta_\gamma(\ell'') + \ldots \tag{424}$$

where $F^{(0)}(T)$ is the contribution to the vibrational free energy from the harmonic part only of Eq. (420). Explicit expressions for the correlation function $\langle u_\alpha(\ell)u_\beta(\ell')\rangle$ have been obtained earlier in this chapter, Eqs. (274a) and (276a), in the harmonic approximation. Another useful expression for them can also be derived by an easy generalization of Eq. (331) for the mean square displacements

$$\langle u_\alpha(\ell)u_\beta(\ell')\rangle = \frac{\hbar}{2m}\left[\tilde{D}^{-\frac{1}{2}}\coth\left(\hbar\tilde{D}^{\frac{1}{2}}/2k_BT\right)\right]_{\ell\alpha,\ell'\beta} \tag{425}$$

Finally, the total free energy of the crystal is

$$\mathcal{F}(T) = \Phi_S + F(T) \tag{426}$$

At each temperature T, we can obtain the new equilibrium positions of all the atoms by minimizing the free energy $\mathcal{F}(T)$ as a function of the $\{\varepsilon_{\nu\lambda}\}$ and $\{d_\alpha(\ell)\}$.

The bulk equilibrium distances are obtained from

$$\frac{\partial \mathcal{F}(T)}{\partial \varepsilon_{\nu\lambda}} = \sum_{\ell} \Phi_\nu(\ell)x_\lambda(\ell) + V\sum_{\beta\mu} C_{\nu\lambda\beta\mu}\varepsilon_{\beta\mu} + F_{\nu\lambda}(T) = 0 \tag{427}$$

where the volume of the crystal is V, the $\{C_{\nu\lambda\beta\mu}\}$ are the elastic moduli

$$C_{\nu\lambda\beta\mu} = \frac{1}{V} \sum_{\ell\ell'} \Phi_{\nu\beta}(\ell\ell')x_\lambda(\ell)x_\mu(\ell') \tag{428}$$

and

$$F_{\nu\lambda}(T) = \frac{1}{2} \sum_{\substack{\ell\ell'\ell'' \\ \alpha\beta}} \langle u_\alpha(\ell)u_\beta(\ell')\rangle \Phi_{\alpha\beta\nu}(\ell\ell'\ell'')x_\lambda(\ell'') \tag{429}$$

Similarly, the minimization of $F(T)$ with respect to $d_\alpha(\ell)$ gives

$$\begin{aligned} \Phi_\alpha(\ell) &+ \sum_{\ell'\beta} \Phi_{\alpha\beta}(\ell\ell')\eta_\beta(\ell') \\ &+ \frac{1}{2} \sum_{\ell'\ell''\beta\gamma} \Phi_{\alpha\beta\gamma}(\ell\ell'\ell'')\eta_\beta(\ell')\eta_\gamma(\ell'') \\ &+ F_\alpha(\ell|T) = 0 \end{aligned} \tag{430}$$

where

$$F_\alpha(\ell|T) = \frac{1}{2} \sum_{\substack{\ell'\ell'' \\ \beta\gamma}} \langle u_\beta(\ell')u_\gamma(\ell'')\rangle \Phi_{\beta\gamma\alpha}(\ell'\ell''\ell) \tag{431}$$

Equations (427) and (430) provide us with the new equilibrium positions at a given temperature T of all the atoms, both in the bulk and in the vicinity of the surface.

Static Relaxation

When the effects of anharmonicity and temperature are not included, there is no bulk expansion, but there can still be a static surface relaxation; one can calculate the latter from the form that Eq. (430) takes in this limit

$$\Phi_\alpha(\ell) + \sum_{\ell'\beta} \Phi_{\alpha\beta}(\ell\ell')d_\beta(\ell') = 0 \qquad (432)$$

Gazis and Wallis (163) have solved Eq. (432) for a semi-infinite linear chain with nearest and next-nearest neighbor interactions. A similar calculation has been carried out for a bcc lattice with a (001) surface by Clark et al. (164) using a Lennard-Jones interaction between nearest and next-nearest neighbors. In both cases, the static displacements were found to vary as $\exp(-q\ell_3)$ for a surface at $\ell_3 = 0$, where ℓ_3 is an integer ≥ 0 labeling the lattice planes, q has the form $q = q_0 + i\pi\theta$, q_0 is a real positive number of the order of unity, and θ is 0 or 1 depending on the force constants.

A number of calculations of surface static relaxations have been reported based on numerical solutions using models involving Lennard-Jones or Morse potential interactions between pairs of atoms. The results (165-179) typically show an expansion outward of the crystal at the surface. The increase in the interlayer spacing is in general of the order of a few percent at the surface and decreases rapidly towards the interior. It is also possible to obtain tangential relaxation in the surface planes (170) as well as normal relaxation.

The static relaxation near surface steps and kinks was also studied recently (180). However, these calculations do not take into account the electronic rearrangements near the surface which in general create surface dipole layers and then induce a contraction of the static surface interlayer spacings (181-186). A recent study (186) takes into account both effects (unbalanced linear forces $\{\Phi_\alpha(\ell)\}$ and electronic rearrangements) and finds for a specific example that these two effects are of the same order of magnitude, but with opposite signs.

Thermal Expansion

The change of the equilibrium spacing with temperature leads to the phenomenon of thermal expansion, and we may anticipate from the above discussion that the latter may also be different near a surface than it is in the bulk.

Experimental values of surface thermal expansion can be obtained from the temperature shifts of Bragg peaks in low-energy electron diffraction, as discussed by Wilson and Bastow (187). The positions of normal-incidence Bragg peaks are specified in the kinematic or single-scattering approximation by the equation

$$|\vec{Q}| = 4\pi/\lambda = n\pi/a_\perp \tag{433}$$

where $\vec{Q}$ is the scattering vector (the difference of the scattered and incident wave vectors of the electron), λ is the wavelength of the electron, n is an integer, and $a_\perp$ is the spacing between equivalent layers in the direction normal to the surface. The wavelength λ is related to the accelerating potential E_p and the inner potential V_i by the equation

$$\lambda = [150.4/E_p'(\text{eV})]^{\frac{1}{2}} \qquad (434)$$

where $E_p' = E_p + V_i$. The thermal expansion is given by

$$\alpha = \frac{1}{a_\perp}\frac{da_\perp}{dT} = -\frac{1}{2}\frac{1}{E_p'}\frac{dE_p'}{dT} \qquad (435)$$

For low values of E_p, the scattering is mainly from the surface and the surface thermal expansion is obtained, whereas for high values of E_p, which lead to deep penetration of the electrons into the crystal, the bulk thermal expansion is obtained.

Experimental values of the surface thermal expansion have been determined for several systems. Gelatt et al. (188) have reported results for the ratio α_s/α_b (the subscripts s and b refer to the surface and bulk, respectively) for Ag(111) and Ni(111) which are in the range of one to two. Wilson and Bastow (187) have studied the (100) surfaces of Cr and Mo and found values of α_s/α_b of the order of two or three. Ignatiev and Rhodin (189) have investigated the (111) surface of Xe and obtained α_s/α_b values of about four or five.

The theoretical treatment of surface thermal expansion can be developed on several levels. A simple discussion has been given by Wilson and Bastow (187) based on the Grüneisen expression for thermal expansion which can be written in the approximate form

$$\alpha = 2\gamma C_v/3Ea_0^2 \qquad (436)$$

where γ is the Grüneisen constant, C_v is the specific heat per atom at constant volume, E is Young's modulus, and a_0 is the lattice constant. Considering only high temperatures, we have $C_v = 3k_B$, where k_B is Boltzmann's constant. The quantity Ea_0^2 can be regarded as a force constant related to the Debye frequency ω_D and the Debye temperature Θ_D by the equations

$$Ea_0^2 = M\omega_D^2 = M(k_B\Theta_D/\hbar)^2 \tag{437}$$

where M is the atomic mass. Equation (436) now takes the form

$$\alpha = \frac{k_B\gamma}{a_0^2 M}\left(\frac{\hbar}{k_B\Theta_D}\right)^2 \tag{438}$$

If γ is the same at the surface as in the bulk, then

$$\frac{\alpha_s}{\alpha_b} = \left(\frac{\Theta_{Db}}{\Theta_{Ds}}\right)^2 = \frac{\langle u_s^2\rangle}{\langle u_b^2\rangle} \tag{439}$$

where $\langle u_s^2\rangle$ and $\langle u_b^2\rangle$ are the mean square displacements of surface and bulk atoms, respectively. Equation (439) appears to be in rough agreement with available experimental data.

In a more refined theory, one would calculate the thermal expansion by minimizing the Helmholtz free energy with respect to variations in the interlayer spacings. Kenner and Allen (179) have done this and obtained the following expression for the change in the thermal expansion of the ith layer relative to the bulk value

$$\Delta\alpha_i = k_B \sum_j a_{oi}^{-1} (\overleftrightarrow{\Phi}^{-1})_{ij} a_{oj}^{-1} \sum_p \gamma_i(\omega_p)$$

$$\cdot \left(\hbar\omega_p / 2k_B T\right)^2 \operatorname{cosech}^2\left(\hbar\omega_p / 2k_B T\right) \qquad (440)$$

Here $\overleftrightarrow{\Phi}$ is the potential energy matrix, ω_p is the pth normal mode frequency, $\gamma_j(\omega_p)$ is the Grüneisen parameter defined by

$$\gamma_i(\omega) = -a_i \frac{\partial \ln \omega}{\partial a_i} \qquad (441)$$

a_i is the ith lattice spacing, and a_{oi} is a reference lattice spacing. Using Lennard-Jones interactions, Kenner and Allen have made calculations for the (100) and (111) surfaces of Ar, Kr, and Xe. They find a peak in α_s/α_b at low temperatures associated with the dispersion of surface modes. In the temperature range of the experiments of Ignatiev and Rhodin (189) on Xe(111), Kenner and Allen obtain the result $\alpha_s/\alpha_b \cong 1.9$, which is significantly smaller than the experimental result quoted above.

The minimization of the Helmholtz free energy has also been used by Dobrzynski and Maradudin (162) to calculate the surface dynamic displacements, Eqs. (427) and (430), and then the surface thermal expansion. Since they retained only third-order anharmonic terms, they were able to perform much of their calculation analytically, in contrast to the numerical methods of Kenner and Allen. Specific calculations of the surface thermal expansion for the (100) surface of α-iron yielded results for α_s/α_b that are somewhat greater than unity at high temperatures, but increase rapidly at temperatures below 50°K. The peak in α_s/α_b was not obtained at low temperatures, because of their calculation of the correlation functions by an expansion around the Einstein approximation, Eq. (337).

Force Constant Changes and Renormalized Phonons

Once the new surface atomic positions are known, one can calculate the renormalized harmonic force constants $\{\hat{\Phi}_{\alpha\beta}(\ell\ell')\}$ from Eq. (422). This was done by Cheng et al. (190,191) for the (001) surface of several bcc crystals, using Lennard-Jones potentials. They found that an outward relaxation of about 5% may decrease the value of the surface force constants by as much as 30%.

Djafari-Rouhani et al. (144,145,192) calculated the surface force constant changes for an adsorbed monolayer of one rare gas on (001) and (111) surfaces of another rare gas solid. They found that, at a given temperature, T_c, the frequency of an acoustical surface phonon, near the Brillouin zone edge, may vanish. One obtains a soft surface phonon, which predicts that the monolayer is unstable at this temperature T_c and will probably rearrange into another surface superstructure. They (140) also used the renormalized surface force constants to evaluate the variation with temperature of the optical surface phonons due to a monolayer of oxygen adsorbed on a (001) Ni surface.

The above calculations (see Refs. 190-193) of the surface force constant changes do not include the electronic rearrangements near the surface. Calculations taking this effect into account are under study (184, 193,194).

SURFACE WAVE ATTENUATION

The attenuation of surface waves is of interest both as a topic in pure physics and as a property which plays an important role in semiconductor devices. Clearly, if a surface wave attenuates too rapidly, it will be relatively useless for transmitting information. There are a number of mechanisms for surface wave attenuation. These include scattering by anharmonic interactions, by defects such as impurities, and by current carriers such as conduction electrons in metals.

A very simple approach to surface wave attenuation has been given by Press and Healy (195) who consider an isotropic solid. They start from the equation specifying the speed of Rayleigh waves c_R in terms of the speeds of longitudinal and transverse bulk waves, c_ℓ and c_t, respectively, Eq. (84)

$$\left(2 - \frac{c_R^2}{c_t^2}\right)^4 - 16\left(1 - \frac{c_R^2}{c_\ell^2}\right)\left(1 - \frac{c_R^2}{c_t^2}\right) = 0 \qquad (442)$$

Press and Healy introduce dissipation by letting each speed become complex, so that

$$c_j \rightarrow c_j\left(1 + i\delta_j\right) \qquad j = R, \ell, t \qquad (443)$$

Substituting Eq. (443) into Eq. (442) and letting $\alpha_j = \omega\delta_j/2c_j$, one obtains the following relation between the attenuation constant for Rayleigh waves, α_R, and those for longitudinal and transverse bulk waves, α_ℓ and α_t

$$\alpha_R = \left\{\left[4\left(1 - \frac{c_R^2}{c_\ell^2}\right) - \left(2 - \frac{c_R^2}{c_t^2}\right)^3\right]\left(\frac{c_t}{c_R}\right)\alpha_t + \left[4\left(1 - \frac{c_R^2}{c_t^2}\right)\left(\frac{c_t^2}{c_\ell^2}\right)\right]\left(\frac{c_\ell}{c_R}\right)\alpha_\ell\right\} \div \left\{4\left(1 - \frac{c_R^2}{c_t^2}\right)\frac{c_t^2}{c_\ell^2} + 4\left(1 - \frac{c_R^2}{c_\ell^2}\right) - \left(2 - \frac{c_R^2}{c_t^2}\right)^3\right\} \qquad (444)$$

For the case where $c_\ell = \sqrt{3}\, c_t$ and $c_R = 0.9194\, c_t$, the result is $\alpha_R = 0.942\, \alpha_t + 0.252\, \alpha_\ell$. Equation (444) seems to be reasonably well obeyed by many noncrystalline solids which in the first approximation are isotropic.

The preceding discussion is applicable only for surface waves whose wavelength is very large compared to a lattice spacing and involves no assumption concerning the mechanism. More detailed calculations based on a specific mechanism have been carried out, particularly in the long-wavelength approximation. An example is the work of King and Sheard (196) on anharmonic scattering. We shall not discuss their work in detail, but instead will focus our attention on a lattice dynamical calculation following the treatment of Maradudin and Mills (197).

We restrict our attention to a monatomic simple cubic lattice which has nearest and next-nearest neighbor central harmonic interactions and nearest neighbor cubic anharmonic interactions. As a measure of the attenuation, we take the damping constant $\Gamma_s(\omega)$ which is related to the proper self-energy $P_s(\omega)$ by the equation

$$\Delta_s(\omega) + i\Gamma_s(\omega) = -\frac{k_B T}{\hbar} \lim_{\varepsilon\to 0} P_s(\omega - i\varepsilon) \qquad (445)$$

where $\Delta_s(\omega)$ is the frequency shift and ω is to be taken equal to the surface mode frequency ω_R. To lowest non-vanishing order in perturbation theory, the proper self-energy can be expressed as

$$P_s(i\omega_\ell) = (\hbar/4\omega_s M) \sum_{\ell\,\ell'\,\ell''} \sum_{m\,m'\,m''} \sum_{\alpha_1\alpha_2\alpha_3} \sum_{\beta_1\beta_2\beta_3}$$

$$\cdot\ \Phi_{\alpha_1\alpha_2\alpha_3}(\ell\,\ell'\,\ell'') \times \Phi_{\beta_1\beta_2\beta_3}(m\,m'\,m'') B_s(\ell\alpha_1) B_s(m\,\beta_1)$$

$$\cdot \sum_{n=-\infty}^{\infty} U_{\alpha_2\beta_2}(\ell' m'; -\omega_n^2)\, U_{\alpha_3\beta_3}[\ell'' m''; -(\omega_\ell - \omega_n)^2] \qquad (446)$$

In this equation, $\omega_\ell = 2\pi\ell k_B T/\hbar$, where ℓ is an integer; M is the mass of an atom; $B_s(\ell\alpha)$ is an eigenvector component of the dynamical matrix including the free surfaces; $\Phi_{\alpha_1\alpha_2\alpha_3}(\ell\,\ell'\,\ell'')$ is the cubic anharmonic coefficient in the cubic anharmonic Hamiltonian

$$H_3 = \frac{1}{6} \sum_{\ell\,\ell'\,\ell''} \sum_{\alpha_1\alpha_2\alpha_3} \Phi_{\alpha_1\alpha_2\alpha_3}(\ell\,\ell'\,\ell'')\, u_{\alpha_1}(\ell)\, u_{\alpha_2}(\ell')\, u_{\alpha_3}(\ell'') \qquad (447)$$

and the Green's function $U_{\alpha\beta}(\ell\,\ell';\omega^2)$ is specified by

$$U_{\alpha\beta}(\ell\ell';\omega^2) = \frac{1}{M}\sum_s \frac{B_s(\ell\alpha)B_s(\ell'\beta)}{\omega^2 - \omega_s^2} \tag{448}$$

The evaluation of the proper self-energy is facilitated if the Green's function $U_{\alpha\beta}(\ell\,\ell';\omega^2)$ and the eigenvector components $B_s(\ell\alpha)$ are expanded in terms of the eigenvector components and eigenfrequencies of the unperturbed crystal without the free surfaces. Thus, we write

$$U_{\alpha\beta}(\ell\,\ell';\omega^2) = \frac{1}{NM}\sum_{\vec{k}_1 j_1}\sum_{\vec{k}_2 j_2} e_\alpha(\vec{k}_1 j_1)e_\beta(\vec{k}_2 j_2) \cdot e^{i\vec{k}_1\cdot\vec{x}(\ell) + i\vec{k}_2\cdot\vec{x}(\ell')} u(\vec{k}_1 j_1;\vec{k}_2 j_2;\omega^2) \tag{449}$$

$$B_s(\ell\alpha) = \frac{1}{N^{\frac{1}{2}}}\sum_{\vec{k}j} e_\alpha(\vec{k}j)e^{i\vec{k}\cdot\vec{x}(\ell)}C_s(\vec{k}j) \tag{450}$$

where N is the number of atoms in the crystal, and $\vec{e}(\vec{k}j)$ is the unit polarization vector for the normal mode of wave vector $\vec{k}$ and branch index j. It should be noted that the perturbed Green's function is a function of ℓ and ℓ' separately, not just of their difference, so a double Fourier series is required in its expansion. The proper self-energy can now be rewritten as

$$P_s(i\omega_\ell) = \frac{72}{\hbar^2\omega_s} \sum_{\vec{k}_1\vec{k}_2\vec{k}_3} \sum_{j_1 j_2 j_3} \sum_{\vec{k}_1'\vec{k}_2'\vec{k}_3'} \sum_{j_1' j_2' j_3'}$$

$$\cdot\ C_s(\vec{k}_1 j_1) C_s(\vec{k}_1' j_1')$$

$$\cdot\ W(\vec{k}_1 j_1;\vec{k}_2 j_2;\vec{k}_3 j_3) W(\vec{k}_1' j_1';\vec{k}_2' j_2';\vec{k}_3' j_3')$$

$$\cdot \sum_{n=-\infty}^{\infty} u(\vec{k}_2 j_2;\vec{k}_2' j_2';-\omega_n^2)\, u\left(\vec{k}_3 j_3;\vec{k}_3' j_3';-(\omega_\ell - \omega_n)^2\right) \tag{451}$$

where

$$W(\vec{k}_1 j_1;\vec{k}_2 j_2;\vec{k}_3 j_3) = \frac{1}{6}\left(\frac{\hbar}{2NM}\right)^{3/2} \sum_{\ell\ell'\ell''} \sum_{\alpha\beta\gamma} \Phi_{\alpha\beta\gamma}(\ell\ell'\ell'')$$

$$\cdot\ e_\alpha(\vec{k}_1 j_1) e_\beta(\vec{k}_2 j_2) e_\gamma(\vec{k}_3 j_3)$$

$$\cdot\ e^{i\vec{k}_1 \cdot \vec{x}(\ell) + i\vec{k}_2 \cdot \vec{x}(\ell') + i\vec{k}_3 \cdot \vec{x}(\ell'')} \tag{452}$$

The evaluation of the sum over n in Eq. (451) is facilitated by defining the spectral density $\bar{u}(\vec{k}j;\vec{k}'\vec{j}';\nu)$

$$u(\vec{k}j;\vec{k}'\vec{j}';\omega^2) = \int_{-\infty}^{\infty} \frac{\bar{u}(\vec{k}j;\vec{k}'\vec{j}';\nu)}{\nu - \omega}\, d\nu \tag{453}$$

With the aid of the theory of contour integration, one can show that

$$\sum_{n=-\infty}^{\infty} \left(\frac{1}{\nu_1 - i\omega_n}\right)\left(\frac{1}{\nu_2 - i(\omega_\ell - \omega_n)}\right) = \frac{\hbar}{k_B T} \frac{[\theta(\nu_1) + \theta(\nu_2)]}{(\nu_1 + \nu_2 - i\omega_\ell)} \tag{454}$$

where

$$\theta(\nu) = \tfrac{1}{2} \coth(\hbar\nu/2k_B T) \tag{455}$$

The damping constant $\Gamma_s(\omega)$ can now be obtained from Eqs. (445), (451), and (455)

$$\Gamma_s(\omega) = \frac{72\pi}{\hbar^2\omega_s} \sum_{\vec{k}_1\vec{k}_2\vec{k}_3} \sum_{j_1j_2j_3} \sum_{\vec{k}_1'\vec{k}_2'\vec{k}_3'} \sum_{j_1'j_2'j_3'}$$

$$\cdot\ C_s(\vec{k}_1 j_1) C_s(\vec{k}_1' j_1')$$

$$\cdot\ W(\vec{k}_1 j_1; \vec{k}_2 j_2; \vec{k}_3 j_3) W(\vec{k}_1' j_1'; \vec{k}_2' j_2'; \vec{k}_3' j_3')$$

$$\cdot \int_{-\infty}^{\infty} d\nu_1 \int_{-\infty}^{\infty} d\nu_2\ \bar{u}(\vec{k}_2 j_2; \vec{k}_2' j_2'; \nu_1) \bar{u}(\vec{k}_3 j_3; \vec{k}_3' j_3'; \nu_2)$$

$$\cdot\ [\theta(\nu_1) + \theta(\nu_2)] \delta(\omega_s - \nu_1 - \nu_2) \tag{456}$$

The result expressed by Eq. (456) is quite general and specifies the damping constant for the sth normal mode. We now specialize to the case where the normal mode is a Rayleigh surface wave. We simplify the problem by noting that the wavelength of an ultrasonic surface phonon is long compared to the wavelength of thermal surface phonons, even at low temperatures, and that the penetration distance is proportional to the wavelength. Consequently, one expects that the interaction of an ultrasonic surface phonon with thermal surface phonons will be negligible compared to the interaction with thermal bulk phonons. Taking into account only the scattering by thermal bulk phonons and neglecting the effect of the surface on these phonons, we can approximate the spectral density by its value $\bar{u}_0(\vec{k}j;\vec{k}'j';\omega)$ for the perfect crystal

$$\bar{u}_0(\vec{k}j;\vec{k}'j';\omega) = \Delta(\vec{k} + \vec{k}')\delta_{jj'}\frac{\{\delta[\omega + \omega_j(\vec{k})] - \delta[\omega - \omega_j(\vec{k})]\}}{2\omega_j(\vec{k})} \qquad (457)$$

where $\Delta(\vec{k})$ equals unity when $\vec{k}$ equals the reciprocal lattice vector and vanishes otherwise, and $\omega_j(\vec{k})$ is the frequency of the normal mode of wave vector $\vec{k}$ and branch index j for the perfect crystal. Substitution of Eq. (457) into Eq. (456) and integration over ν_1 and ν_2 yields

$$\Gamma_s(\omega) = \frac{18\pi}{\hbar^2\omega_s}\sum_{\vec{k}j}\sum_{\vec{k}'j'}\sum_{\vec{k}_1j_1}\sum_{\vec{k}_2j_2} \sigma_s(\vec{k}j)\sigma_s(\vec{k}'j')[\omega_j(\vec{k})\omega_{j'}(\vec{k}')]^{\frac{1}{2}}$$

$$\cdot\ V(\vec{k}j;\vec{k}_1j_1;\vec{k}_2j_2)V^*(-\vec{k}'j';\vec{k}_1j_1;\vec{k}_2j_2)$$

$$\cdot \left\{ \left[n(\vec{k}_1 j_1) + n(\vec{k}_2 j_2) + 1 \right] \delta\left(\omega_s - \omega_{j_1}(\vec{k}_1) - \omega_{j_2}(\vec{k}_2) \right) \right.$$

$$+ \left[n(\vec{k}_1 j_1) - n(\vec{k}_2 j_2) \right] \left[\delta\left(\omega_s + \omega_{j_1}(\vec{k}_1) - \omega_{j_2}(\vec{k}_2) \right) \right.$$

$$\left. \left. - \delta\left(\omega_s - \omega_{j_1}(\vec{k}_1) + \omega_{j_2}(\vec{k}_2) \right) \right] \right\} \tag{458}$$

where $n(\vec{k}j) \equiv n\left(\omega_j(\vec{k})\right) = \theta\left(\omega_j(\vec{k})\right) - \frac{1}{2}$ and where

$$V(\vec{k}_1 j_1; \vec{k}_2 j_2; \vec{k}_3 j_3) = \left[\omega_{j_1}(\vec{k}_1)\, \omega_{j_2}(\vec{k}_2)\, \omega_{j_3}(\vec{k}_2) \right]^{-\frac{1}{2}}$$

$$\cdot\, W(\vec{k}_1 j_1; \vec{k}_2 j_2; \vec{k}_3 j_3) \tag{459}$$

We now address our attention to the calculation of the quantity $C_s(\vec{k}j)C_s(\vec{k}'j')$ for a Rayleigh wave. From Eq. (448) we see that we can write

$$\frac{B_s(\ell\alpha)B_s(\ell'\beta)}{2\omega_s} = M \operatorname{Res}_{\omega_s}\{U_{\alpha\beta}(\ell\ell';\omega^2)\} \tag{460}$$

where $\operatorname{Res}_{\omega_s}\{U_{\alpha\beta}(\ell\ell';\omega^2)\}$ stands for the residue of $U_{\alpha\beta}(\ell\ell';\omega^2)$ at the simple pole $\omega = \omega_s$. Utilizing Eqs. (449) and (450), we obtain

$$\frac{C_s(\vec{k}j)C_s(\vec{k}'j')}{2\omega_s} = \operatorname{Res}_{\omega_s}\{u(\vec{k}j;\vec{k}'j';\omega^2)\} \tag{461}$$

The double Fourier-transformed Green's function can be written in the form

$$u(\vec{k}j;\vec{k}'j';\omega^2) = u_0(\vec{k}j;\vec{k}'j';\omega^2) + \Delta u(\vec{k}j;\vec{k}'j';\omega^2) \quad (462)$$

where $u_0(\vec{k}j;\vec{k}'j';\omega^2)$ is the perfect crystal contribution given by

$$u_0(\vec{k}j;\vec{k}'j';\omega^2) = \frac{\Delta(\vec{k} + \vec{k}')\delta_{jj'}}{\omega^2 - \omega^2(\vec{k}j)} \quad (463)$$

and $\Delta u(\vec{k}j;\vec{k}'j';\omega^2)$ is the contribution due to the surface.

We now specialize to the model of interest and create a pair of (001) free surfaces by setting to zero all interactions coupling atoms in the plane $\ell_3 = 1$ with those in the plane $\ell_3 = 0$. For each surface atom a total of five "bonds" are cut. The surface correction $\Delta u(\vec{k}j;\vec{k}'j';\omega^2)$ can then be expressed as

$$\Delta u(\vec{k}j;\vec{k}'j';\omega^2) = \Delta(k_1 + k_1')\Delta(k_2 + k_2')$$

$$\cdot \sum_{\ell\ell'} \frac{v_\ell(\vec{k}j)v_{\ell'}(\vec{k}'j')}{(\omega^2 - \omega_j^2(\vec{k}))(\omega_{j'}^2 - \omega^2(\vec{k}'))}$$

$$\cdot \left[\overleftrightarrow{I} - \overleftrightarrow{M}(k_1k_2;\omega^2)\right]^{-1}_{\ell\ell'} \quad (464)$$

where ℓ and ℓ' take on only the five values corresponding to the bonds broken in creating the surface, $\overleftrightarrow{I}$ is the 5 x 5 unit matrix, and $\overleftrightarrow{M}(k_1k_2;\omega^2)$ is a 5 x 5 matrix given by

$$M_{\ell\ell'}(k_1k_2;\omega^2) = \sum_{k_3j} \frac{v_\ell(-\vec{k}j)v_{\ell'}(\vec{k}j)}{\omega^2 - \omega_j^2(\vec{k})} \qquad (465)$$

$$v_\ell(\vec{k}j) = \delta_{\ell_3,-1} \frac{2[\phi''(r_\ell)]^{\frac{1}{2}}}{(LM)^{\frac{1}{2}} r_\ell} [\vec{x}(\ell) \cdot \vec{e}(\vec{k}j)]$$

$$\cdot\, e^{(i/2)\vec{k}\cdot\vec{x}(\ell)} \sin \tfrac{1}{2}\, \vec{k}\cdot\vec{x}(\ell) \qquad (466)$$

Here $\phi''(r_\ell)$ is the second derivative of the interaction potential of two atoms separated by the distance r_ℓ and L is the number of atoms on a cube edge. The evaluation of Eq. (461) is rather tedious, and the reader is referred to the paper by Maradudin and Mills (197) for details. In the long-wavelength limit, which is all that is required at low temperatures, one obtains

$$\frac{C_s(\vec{k}j)C_s(\vec{k}'j')}{2\omega_s} = -\frac{c_R c_t^4}{La_0\omega_R^2} \frac{(0.39332)}{(1+\sqrt{3})}$$

$$\cdot \frac{\Delta(q_1 + k_1')\Delta(q_2 + k_2')}{\left[\left(c_j^2 - c_R^2\right)r^2 + c_j^2 k_3^2\right]\left[\left(c_{j'}^2 - c_R^2\right)r^2 + c_{j'}^2 k_3'^2\right]}$$

$$\times \left\{\left[e_1(\vec{k}'j\,)k'_3 + e_3(\vec{k}'j')9_1\right]q_1\right.$$

$$+ \left[e_2(\vec{k}j)k_3 + e_3(\vec{k}j)9_2\right]q_2\Big\}$$

$$\times \Big\{\left[e_1(\vec{k}'j')k_3' + e_3(\vec{k}'j')k_1'\right]q_1$$

$$+ \left[e_2(\vec{k}'j')k_3' + e_3(\vec{k}'j')k_2'\right]q_2\Big\} \qquad (467)$$

where a_0 is the nearest neighbor separation and $r^2 = 9_1^2 + 9_2^2$.

The anharmonic coefficient $V(\vec{k}_1j_1;\vec{k}_2j_2;\vec{k}_3j_3)$ in principle is affected by the surface. However, if the interaction potential $\phi(r)$ drops off rapidly with increasing r, the effect of the surface on the anharmonic coefficients gives a negligible contribution to the damping constant for deeply penetrating Rayleigh waves. We shall include only nearest neighbor contributions to the anharmonic coefficients and shall neglect the effect of the surface. We also retain only the contribution from $\phi'''(r)$ and neglect the contributions from $\phi''(r)$ and $\phi'(r)$. Then we can write

$$V(\vec{k}j;\vec{k}_1j_1;\vec{k}_2j_2) \cdot V^*(-\vec{k}'j';\vec{k}_1j_1;\vec{k}_2j_2)$$

$$= (4/9N)(\hbar/2M)^3[\phi'''(a_0)]^2$$

$$\cdot \Delta(\vec{k} + \vec{k}')\Delta(\vec{k} + \vec{k}_1 + \vec{k}_2)$$

$$\times \sum_{\ell\ell'} e^{-\frac{i}{2}\vec{G}\cdot[\vec{x}(\ell) - \vec{x}(\ell')]} V_\ell(\vec{k}j)V_\ell(\vec{k}j')$$

$$\times\ V_{\ell}(\vec{k}_1 j_1) V_{\ell'}(\vec{k}_1 j_1) V_{\ell'}(\vec{k}_2 j_2) V_{\ell'}(\vec{k}_2 j_2) \qquad (468)$$

where $V_{\ell}(\vec{k}j) = \frac{1}{a_0}\{\vec{x}(e) \cdot \vec{e}(\vec{k}j)/[\omega(\vec{k}j)]^{\frac{1}{2}}\}\sin \frac{1}{2}\ \vec{x}(e) \cdot \vec{k}$, $\vec{G}$ is a reciprocal lattice vector defined by $\vec{k} + \vec{k}_1 + \vec{k}_2 = \vec{G}$ and $\vec{x}(e)a_0 = (\ell_1, \ell_2, \ell_3)$.

One now has the ingredients at hand to evaluate the damping constant. Substituting Eqs. (467) and (468) into Eq. (458) and working in the low-temperature and long-wavelength limits, one obtains after considerable tedious manipulation

$$\Gamma_R = \frac{0.3971}{(2\pi)^3}\ \zeta(4)\left[0.0521 + 0.337\left(\frac{c_t}{c_\ell}\right)^8\right][\phi'''(a_0)]^2$$

$$\frac{\omega_R (k_B T)^4}{\rho^3 c_R^2 c_t^7 \hbar^3} \qquad (469)$$

We note that Γ_R has the same ωT^4 dependence derived by Landau and Rumer (198) for the damping of bulk waves. However, the magnitude of the damping for Rayleigh waves is significantly larger than for bulk waves.

Recent experimental work has verified the frequency and temperature dependences predicted by Eq. (469). Salzmann, Plieninger, and Dransfeld (199) and Budreau and Carr (200) have carried out experimental studies of Rayleigh wave damping on quartz surfaces. They verify the ωT^4 dependence below 40°K. Above 60°K, the damping becomes independent of temperature and varies as ω^2. The latter dependence can be understood in terms of the viscosity theories of Maris (201) and of King and Sheard (202).

We have already noted that Rayleigh waves can be damped as a result of their interaction with impurities or other imperfections. Steg and Klemens (203) have made a calculation of the Rayleigh-wave damping constant for impurity scattering for the case of an isotropic continuum. Their result exhibits an ω^5 dependence on frequency, but no explicit temperature dependence. An alternate calculation of this damping is described in the next section.

A CONTINUUM APPROACH TO SURFACE LATTICE DYNAMICS

The dynamical theory of semi-infinite crystals and of crystal slabs presented in the preceding sections of this chapter has been based on lattice theory: the media studied have been described by doubly periodic arrays of interacting atoms or ions filling a half-space or constituting a crystal slab of finite thickness. In this section we present a discussion of dynamical properties of planar, stress-free surfaces on perfect and imperfect solids, based on continuum theory, in which the emphasis is on obtaining essentially analytic results. The latter requirement means that much, but not all, of the work to be described has been carried out for elastically isotropic media. The consequent loss of generality is more than compensated, however, by the explicit nature of the results obtained.

The work presented here is intended to complement, rather than to replace, results for dynamical properties of solid surfaces obtained by purely numerical calculations in which the normal mode frequencies and the corresponding unit eigenvectors of a semi-infinite crystal are obtained by solving the equations of motion of such a crystal by representing it as a slab of a finite number of atomic layers. The explicit, analytic results obtained display their dependences on the parameters of the problem, such as temperature, frequency, and distance into the solid from the surface, in a way that purely numerical results cannot, and can help to interpret the results of the usually more detailed numerical calculations. In addition, certain kinds of calculations, e.g., the determination of the attenuation of Rayleigh surface waves by point defects, and the study of effects associated with interfaces, appear to be more easily carried out on the basis of continuum theory at the present time than by lattice theory.

The continuum approach to the study of the dynamical properties of planar, stress-free surfaces will be illustrated here by its application to three problems: (a) surface atom mean square displacements; (b) the scattering of Rayleigh surface waves by point defects; and (c) the surface contribution to the low-temperature specific heat of a semi-infinite solid. The dynamical Green's functions for a semi-infinite elastic medium bounded by a planar, stress-free surface play a central role in the first two of these illustrations. A different, but related, type of Green's function is found to be helpful in obtaining the surface contributions to the low-temperature specific heat of a semi-infinite solid, and will be discussed below.

The three problems selected for discussion hardly exhaust the kinds of problems in the dynamical theory of solid surfaces that can be studied essentially analytically by the use of continuum theory. They are, however, representative and paradigmatic. A brief discussion of other problems which have been, are being, or can be studied from this point of view, for the reasons indicated above, is presented at the end of this section.

Surface Atom Mean Square Displacements

The dynamical Green's functions for a semi-infinite elastic medium bounded by a planar, stress-free surface enable us to obtain almost immediately the mean square displacement of an atom in the vicinity of the surface of a semi-infinite crystal, a quantity that is susceptible to experimental study by the techniques of low-energy electron diffraction.

Probably the most detailed calculations of the mean square displacement of an atom in the surface layers of a crystal have been carried out from a discrete lattice point of view. Such calculations and their results have already been described. However, the earliest calculations of this quantity were carried out on the basis of continuum theory (204) which, despite its limitations, is capable of yielding information of a kind which would be difficult to extract from purely numerical, discrete lattice calculations, as we shall see.

We now show how the mean square displacement can be related to the Green's functions and present results for a semi-infinite isotropic elastic continuum, bounded by a planar, stress-free surface.

To express the mean square displacement $\langle u_\alpha^2(\vec{x})\rangle$ in terms of the dynamical Green's functions of elasticity theory we begin with the normal coordinate transformation of Eq. (288)

$$\langle u_\alpha^2(\vec{x})\rangle = \frac{\hbar}{2\rho} \sum_{ss'} \frac{1}{(\omega_s \omega_{s'})^{\frac{1}{2}}} \Big\{ v_\alpha^{(s)}(\vec{x}) v_\alpha^{(s')}(\vec{x}) \langle b_s b_{s'}\rangle + v_\alpha^{(s)}(\vec{x}) v_\alpha^{(s')}(\vec{x})^* \langle b_s b_{s'}^+\rangle + v_\alpha^{(s)}(\vec{x})^* v_\alpha^{(s')}(\vec{x}) \cdot \langle b_s^+ b_{s'}\rangle + v_\alpha^{(s)}(\vec{x})^* v_\alpha^{(s')}(\vec{x})^* \langle b_s^+ b_{s'}^+\rangle \Big\} \tag{470}$$

The statistical averages given by Eq. (297) reduce this expression to

$$\langle u_\alpha^2(\vec{x})\rangle = \frac{\hbar}{2\rho}\sum_s \frac{v_\alpha^{(s)}(\vec{x})v_\alpha^{(s)}(\vec{x})^*}{\omega_s}\left(2n_s + 1\right) \quad (471)$$

We next use the following representation for $2n_s + 1$

$$2n_s + 1 = \frac{2\omega_s}{\beta\hbar}\sum_{n=-\infty}^{\infty}\frac{1}{\Omega_n^2 + \omega_s^2}, \quad \Omega_n = \frac{2\pi n}{\beta\hbar} \quad (472)$$

to rewrite Eq. (471) as

$$\langle u_\alpha^2(\vec{x})\rangle = \frac{k_B T}{\rho}\sum_{n=-\infty}^{\infty}\sum_s \frac{v_\alpha^{(s)}(\vec{x})v_\alpha^{(s)}(\vec{x})^*}{\Omega_n^2 + \omega_s^2} \quad (473)$$

If we compare this result with the bilinear expansion (Eq. 302) of the Green's function $D_{\alpha\beta}(\vec{x},\vec{x}'|\omega)$ we obtain:

$$\langle u_\alpha^2(\vec{x})\rangle = -\frac{k_B T}{\rho}\sum_{n=-\infty}^{\infty} D_{\alpha\alpha}\left(\vec{x},\vec{x}|i\Omega_n\right) \quad (474)$$

The result given by Eq. (474) simplifies into two limiting cases. At the absolute zero of temperature the summation over n can be replaced by integration

$$\langle u_\alpha^2(\vec{x})\rangle = -\frac{\hbar}{2\pi\rho}\int_{-\infty}^{\infty} d\Omega D_{\alpha\alpha}(\vec{x},\vec{x}|i\Omega) \quad (475)$$

In the opposite limit of high temperatures, the only term making a nonzero contribution to the sum in Eq. (474) as $T \to \infty$ is the one for which $n = 0$. Consequently, in this limit we find

$$\langle u_\alpha^2(\vec{x})\rangle = -\frac{k_B T}{\rho} D_{\alpha\alpha}(\vec{x},\vec{x}|0) \tag{476}$$

In what follows we will focus on the high-temperature limit because many low-energy electron diffraction experiments are carried out under conditions where this approximation is valid, and because analytic expressions for $\langle u_\alpha^2(\vec{x})\rangle$ can be obtained in this limit, for an elastically isotropic medium, and for an hexagonal medium with its stress-free surface parallel to the basal plane.

If Eqs. (303) and (305) are used in Eq. (476), the expression for $\langle u_\alpha^2(\vec{x})\rangle$ can be stated in terms of the quantities that are actually calculated

$$\begin{aligned}\langle u_\alpha^2(\vec{x})\rangle &= -\frac{k_B T}{\rho}\int \frac{d^2k_\parallel}{(2\pi)^2} D_{\alpha\alpha}\left(\vec{k}_\parallel 0|x_3x_3\right) \\ &= -\frac{k_B T}{\rho}\sum_{\mu\nu}\int \frac{d^2k_\parallel}{(2\pi)^2} S_{\alpha\mu}^{-1}(\hat{k}_\parallel) d_{\mu\nu}\left(k_\parallel 0|x_3x_3\right) S_{\nu\alpha}(\hat{k}_\parallel)\end{aligned} \tag{477}$$

These expressions show the expected result, i.e., that $\langle u_\alpha^2(\vec{x})\rangle$ depends only on the coordinate normal to the surface.

The Fourier coefficients $\{d_{\mu\nu}\left(k_\parallel 0|x_3x_3'\right)\}$ are of interest in their own right for a variety of physical applications (205), in addition to their utility in calculations of the mean square displacement of a surface atom in the classical limit. For an isotropic elastic medium occupying the upper half-space $x_3 > 0$ the nonzero coefficients, obtained by taking the limit as $\omega \to 0$ in the results given by Eqs. (307) are

$$d_{11}(k_\parallel 0|x_3x_3') = -\frac{1}{4k_\parallel c_t^2c_\ell^2}\left\{e^{-k_\parallel|x_3 - x_3'|}\left[\left(c_\ell^2 + c_t^2\right)\right.\right.$$

$$\left.\left. - \left(c_\ell^2 - c_t^2\right)k_\parallel|x_3 - x_3'|\right]\right\}$$

$$- \frac{e^{-k_\parallel(x_3 + x_3')}}{4k_\parallel c_t^2c_\ell^2}\left\{\frac{c_\ell^4 + c_t^4}{c_\ell^2 - c_t^2}\right.$$

$$- \left(c_\ell^2 + c_t^2\right)k_\parallel(x_3 + x_3')$$

$$\left. + 2\left(c_\ell^2 - c_t^2\right)k_\parallel^2\ x_3x_3'\right\} \tag{478a}$$

$$d_{13}(k_\parallel 0|x_3x_3') = \frac{i}{4k_\parallel}\frac{c_\ell^2 - c_t^2}{c_t^2c_\ell^2}\, e^{-k_\parallel|x_3 - x_3'|}k_\parallel(x_3 - x_3')$$

$$- \frac{i}{4k_\parallel}\, e^{-k_\parallel(x_3 + x_3')}$$

$$\times \left\{ \frac{2}{c_\ell^2 - c_t^2} - \frac{c_\ell^2 + c_t^2}{c_t^2 c_\ell^2} k_\parallel (x_3 - x_3') - \frac{2\left(c_\ell^2 - c_t^2\right)}{c_t^2 c_\ell^2} k_\parallel{}^2 \; x_3 x_3' \right\} \qquad (478b)$$

$$d_{22}(k_\parallel 0|x_3x_3') = -\frac{1}{2k_\parallel c_t^2} e^{-k_\parallel |x_3 - x_3'|} - \frac{1}{2k_\parallel c_t^2} e^{-k_\parallel (x_3 + x_3')} \qquad (478c)$$

$$d_{31}(k_\parallel|x_3x_3') = \frac{i}{4k_\parallel} \frac{c_\ell^2 - c_t^2}{c_t^2 c_\ell^2} e^{-k_\parallel |x_3 - x_3'|} k_\parallel (x_3 - x_3') + \frac{i}{4k_\parallel} e^{-k_\parallel (x_3 + x_3')}$$

$$\times \left\{ \frac{2}{c_\ell^2 - c_t^2} + \frac{c_\ell^2 + c_t^2}{c_t^2 c_\ell^2} k_\parallel (x_3 - x_3') - \frac{2\left(c_\ell^2 - c_t^2\right)}{c_t^2 c_\ell^2} k_\parallel{}^2 \; x_3 x_3' \right\} \qquad (478d)$$

$$d_{33}(k_\parallel 0|x_3x_3') = -\frac{1}{4k_\parallel c_t^2c_\ell^2}\Bigg\{e^{-k_\parallel|x_3-x_3'|}\Big[\left(c_\ell^2+c_t^2\right) + \left(c_\ell^2-c_t^2\right)k_\parallel|x_3-x_3'|\Big]\Bigg\}$$

$$-\frac{e^{-k_\parallel(x_3+x_3')}}{4k_\parallel c_t^2c_\ell^2}\left\{\frac{c_\ell^4+c_t^4}{c_\ell^2-c_t^2} + \left(c_\ell^2+c_t^2\right)k_\parallel(x_3+x_3') + 2\left(c_\ell^2-c_t^2\right)k_\parallel^2\, x_3x_3'\right\} \quad (478e)$$

In each case the first term, which is a function of x_3 and x_3' only through their difference, is the Fourier coefficient $d_{\mu\nu}^{(\infty)}(k_\parallel 0|x_3x_3')$, which is obtained for an infinitely extended medium. The second term, which depends on x_3 and x_3' separately, is the contribution $d_{\mu\nu}^{(S)}(k_\parallel 0|x_3x_3')$ reflecting the presence of a stress-free surface at the plane $x_3 = 0$.

Because the Green's function $D_{\alpha\beta}(k_\parallel\omega|x_3x_3')$ can be written as the sum of the Green's function $D_{\alpha\beta}^{(\infty)}(k_\parallel\omega|x_3x_3')$ for an infinitely extended elastic medium and a contribution which causes the sum to satisfy the corresponding boundary conditions, it follows that the mean square displacement $\langle u_\alpha^2(\vec{x})\rangle$ can be decomposed in the same way

$$\langle u_\alpha^2(\vec{x})\rangle = \langle u_\alpha^2(\vec{x})\rangle_B + \langle u_\alpha^2(\vec{x})\rangle_S \tag{479}$$

The first term on the right-hand side is the bulk contribution to the mean square displacement, which should be independent of position, while the second term is the contribution associated with the presence of a surface and is expected to go to zero with increasing distance into the medium from the surface.

When the expressions for $\left\{d_{\mu\nu}^{(\infty)}(k_\parallel 0|x_3 x_3')\right\}$ from Eqs. (478) are substituted into Eq. (477), the integral over $\vec{k}_\parallel$ diverges unless one imposes a cutoff k_D on the magnitude of $\vec{k}_\parallel$, where k_D is of the order of the reciprocal of a typical interatomic spacing in a crystal. Such a cutoff arises naturally in a lattice theory, in which the corresponding integral extends over only the two-dimensional first Brillouin zone for the two-dimensional Bravais lattice underlying the semi-infinite crystal, and is associated with the existence of a minimum length in a crystal, the shortest primitive translation vector. No such length exists in a continuum theory, and a cutoff on the magnitude of $\vec{k}_\parallel$ has to be imposed in an ad hoc fashion.

The results for $\langle u_\alpha^2(\vec{x})\rangle_B$ obtained in this fashion are

$$\langle u_1^2(\vec{x})\rangle_B = \langle u_2^2(\vec{x})\rangle_B = \frac{k_B T}{8\pi\rho}\left[\frac{1}{2}\left(\frac{3}{c_t^2} + \frac{1}{c_\ell^2}\right)\right]k_D \tag{480a}$$

$$\langle u_3^2(\vec{x})\rangle_B = \frac{k_B T}{8\pi\rho}\left[\frac{1}{c_t^2} + \frac{1}{c_\ell^2}\right]k_D \tag{480b}$$

These results display the surprising feature that $\langle u_3^2(\vec{x})\rangle_B \neq \langle u_1^2(\vec{x})\rangle_B = \langle u_2^2(\vec{x})\rangle_B$, i.e., the bulk contribution to the mean square displacement in an isotropic elastic medium is not isotropic. This is a standard difficulty in the theory of semi-infinite elastic media. In the present case this anisotropy is due to the introduction of a Debye cutoff to the two-dimensional wave vector $\vec{k}_{\parallel}$ rather than to the full three-dimensional wave vector $\vec{k}$. In effect, one is integrating throughout the volume of a cylinder in wave vector space rather than a sphere. One can restore the bulk isotropy by introducing an integration throughout a Debye sphere rather than throughout a cylinder. This can be done in the following way.

The Fourier coefficient $d^{(\infty)}_{\mu\nu}(k_{\parallel}0|x_3x_3')$ for an infinitely extended isotropic elastic medium can be expressed in the form of a Fourier integral

$$d^{(\infty)}_{11}(k_{\parallel}0|x_3x_3') = -\frac{e^{-k_{\parallel}|x_3 - x_3'|}}{4k_{\parallel}c_t^2c_\ell^2}\left[\left(c_\ell^2 + c_t^2\right) - \left(c_\ell^2 - c_t^2\right)k_{\parallel}|x_3 - x_3'|\right]$$

$$= -\frac{1}{c_t^2c_\ell^2}\int_{-\infty}^{\infty}\frac{dk_3}{2\pi}\, e^{ik_3(x_3 - x_3')} \cdot \frac{c_\ell^2k_{\parallel}^2 + c_t^2k_{\parallel}^2}{\left(k_{\parallel}^2 + k_3^2\right)^2} \tag{481a}$$

$$d^{(\infty)}_{13}(k_{\parallel}0|x_3x_3') = \frac{i}{4k_{\parallel}}\frac{c_\ell^2 - c_t^2}{c_t^2c_\ell^2}\, e^{-k_{\parallel}|x_3 - x_3'|}\, k_{\parallel}(x_3 - x_3')$$

$$= \frac{c_\ell^2 - c_t^2}{c_t^2 c_\ell^2} \int_{-\infty}^{\infty} \frac{dk_3}{2\pi} e^{ik_3(x_3 - x_3')}$$

$$\frac{k_\parallel k_3}{\left(k_\parallel^2 + k_3^2\right)^2} = d_{31}^{(\infty)}(k_\parallel 0|x_3 x_3') \qquad (481b)$$

$$d_{22}^{(\infty)}(k_\parallel 0|x_3 x_3') = \frac{1}{2k_\parallel c_t^2} e^{-k_\parallel |x_3 - x_3'|}$$

$$= -\frac{1}{c_t^2}\int \frac{dk_3}{2\pi} e^{ik_3(x_3 - x_3')} \frac{1}{\left(k_\parallel^2 + k_3^2\right)}$$

(481c)

$$d_{33}^{(\infty)}(k_\parallel 0|x_3 x_3') = -\frac{e^{-k_\parallel |x_3 - x_3'|}}{4k_\parallel c_t^2 c_\ell^2}\Big[\left(c_\ell^2 + c_t^2\right)$$

$$+ \left(c_\ell^2 - c_t^2\right) k_\parallel |x_3 - x_3'|\Big]$$

$$= -\frac{1}{c_t^2 c_\ell^2}\int_{-\infty}^{\infty} \frac{dk_3}{2\pi} e^{ik_3(x_3 - x_3')}$$

$$\frac{c_\ell^2 k_\parallel^2 + c_t^2 k_3^2}{\left(k_\parallel^2 + k_3^2\right)^2} \qquad (481d)$$

When Eqs. (481) are substituted into Eq. (477) and the resulting integration over the three-dimensional wave vector $\vec{k}$ is restricted to the volume of a Debye sphere of radius k_D, the isotropic result is obtained for the bulk contribution to the mean square displacement

$$\langle u_1^2(\vec{x})\rangle_B = \langle u_2^2(\vec{x})\rangle_B = \langle u_3^2(\vec{x})\rangle_B = \frac{k_B T}{6\pi^2 \rho}\left(\frac{2}{c_t^2} + \frac{1}{c_\ell^2}\right)k_D \tag{482}$$

The surface contribution to the mean square displacement is intrinsically anisotropic because of the lowering of the symmetry of the isotropic medium by the introduction of a surface. Consequently, no such symmetrization procedure is required in obtaining $\langle u_\alpha^2(\vec{x})\rangle_S$. The following results are obtained (Ref. 206)

$$\langle u_1^2(\vec{x})\rangle_S = \langle u_2^2(\vec{x})\rangle_S = \frac{k_B T}{8\pi\rho c_t^2}\left\{\frac{1}{2}\left(1 + \frac{c_t^2}{c_\ell^2}\right)\right.$$

$$\cdot\left[\frac{(1 + 2k_D x_3)e^{-2k_D x_3} - 1}{2x_3}\right]$$

$$+\left(1 - \frac{c_t^2}{c_\ell^2}\right)\left[\frac{1 - \left(1 + 2k_D x_3 + 2k_D^2 x_3^2\right)e^{-2k_D x_3}}{4x_3}\right]$$

$$+ \left[1 + \frac{1}{2c_\ell^2}\left(\frac{c_\ell^4 + c_t^4}{c_\ell^2 - c_t^2}\right)\right]\left[\frac{1 - e^{-2k_D x_3}}{2x_3}\right]\Bigg\}$$

$$\xrightarrow[x_3 \to \infty]{} \frac{k_B T}{32\pi\rho c_t^2} \frac{3c_\ell^4 - 4c_\ell^2 c_t^2 + 3c_t^4}{c_\ell^2\left(c_\ell^2 - c_t^2\right)} \frac{1}{x_3} \tag{483a}$$

$$\langle u_3^2(\vec{x})\rangle_S = \frac{k_B T}{8\pi\rho c_t^2}\Bigg\{\left(1 + \frac{c_t^2}{c_\ell^2}\right)\left[\frac{1 - \left(1 + 2k_D x_3\right)e^{-2k_D x_3}}{2x_3}\right]$$

$$+ 2\left(1 - \frac{c_t^2}{c_\ell^2}\right)\left[\frac{1 - \left(1 + 2k_D x_3 + 2k_D^2 x_3^2\right)e^{-2k_D x_3}}{4x_3}\right]$$

$$+ \frac{c_\ell^4 + c_t^4}{c_\ell^2\left(c_\ell^2 - c_t^2\right)}\left[\frac{1 - e^{-2k_D x_3}}{2x_3}\right]\Bigg\}$$

$$\xrightarrow[x_3 \to \infty]{} \frac{k_B T}{16\pi\rho c_t^2} \frac{3c_\ell^4 - 2c_\ell^2 c_t^2 + c_t^4}{c_\ell^2\left(c_\ell^2 - c_t^2\right)} \frac{1}{x_3} \tag{483b}$$

The Debye radius k_D appearing in Eqs. (483) is that appropriate to the two-dimensional first Brillouin zone for the Bravais lattice describing the periodicity parallel to the surface of the semi-infinite crystal being approximated by a continuum. It is therefore different in general from the Debye radius k_D appearing in Eqs. (482), which is that corresponding to the three-dimensional first Brillouin zone for the Bravais lattice describing the periodicity of the bulk of the crystal. For simplicity, we have assumed the two values of k_D to be equal in presenting the results in Eqs. (482) and (483).

Dennis et al. (207) used another value of k_D in their calculation. At any given value of x_3, $\langle u_3^2(\vec{x})\rangle_S$ is always larger than $\langle u_1^2(\vec{x})\rangle_S = \langle u_2^2(\vec{x})\rangle_S$ for all values of c_t/c_ℓ compatible with elastic stability, i.e., $0 < c_t^2/c_\ell^2 < 3/4$.

From the results given by Eqs. (483) we see that each component of the mean square displacement decreases exponentially rapidly over a distance of a very few interatomic spacings into the solid from the surface $\left(x_3 \sim (2k_D)^{-1}\right)$, and then decreases much more slowly, as x_3^{-1}. the latter dependence (208) appears to have been obtained first by Mindlin (209) in calculations of displacement correlation functions for a semi-infinite, isotropic ela tic medium. It has also been demonstrated recently (146 210) on the basic of a simple lattice dynamical model, but it seems to be a result which would be difficult to extract from a purely numerical calculation based either on a lattice dynamical Green's function approach (98) or on a representation of a semi-infinite crystal by a slab of a finite number of layers. We also note that the large x_3 behavior of $\langle u_\alpha^2(\vec{x})\rangle_S$ is independent of the Debye radius k_D, and can thus be obtained by letting $k_D \to \infty$. In other words, this behavior can be obtained purely from elasticity theory, in which such a cutoff does not exist. However, a finite value of $\langle u_\alpha^2(\vec{x})\rangle_S$ at $x_3 = 0$ is obtained only if we retain a finite value of k_D.

The methods of this section have also been used to obtain analytic expressions for the mean square displacement in an hexagonal medium with a stress-free planar surface parallel to its basal plane (211). The results obtained are qualitatively the same as those described here.

Attenuation of Rayleigh Waves by Point Defects

The development of electronic devices using surface elastic waves has stimulated considerable interest in the fundamental properties of these waves. Important among these properties is the mean free path, or the inverse attenuation length, of surface waves.

The mechanisms attenuating elastic surface waves can be divided into intrinsic and extrinsic mechanisms. Chief among the intrinsic mechanisms in insulating crystals is the scattering of Rayleigh waves by thermally excited phonons in the substrate through the cubic anharmonic terms in the potential energy of the crystal. We have already shown that at least for insulators, there is a good understanding of the intrinsic processes responsible for the attenuation of Rayleigh waves.

In recent years, therefore, there has been considerable interest in extrinsic mechanisms for attenuating Rayleigh waves, particularly those extrinsic mechanisms which are specific to the near vicinity of the surface. Thus, the effects of surface roughness on the attenuation of Rayleigh waves have been studied (77), as have also the effects of mass density fluctuations in the surface of the solid (212).

The effects of point defects on the damping of Rayleigh waves was investigated theoretically by Steg and Klemens (203) for the case of a point-mass defect having a mass change Δm. By the use of perturbation theory they found that the attenuation rate is proportional to $(\Delta m)^2 \cdot \omega_R^5$. Subsequently, Sakuma (213) reexamined the problem by a nonperturbative approach, confirmed the ω_R^5 dependence of the attenuation rate at sufficiently low frequencies, and examined the possibility of resonant scattering of Rayleigh waves by such point defects.

The dynamic Green's functions can be used in a non-perturbative calculation of the attenuation of Rayleigh waves due to their scattering by point-mass defects (214), a procedure which differs in nature from the earlier treatments of this problem, and which also yields more physical results for the resonant scattering of these waves than those obtained by Sakuma.

We illustrate this application of these continuum Green's functions by considering the scattering of a Rayleigh wave by a point-mass defect situated at the point $\vec{x}_0 = (0,0,x_{03})$ in a semi-infinite elastic medium occupying the half-space $x_3 > 0$. The equations of motion of the medium in the presence of the defect are

$$-\rho \frac{\partial^2}{\partial t^2} u_\alpha(\vec{x},t) + \sum_{\beta\mu\nu} C_{\alpha\beta\mu\nu} \frac{\partial^2}{\partial x_\beta \partial x_\nu} u_\mu(\vec{x},t) = \Delta m \delta(\vec{x} - \vec{x}_0) \frac{\partial^2}{\partial t^2} u_\alpha(\vec{x},t) \tag{484}$$

and are subject to the boundary conditions of Eq. (279). With the aid of the Green's function defined by Eq. (299) we can convert Eq. (484) into an integral equation

$$u_\alpha(\vec{x},t) = u_\alpha^{(0)}(\vec{x},t) + \frac{\Delta m}{\rho} \sum_\beta \int dt' D_{\alpha\beta}(\vec{x},\vec{x}_0;t - t') \frac{\partial^2}{\partial t'^2} u_\beta(\vec{x}_0,t') \tag{485}$$

where $u_\alpha^{(0)}(\vec{x},t)$ is the displacement field of the incident

Rayleigh wave propagating along the surface $x_3 = 0$ of the semi-infinite medium. With the Fourier decompositions $u_\alpha(\vec{x},t) = u_\alpha(\vec{x}|\omega)\exp(-i\omega t)$ and $u_\alpha^{(0)}(\vec{x},t) = u_\alpha^{(0)}(\vec{x}|\omega)\exp(-i\omega t)$, we readily find that the amplitude of the scattered displacement field is given by

$$u_\alpha^{(S)}(\vec{x}|\omega) = u_\alpha(\vec{x}|\omega) - u_\alpha^{(0)}(\vec{x}|\omega)$$
$$= -\frac{\Delta m\omega^2}{\rho}\sum_{\beta\gamma} D_{\alpha\beta}(\vec{x},\vec{x}_0|\omega)T_{\beta\gamma}(\vec{x}_0|\omega)u_\gamma^{(0)}(\vec{x}_0|\omega) \tag{486}$$

where the scattering matrix $T_{\alpha\beta}(\vec{x}_0|\omega)$ is the 3 x 3 matrix given by

$$T_{\alpha\beta}(\vec{x}_0|\omega) = \left(\overleftrightarrow{I} + \frac{\Delta m\omega^2}{\rho}\overleftrightarrow{D}(\vec{x}_0,\vec{x}_0|\omega)\right)^{-1}_{\alpha\beta} \tag{487}$$

In the case of solids for which $D_{\alpha\beta}(\vec{x}_0,\vec{x}_0|\omega)$ is diagonal, e.g., isotropic, cubic, hexagonal solids, $T_{\alpha\beta}(\vec{x}_0|\omega)$ takes the simple form

$$T_{\alpha\beta}(\vec{x}_0|\omega) = \frac{\delta_{\alpha\beta}}{1 + \frac{\Delta m\omega^2}{\rho}D_{\alpha\alpha}(\vec{x}_0,\vec{x}_0|\omega)} \equiv \frac{\delta_{\alpha\beta}}{D_\alpha(\vec{x}_0|\omega)} \tag{488}$$

In scattering problems ω must be given a small positive imaginary part so that the contributions of various poles in the expressions for scattered amplitudes can be evaluated correctly. The functions $D_\alpha(\vec{x}_0|\omega)$ in this case have complex zeros in the low-frequency limit when Δm is positive (a heavy defect), and are therefore referred to as "resonance denominators."

Using the explicit forms of the $D_{\alpha\beta}(\vec{x},\vec{x}'|\omega)$ for an isotropic solid given by Eqs. (307) and (308), the asymptotic behavior of the scattered displacement field $u_\alpha^{(s)}(\vec{x},t) = u_\alpha^{(s)}(\vec{x}|\omega) \times \exp(-i\omega t)$ for $\vec{x}$ far from $\vec{x}_0$ can be evaluated by the method of stationary phase. The result can be written as the sum of four contributions

$$\vec{u}^{(s)}(\vec{x},t) = e^{-i\omega t}\Big\{\vec{u}^{(\ell)}(\vec{x}|\omega) + \vec{u}^{(tp)}(\vec{x}|\omega) + \vec{u}^{(ts)}(\vec{x}|\omega) + \vec{u}^{(R)}(\vec{x}|\omega)\Big\} \tag{489}$$

In Eq. (489), $\vec{u}^{(\ell)}(\vec{x}|\omega)$ is the amplitude describing the scattering of the incident Rayleigh wave into bulk longitudinal waves; $\vec{u}^{(tp)}(\vec{x}|\omega)\left(\vec{u}^{(ts)}(\vec{x}|\omega)\right)$ is the amplitude for scattering into bulk transverse waves of p-polarization (s-polarization); and $\vec{u}^{(R)}(\vec{x}|\omega)$ is the amplitude for scattering into other Rayleigh waves.

With these results in hand it is straightforward to obtain the energy crossing unit area in unit time, averaged over the period of the motion, for both the scattered and the incident displacement fields. This is given by the real part of the complex elastic Poynting vector (77)

$$\zeta_\alpha^c = -\tfrac{1}{2}\sum_{\beta\mu\nu} C_{\alpha\beta\mu\nu}\, \dot{u}_\beta^* \frac{\partial u_\mu}{\partial x_\nu} \tag{490}$$

We restrict ourselves to the case of point defects localized at the surface of the solid $\left(x_{03} = 0\right)$. We assume an area density of defects n_s, and assume further that the incident Rayleigh wave propagates in the positive x_1-direction. The total energy stored per unit time in the incident Rayleigh wave is

$$\frac{dE^{(0)}}{dt} = L_2 \int_0^\infty dx_3 \zeta_1^c(\text{inc.}) \tag{491}$$

where L_2 is the length of the sample in the x_2-direction (i.e., in the direction perpendicular to the direction of propagation of the incident Rayleigh wave). If the length of the sample in the x_1-direction is L_1, there is a total of $n_s L_1 L_2$ defects on the surface. Then if we denote by $dE^{(s)}/dt$ the total energy stored per unit time in the displacement field scattered from a single impurity, the total energy per unit time stored in the displacement field scattered from all the point defects is

$$\frac{dE^{(T)}}{dt} = n_s L_1 L_2 \frac{dE^{(S)}}{dt} \tag{492}$$

We thus obtain the relation

$$\frac{dE^{(T)}}{dt} = \frac{L_1}{\ell} \frac{dE^{(0)}}{dt} \tag{493}$$

which defines the length ℓ. L_1 is the distance traveled by the Rayleigh wave as it passes over the surface. The ratio $(dE^{(T)}/dt)/L_1(dE^{(0)}/dt)$ is the energy lost per unit distance traveled by the Rayleigh wave. This is the inverse of the mean free path of the wave. Thus, from Eq. (493) we see that the length ℓ is the mean free path of the Rayleigh wave. It is given explicitly by

$$\frac{1}{\ell} = n_S \left[\int_0^\infty dx_3 \zeta_1^c(\text{inc.})\right]^{-1} \frac{dE^{(S)}}{dt} \tag{494}$$

The determination of $dE^{(S)}/dt$ from Eqs. (489) and (490) and of $\zeta_1^c(\text{inc.})$ is described in Ref. (214). Because the scattered displacement field $\vec{u}^{(S)}(\vec{x},t)$ is the sum of four contributions, as was indicated in Eq. (489), it follows from this fact and Eq. (490) that $dE^{(S)}/dt$ can be expressed as the sum of four contributions, one for each of the four possible scattering processes

$$\frac{dE^{(S)}}{dt} = \left(\frac{dE^{(S)}}{dt}\right)_{\ell} + \left(\frac{dE^{(S)}}{dt}\right)_{tp} + \left(\frac{dE^{(S)}}{dt}\right)_{ts} + \left(\frac{dE^{(S)}}{dt}\right)_{R} \tag{495}$$

This result combined with Eq. (494) immediately yields the result that the inverse attenuation length is itself the sum of four contributions, one for each of the possible scattering processes

$$\frac{1}{\ell} = \frac{1}{\ell_{\ell}} + \frac{1}{\ell_{tp}} + \frac{1}{\ell_{ts}} + \frac{1}{\ell_{R}} \tag{496}$$

It is shown in Ref. (214) that each of the four terms on the right-hand side of this equation has the functional form

$$\frac{1}{\ell_i} = n_S\left(\frac{\Delta m}{\rho}\right)^2 \omega^5 \left\{\frac{A_i}{|D_1(\vec{0}|\omega)|^2} + \frac{B_i}{|D_3(\vec{0}|\omega)|^2}\right\} \tag{497}$$

where A_i and B_i are constants which in general have to be obtained numerically.

This result is in agreement with those of Steg and Klemens (203) and Sakuma (213) in predicting an inverse attenuation length proportional to the square of the change in mass introduced by the defect and to the fifth power of the frequency of the Rayleigh wave, when the point defects are localized at the surface. The ω^5 dependence has a simple physical origin. It is well known that the cross-section for scattering of bulk phonons by a defect whose dimensions are small compared to the wavelength of the phonons is proportional to ω^4 (215). A bulk phonon has its energy density spread uniformly throughout the volume V of the solid, and the fraction f of the energy density that interacts with the defect is V_D/V, where V_D is the volume of the defect. The ω^4 dependence comes from the combined effect of the matrix element for scattering and the density of final states, and the attenuation constant is proportional to $\omega^4 f$. For a Rayleigh wave that scatters from a surface defect of finite size (small compared with the wavelength of the Rayleigh wave), one has $f \approx V_D\omega/c_R S$, where S is the area of the surface exposed to the Rayleigh wave, ω is its frequency, and c_R is its speed. The length c_R/ω is a rough measure of the penetration depth of the Rayleigh wave into the substrate. The attenuation constant is again proportional to $\omega^4 f$, and the ω^5 dependence is obtained.

In contrast with the result obtained by Sakuma, who obtained resonant scattering when the mass defect corresponds to an impurity atom lighter than the atom of the host crystal it replaces, the result given by Eq. (497) predicts resonant scattering when the impurity atom is heavier than the atom it replaces. The latter result is analogous to that obtained for the scattering of bulk waves by mass defects (215), and appears physically reasonable in the present context.

To obtain an estimate of the magnitude of ℓ^{-1} we consider silicon with $\rho = 2.5$ gm cm^{-3}, $c_R = 4.9 \times 10^5$ cm sec^{-1},

$c_t = 5.3 \times 10^5$ cm sec^{-1}, and $c_\ell = 9.5 \times 10^5$ cm sec^{-1}. If we assume that the defects are localized at the surface with $n_S = 10^{12}$ cm^{-2}, and $\Delta m = 10^{-23}$ gm, we obtain the following estimates for $\omega = 10^{10}$ Hz: $\ell_{(\ell)} = 3.1 \times 10^{15}$ cm, $\ell_{(tp)} = 2.1 \times 10^{15}$ cm, $\ell_{(ts)} = 2.1 \times 10^{15}$ cm, and $\ell_R = 0.77 \times 10^{15}$ cm. If the frequency is increased by a factor of 1000, to 10^{13} Hz, the preceding values are reduced to 3.1 cm, 2.1 cm, 2.1 cm, and 0.77 cm, respectively. The latter lengths are moderately short, but the frequency is much higher than those used in surface wave devices. In typical situations, therefore, it seems likely that scattering of Rayleigh waves by surface roughness (77) will be a more significant extrinsic mechanism in determining the attenuation length than will scattering by mass defects.

The Surface Contribution to the Low-Temperature Specific Heat of a Solid

We have already presented a discussion of the calculation of the surface contribution to the thermodynamic functions of a crystal by purely numerical methods. Such methods yield this contribution over the complete range of temperatures, in the harmonic approximation, for which it is nonvanishing. However, the results are obtained purely numerically and must therefore be presented either graphically or in tabular form, in neither of which forms is information about the analytic dependence of the results on temperature and the parameters characterizing the dynamical properties of the crystal readily obtainable.

There is one limit, however, in which an analytic result is obtainable, and we will discuss the surface contribution to the specific heat of a crystal in this limit.

In the limit of very low temperatures the specific heat at constant volume of a finite solid contains a contribution associated with the surface of the solid which has the form

$$\Delta C_v(T) = BST^2 + o(T^2) \qquad (498)$$

In this expression S is the surface area of the solid and T is the absolute temperature. The proportionality of $\Delta C_v(T)$ to S and T^2 is characteristic of two-dimensional dynamical systems at low temperatures (the latter dependence is the analog for two-dimensional crystals of the well-known "Debye T^3 law" for the specific heat of a three-dimensional crystal at low temperatures (216). In the case that the surface is stress-free the coefficient B is positive because the relaxation of stresses at the boundary depresses the normal mode frequencies of the solid, according to Rayleigh's theorems (9), thus enhancing the density of normal-mode frequencies at its low-frequency end, and increasing the specific heat at low temperatures thereby. In this subsection we obtain an explicit expression for the coefficient B by means of which it can be calculated for any solid.

The interest in knowing B arises from two circumstances First, since only vibration modes of very long wavelength contribute to the specific heat at very low temperatures, the coefficient B can be calculated on the basis of elasticity theory. Its value is therefore a function only of the elastic moduli and the mass density of the medium. In other words, it is model independent and exact: no information about the structure of the primitive unit cell of the crystal nor about the nature and range of the interatomic forces in the crystal is required for its determination. Second, up to the present time it has not been proved to be possible to obtain the coefficient B, or even the T^2 temperature dependence, accurately from purely numerical calculations in which the normal-mode frequencies are determined for a semi-infinite slab of the crystal being studied, consisting of a comparatively small number of atomic layers (of the order of 20) (42). The difficulty of doing this lies in the fact that as the temperature is lowered the wavelength of the phonons making the dominant contribution to the specific heat increases. At some temperature this wavelength becomes long enough that the fact that the solid is a slab, bounded by two stress-free surfaces, and not a semi-infinite medium, bounded by a single stress-free surface, begins to be reflected in the result for the specific heat. At the present time it does not appear to be feasible to make the crystal slabs thick enough to enable the specific heat to be obtained at sufficiently low temperatures that the contribution proportional to T^2 can be obtained accurately. An independent calculation of the coefficient B, therefore, gives the correct limit to which the purely numerical results should be extrapolated in the limit of vanishing temperature.

The specific heat of an arbitrary crystal is given in the harmonic approximation by

$$C_v(T) = k_B \sum_s \frac{\left(\frac{1}{2}\beta\hbar\omega_s\right)^2}{\sinh^2 \frac{1}{2}\beta\hbar\omega_s} = k_B \sum_{n=1}^{\infty} n \sum_s \left(\beta\hbar\omega_s\right)^2 e^{-n\beta\hbar\omega_s} \tag{499}$$

Here $\beta = \left(k_B T\right)^{-1}$, ω_s is the frequency of the sth normal mode, and the sum on s runs over all the normal modes of vibration of the crystal. The second form of this expression is valid at all temperatures. However, it is particularly useful in the limit of low temperatures, where β is large.

The sum on s in Eq. (499) can be evaluated if we introduce the function F(z) of the complex variable z

$$F(z) = \sum_s \frac{1}{z^2 - \omega_s^2} \tag{500}$$

This function has simple poles at $z = \pm\,\omega_s$, $s = 1,2,\ldots$ and the residues at these poles are $\pm\left(1/2\omega_s\right)$, respectively. Because only positive values of ω_s appear on the right-hand side of Eq. (499), we can rewrite the sum on s in this equation as a contour integral

$$C_v(T) = \frac{k_B(\beta\hbar)^2}{\pi i} \sum_{n=1}^{\infty} n \int_{C_1} z^3 e^{-n\beta\hbar z} F(z)\,dz \tag{501}$$

where the contour C_1 is shown in Fig. 2.5.

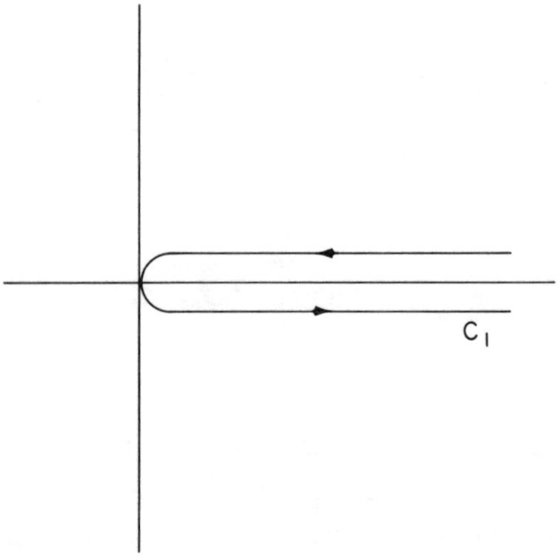

Fig. 2.5.--The integration contour C_1 (in the z-plane) which can be used in the evaluation of additive functions of normal mode frequencies.

The result given by Eq. (501) applies equally to infinite and semi-infinite crystals. If we denote quantities associated with an infinite crystal by a superscript or subscript zero, the difference between the specific heats of a semi-infinite and an infinite crystal possessing the same number of degrees of freedom, i.e., the surface contribution to the specific heat, can be written in the form

$$\Delta C_v(T) = \frac{k_B(\beta\hbar)^2}{\pi i} \sum_{n=1}^{\infty} n \int_{C_1} z^3 e^{-n\beta\hbar}[F(z) - F^{(0)}(z)]dz \tag{502}$$

where

$$F^{(0)}(z) = \sum_s \frac{1}{z^2 - \omega_{0s}^2} \tag{503}$$

and the $\{\omega_{0s}\}$ are the normal mode frequencies of the infinitely extended crystal.

The presence of the descending exponential in the integrand on the right-hand side of Eq. (502) enables us to deform the integration contour C_1 in this equation into the contour C_2 shown in Fig. 2.6 and know that the contribution from the semicircular portion vanishes in the limit as its radius increases indefinitely. We are thus left with the contribution from the integration down the imaginary axis, which we write as

$$\Delta C_v(T) = \frac{-2k_B(\beta\hbar)^2}{\pi} \sum_{n=1}^{\infty} n \int_0^{\infty} \Omega(y)y^3 \sin n\beta\hbar y \, dy \tag{504}$$

where we have put

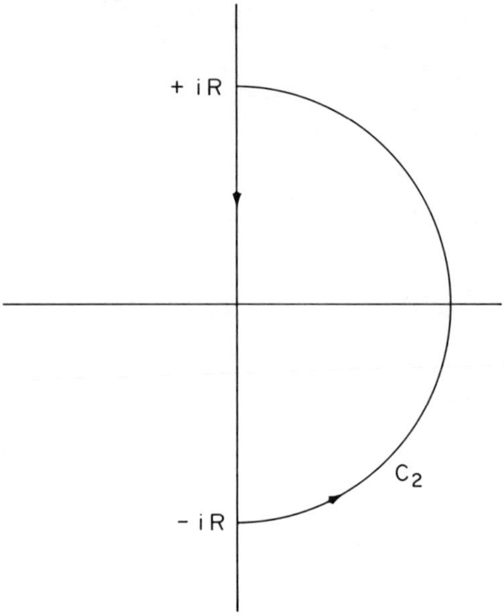

Fig. 2.6.--The integration contour C_2 (in the z-plane) which can be used in the evaluation of additive functions of normal mode frequencies.

$$\Omega(y) = - [F(iy) - F^{(0)}(iy)] \tag{505}$$

and have used the fact, evident from Eqs. (500), (503), and (505) that $\Omega(y)$ is an even function of y.

Because n is greater than or equal to unity, and β is large in the limit of low temperatures, to obtain the surface contribution to the low-temperature specific heat of a crystal we need only the asymptotic behavior of the integral in Eq. (504) in the limit as $n\beta\hbar \to +\infty$.

It is known from the theory of the asymptotic behavior of Fourier integrals (217) that if the only singularity of the function $\Omega(y)$ is a logarithmic dependence on $|y|$ in the limit as $|y| \to 0$, i.e., if

$$\Omega(y) = - A\ell n|y| + o(\ell n|y|), \quad |y| \to 0 \tag{506}$$

then the dominant term in the expansion of the integral in Eq. (504) in the limit of large $n\beta\hbar$ is given by

$$\int_0^\infty \Omega(y)y^3 \sin n\beta\hbar y \, dy = \frac{-3\pi A}{(n\beta\hbar)^4} + o\left((n\beta\hbar)^{-4}\right)$$

$$n\beta\hbar \to +\infty \tag{507}$$

When this result is substituted into Eq. (504) and the sum on n is carried out, we obtain for the surface contribution to the specific heat of a solid

$$\Delta C_v(T) = 6A\ \zeta(3)\ k_B\left(\frac{k_B T}{\hbar}\right)^2 + o(T^2) \tag{508}$$

in the limit as $T \to 0$, where $\zeta(x)$ is the Riemann zeta function.

It is not difficult to show that for real y the function $\Omega(y)$ can be nonanalytic only at $y = 0$ (218). The problem of obtaining the surface contribution to the low-temperature specific heat of a solid reduces, therefore, to showing that the function $\Omega(y)$ associated with the introduction of a free surface into a solid has the asymptotic form given by Eq. (506) in the limit as $|y| \to 0$, and obtaining the coefficient A.

From its definition, Eq. (505), and the results expressed by Eqs. (500) and (503), it follows that the function $\Omega(y)$ can be expressed in terms of the dynamical Green's functions

$$\Omega(y) = -\sum_\alpha \int d^2x_\parallel \int_0^\infty dx_3 \Big[D_{\alpha\alpha}(\vec{x},\vec{x}\,|\,iy) - D_{\alpha\alpha}^{(\infty)}(\vec{x},\vec{x}\,|\,iy)\Big] \tag{509a}$$

$$= -\frac{S}{4\pi^2}\sum_{\alpha}\int_0^{\infty} dx_3 \int d^2k_{\parallel}\,[D_{\alpha\alpha}(\vec{k}_{\parallel}\,iy|x_3x_3) - D^{(\infty)}_{\alpha\alpha}(\vec{k}_{\parallel}\,iy|x_3x_3)] \tag{509b}$$

where we have used Eq. (303) in writing the latter expression and S is the area of the surface of the solid.

In carrying out the integration over $\vec{k}_{\parallel}$ it is convenient to introduce two-dimensional polar coordinates. In doing so we impose a cutoff on the integral over the magnitude of $\vec{k}_{\parallel}$, $0 \le k_{\parallel} \le k_D$, where k_D is of the order of the reciprocal of a nearest neighbor separation in a crystal. We will see that the result we seek is independent of the value chosen for k_D, provided that it remains finite.

We now make the changes of variables in the integral on the right-hand side of Eq. (509b)

$$k_{\parallel} = \frac{u|y|}{c_t}\,,\quad k_{\alpha} = k_{\parallel}\hat{k}_{\alpha}\,,\quad \alpha = 1,2$$

$$\hat{k}_1 = \cos\theta,\quad \hat{k}_2 = \sin\theta$$

$$x_3 = \frac{zc_t}{u|y|} \tag{510}$$

where c_t is some conveniently chosen speed of sound for the elastic medium. For isotropic media it can be taken to be the speed of sound for transverse waves; for cubic crystals a convenient choice might be $c_t = (c_{44}/\rho)^{\frac{1}{2}}$, where c_{44} is one of the shear elastic moduli in the contracted Voigt notation. We also define a new Green's function $D_{\alpha\beta}(u\hat{k}_{\parallel}|zz')$ by

$$D_{\alpha\beta}\left(\frac{u|y|\hat{k}_1}{c_t}, \frac{u|y|\hat{k}_2}{c_t}, iy\left|\frac{zc_t}{u|y|}, \frac{z'c_t}{u|y|}\right.\right) \equiv \frac{D_{\alpha\beta}(u\hat{k}_{\parallel}|zz')}{c_t u|y|} \tag{511}$$

By making the changes of variables (510) in the differential equations (304a) satisfied by $D_{\alpha\beta}(\vec{k}_{\parallel}\omega|x_3x_3')$ and in the accompanying boundary conditions (304b) and making use of the definition (511) we find that the new Green's functions $D_{\alpha\beta}(u\hat{k}_{\parallel}|zz')$ are the solutions of the equations

$$\sum_{\mu}\left\{-\frac{\delta_{\alpha\mu}}{u^2} + \sum_{\beta\nu}\left(\frac{C_{\alpha\beta\mu\nu}}{\rho c_t^2}\right)\left[(1-\delta_{\beta 3})i\hat{k}_{\beta} + \delta_{\beta 3}\frac{d}{dz}\right]\right.$$
$$\left.\left[(1-\delta_{\nu 3})i\hat{k}_{\nu} + \delta_{\nu 3}\frac{d}{dz}\right]\right\}D_{\mu\gamma}(u\hat{k}_{\parallel}|zz')$$
$$= \delta_{\alpha\gamma}\delta(z-z'), \quad z,z' > 0, \ \alpha,\gamma = 1,2,3 \tag{512a}$$

subject to the boundary conditions

$$\sum_{\mu\nu}\left(\frac{C_{\alpha 3\mu\nu}}{\rho c_t^2}\right)\left[(1-\delta_{\nu 3})i\hat{k}_{\nu} + \delta_{\nu 3}\frac{d}{dz}\right]D_{\mu\gamma}(u\hat{k}_{\parallel}|zz')\Big|_{z=0} = 0,$$
$$z' > 0, \ \alpha,\gamma = 1,2,3 \tag{512b}$$

at the surface $z = 0$, and outgoing or exponentially decaying wave conditions at $z = +\infty$. The Green's functions $D^{(\infty)}_{\alpha\beta}(u\hat{k}_{\parallel}|zz')$ corresponding to an infinitely extended medium are obtained as the solutions of Eqs. (512a) subject to outgoing or exponentially decaying boundary conditions at $|z| \to \infty$. It is clear from the forms of the equations (512a) and (512b) that both $D_{\alpha\beta}(u\hat{k}_{\parallel}|zz')$ and $D^{(\infty)}_{\alpha\beta}(u\hat{k}_{\parallel}|zz')$ are independent of y.

The use of Eqs. (510) and (511) enables us to rewrite Eq. (509b) in the form

$$\Omega(y) = -\frac{S}{4\pi^2 c_t^2}\int_0^{c_t k_D/|y|} \frac{du}{u} F(u) \qquad (513a)$$

where

$$F(u) = \sum_{\alpha}\int_0^{2\pi} d\theta \int_0^{\infty} dz \left[D_{\alpha\alpha}(u\hat{k}_{\parallel}|zz) - D^{(\infty)}_{\alpha\alpha}(u\hat{k}_{\parallel}|zz)\right] \qquad (513b)$$

The only dependence of the right-hand side of Eq. (513a) on y is through the upper limit of integration. It is readily seen that the singular behavior of $\Omega(y)$ in the limit of small $|y|$ is determined by the large u behavior of the function F(u). The latter can be obtained as follows.

Because u enters Eq. (512a) for $D_{\alpha\beta}(u\hat{k}_{\parallel}|zz')$ only in the term $-\delta_{\alpha\mu}/u^2$ on the left-hand side, and this term is small when u is large, we introduce a new Green's function $G_{\alpha\beta}(\hat{k}_{\parallel}|zz')$, independent of u, as the solution of the equation

$$\sum_{\beta\mu\nu}\left(\frac{C_{\alpha\beta\mu\nu}}{\rho c_t^2}\right)\left[(1-\delta_{\beta 3})i\hat{k}_{\beta} + \delta_{\beta 3}\frac{d}{dz}\right]\cdot\left[(1-\delta_{\nu 3})i\hat{k}_{\nu} + \delta_{\nu 3}\frac{d}{dz}\right]G_{\mu\gamma}(\hat{k}_{\parallel}|zz')$$

$$= \delta_{\alpha\gamma}\delta(z - z'),$$

$$z, z' > 0, \quad \alpha, \gamma = 1,2,3 \quad (514a)$$

subject to the boundary conditions

$$\sum_{\mu\nu} \left(\frac{C_{\alpha 3\mu\nu}}{\rho c_t^2}\right) \left[(1 - \delta_{\nu 3}) i\hat{k}_\nu + \delta_{\nu 3} \frac{d}{dz}\right] G_{\mu\gamma}(\hat{k}_\parallel | zz')\bigg|_{z=0} = 0,$$

$$z' > 0, \quad \alpha, \gamma = 1,2,3 \quad (514b)$$

at the surface z = 0, and outgoing or exponentially decaying wave conditions at $z = +\infty$. It can be shown, from a comparison of Eqs. (514) and (304) that the Green's function $G_{\alpha\beta}(\hat{k}_\parallel | zz')$ and the static Green's function $D_{\alpha\beta}(\hat{k}_\parallel 0 | x_3 x_3')$ are simply related

$$G_{\alpha\beta}(\hat{k}_\parallel | zz') = c_t^2 k_\parallel D_{\alpha\beta}\left(\vec{k}_\parallel 0 \middle| \frac{z}{k_\parallel}, \frac{z'}{k_\parallel}\right) \quad (515)$$

The differential equation (512a) for $D_{\alpha\beta}(u\hat{k}_\parallel | zz')$ can now be rewritten as an integral equation with the aid of this new Green's function

$$D_{\alpha\beta}(u\hat{k}_\parallel | zz') = G_{\alpha\beta}(\hat{k}_\parallel | zz')$$

$$+ \frac{1}{u^2} \sum_\gamma \int_0^\infty G_{\alpha\gamma}(\hat{k}_\parallel | zz'') D_{\gamma\beta}(u\hat{k}_\parallel | z''z') dz'' \quad (516)$$

Although this equation holds for all u, it is particularly convenient in the limit of large u, when it can be solved by iteration

$$D_{\alpha\beta}(u\hat{k}_{\parallel}|zz') = G_{\alpha\beta}(\hat{k}_{\parallel}|zz') + \frac{1}{u^2}\sum_{\gamma}\int_0^{\infty} G_{\alpha\gamma}(k_{\parallel}|zz'')G_{\gamma\beta}(\hat{k}_{\parallel}|z''z')\,dz'' + 0(u^{-4}) \tag{517}$$

It follows from this result that in the limit of large u the function F(u) defined by Eq. (513b) has an expansion in inverse powers of u^2

$$F(u) \sim F_0 + \frac{F_2}{u^2} + \frac{F_4}{u^4} + \ldots, \quad u \to \infty \tag{518}$$

where the $\{F_{2n}\}$ are constant coefficients. In particular, we have that

$$F_0 = \sum_{\alpha}\int_0^{2\pi} d\theta \int_0^{\infty} dz\left[G_{\alpha\alpha}(\hat{k}_{\parallel}|zz) - G_{\alpha\alpha}^{(\infty)}(\hat{k}_{\parallel}|zz)\right] \tag{519}$$

where $G_{\alpha}^{(\infty)}(\hat{k}_{\parallel}|zz')$ is the solution of Eq. (514a) which satisfies outgoing or exponentially decaying wave conditions at $z = \pm\infty$.

From Eqs. (513a) and (518) it is straightforward to show that the only singular contribution to the function $\Omega(y)$ in the limit of small $|y|$ is given by

$$\Omega(y) \sim \frac{SF_0}{4\pi^2 c_t^2}\,\ell n|y| + o(\ell n|y|), \quad |y| \to 0 \tag{520}$$

From Eqs. (506) and (508) we see that this result implies that the surface contribution to the low-temperature specific heat is given by

$$\Delta C_v(T) \sim - \frac{3SF_0}{2\pi^2 c_t^2} \zeta(3) k_B \left(\frac{k_B T}{\hbar}\right)^2 + o(T^2) \qquad (521)$$

This result, together with Eqs. (515) and (519), solves the problem of obtaining the surface contribution to the low-temperature specific heat of an arbitrary solid bounded by a single stress-free planar surface, and displays the functional dependence of this contribution on the surface area and temperature in an explicit fashion. A definite prescription for the determination of the numerical coefficient F_0 is given.

The reduced Green's functions $G_{\alpha\beta}(\hat{k}_\parallel|zz')$ have been obtained for an isotropic elastic medium (81), a hexagonal medium with its stress-free surface parallel to the basal plane (81), and a cubic medium with its stress-free surface parallel to a (001) surface (219). For the hexagonal medium the following result for $\Delta C_v(T)$ has been obtained (78,81)

$$\Delta C_v(T) = 6\pi \frac{k_B^3}{h^2} \zeta(3) \left\{ \frac{\rho}{c_{11} - c_{12}} + \frac{\rho}{2\left(\frac{c_{11}c_{33}}{c_{13}} - c_{13}\right)} \times \frac{R - P}{\Delta} \right\} ST^2 + o(T^2) \qquad (522a)$$

where

$$\Delta = 2\left(\frac{c_{11}}{c_{33}}\right)^{\frac{1}{2}} + \frac{c_{11}}{c_{44}} - \frac{c_{13}}{c_{33}}\left(2 + \frac{c_{13}}{c_{44}}\right) \qquad (522b)$$

$$R = 4\left(\frac{c_{33}}{c_{11}}\right)^{\frac{1}{2}}\left[\frac{c_{13}}{c_{33}} + \frac{c_{11}}{c_{13}} - \frac{2c_{11}}{c_{33}} - \frac{c_{11}}{c_{44}}\left(\frac{c_{11}}{c_{13}} - \frac{c_{13}}{c_{33}}\right)\right] \tag{522c}$$

$$P = 6\left(1 - \frac{c_{11}}{c_{13}}\right) - 2\,\frac{c_{13}}{c_{33}} - \frac{c_{11}}{c_{44}}\,\frac{c_{33}}{c_{13}}\left(1 + \frac{c_{11}}{c_{44}}\right)$$
$$+ 2\,\frac{c_{11}}{c_{44}} + 2\,\frac{c_{13}}{c_{33}}\left(\frac{c_{13}}{c_{11}} - \frac{c_{13}}{c_{44}}\right)$$
$$+ \frac{c_{13}}{c_{44}}\left[2\,\frac{c_{11}}{c_{44}} + \frac{c_{13}}{c_{33}}\left(\frac{c_{13}}{c_{11}} - \frac{c_{13}}{c_{44}}\right)\right] \tag{522d}$$

An isotropic medium can be regarded as a special case of a hexagonal medium, which is obtained by setting $c_{13} = c_{12} = c_{11} - 2c_{44}$ and $c_{33} = c_{11}$. In this limit, Eq. (522a) reduces to

$$\Delta C_v(T) = 3\pi\,\frac{k_B^3}{h^2}\,\zeta(3)\,\frac{2c_t^4 - 3c_t^2 c_\ell^2 + 3c_\ell^4}{c_t^2 c_\ell^2\left(c_\ell^2 - c_t^2\right)}\,ST^2 + o(T^2) \tag{523}$$

which was first obtained by Dupuis, Mazo, and Onsager (220), and which has subsequently been rederived by other authors in other ways (221).

In the case of cubic crystals the integral over θ in the expression (519) for F_0 has to be evaluated numerically. Values of $-F_0$ have been tabulated as functions of c_{11}/c_{44} and c_{12}/c_{44} (224). From these tables one obtains a value of $B = 3.034 \times 10^{-5}$ erg/cm^2(°K)^3 for NaCl, when the 0 °K values of the elastic moduli obtained by Overton and Swim (222) are used. This value is to be compared with the value of $B = 3.80 \times 10^{-5}$ erg/cm^2(°K)^3 obtained for a slab of 15 atomic layers by Chen et al.

(223), from values of $\Delta C_V(T)$ in the temperature range between 5 and 20 °K. Barkman, Anderson, and Brackett (224) experimentally determined the value of B for NaCl to be 4.6×10^{-5} erg/cm^2(°K)^3, which is approximately 50% larger than the theoretical result.

An alternative, Green's function method of obtaining the surface contribution to the low-temperature specific heat of a semi-infinite solid has recently been developed by Garcia-Moliner and his colleagues (82,225), and used to derive the results given by Eq. (522).

Further Applications of the Continuum Approach

The mean square velocity of a point in the vicinity of the surface of an elastically isotropic medium bounded by a stress-free planar surface can also be expressed in terms of the continuum Green's functions

$$\langle v_\alpha^2(\vec{x})\rangle = \frac{k_B T}{6\pi^2 \rho} k_D^3 + \frac{2k_B T}{\rho} \sum_{n=1}^{\infty} \left[\frac{k_D^3}{6\pi^2} + \Omega_n^2 D_{\alpha\alpha}(\vec{x},\vec{x}|i\Omega_n) \right] \tag{524}$$

This is a less interesting dynamical property of a solid surface than the mean square displacement, however. This is because at elevated temperatures the effects of the surface appear first in the leading quantum correction to the classical result, the first term on the right-hand side of Eq. (524).

The mean square displacement of a point in the vicinity of the plane interface between two different elastic media can be calculated on the basis of appropriate generalizations of the Green's functions. Such calculations have been carried out for two different isotropic elastic media in contact (226).

The same Green's functions can also be used to obtain the interface contribution to the low-temperature specific heat of such composite systems (227).

These Green's functions were used also for the calculation of the Stoneley waves for two different hexagonal media in contact along a plane interface which is parallel to each of their basal planes (228).

The static and dynamic Green's functions for an isotropic elastic medium bounded by a pair of parallel, stress-free planar surfaces have been obtained recently (229). It is therefore possible now to obtain the mean square displacement of a point in a crystal slab, as well as the surface contribution to the low-temperature specific heat of a slab, and to determine the effects of the finite thickness of the slab on these properties.

The continuum Green's functions described earlier can be used to obtain dynamical correlation functions for a semi-infinite elastic medium, in addition to static correlation functions such as the mean square displacement and mean square velocity. For example (see Ref. 230)

$$\int_{-\infty}^{\infty} dt e^{i\omega t} \langle u_\alpha(\vec{x},t) u_\beta(\vec{x}',0)\rangle = \frac{i\hbar\{n(\omega)+1\}}{\rho} \cdot \left\{D_{\alpha\beta}(\vec{x},\vec{x}'|\omega + i0) - D_{\alpha\beta}(\vec{x},\vec{x}'|\omega - i0)\right\} \tag{525}$$

This result has been used recently in a calculation of the Brillouin scattering of light from the surface of an opaque medium (230).

The preceding calculations have all been carried out in the harmonic approximation. However, the continuum Green's functions can also be used in calculations of certain anharmonic properties of solid surfaces. They can be used as the basis of a rather direct and simple calculation of second harmonic generation of Rayleigh waves as they propagate along a solid surface, due to the anharmonic terms in the crystal's potential energy (231). Other nonlinear processes involving Rayleigh waves (e.g., the nonlinear mixing of two Rayleigh waves) can also be studied in the same manner.

The simplicity of such calculations, and the attractive explicitness of the results obtained, are likely to result in studies of many other dynamical properties of solid surfaces analytically, by the methods of continuum mechanics.

REFERENCES

1. M. Born and K. Huang, Dynamical Theory of Crystal Lattices (Oxford Univ. Press, London and New York, 1954).

2. W. Ledermann, Proc. Roy. Soc., A182, 362 (1944).

3. J. R. Hardy, Phil. Mag., 7, 315 (1962).

4. A. A. Maradudin, E. W. Montroll, G. H. Weiss, and I. P. Ipatova, Theory of Lattice Dynamics in the Harmonic Approximation (Academic Press, New York, 1971).

5. W. Ludwig and B. Lengeler, Solid State Commun., 2, 83 (1964).

6. See pp. 523-527 and 577-578, Reference 4.

7. F. Seitz, Ann. Math., 37, 17 (1936).

8. E. A. Wood, Bell Syst. Tech. Journal, 43, 541 (1964); Bell Syst. Tech. Publ. Monograph No. 4680 (1964).

9. Lord Rayleigh, Theory of Sound, Vol. I (Dover, New York, 1945).

10. Lord Rayleigh, Proc. London Math Soc., 17, 4 (1887).

11. G. W. Farnell, in Physical Acoustics, Vol. 6, Edited by W. P. Mason and R. N. Thurston (Academic Press, New York, 1970), p. 109.

12. B. A. Auld, Acoustic Fields and Waves in Solids (John Wiley, New York, 1973).

13. E. Dieulesaint and D. Royer, in Handbook of Surfaces and Interfaces, Vol. 2, Edited by L. Dobrzynski (Garland, New York, 1975), p. 109.

14. See, for example, L. D. Landau, and E. M. Lifshitz, Theory of Elasticity (Pergamon Press, New York, 1970), p. 102.

15. D. C. Gazis, R. Herman, and R. F. Wallis, Phys. Rev., 119, 533 (1960).

16. T. C. Lim and G. W. Farnell, J. Appl. Phys., 39, 4319 (1968); J. Acoust. Soc. Am., 45, 845 (1969).

17. F. R. Rollins, T. C. Lim, and G. W. Farnell, Appl. Phys. Lett., 12, 236 (1968).

18. H. Lamb, Phil. Trans. Roy. Soc. London, Ser. A, 203, 1 (1904).

19. See, for example, I. Tolstoy, Wave Propagation (McGraw-Hill, New York, 1973), pp. 221-226.

20. R. Stoneley, Proc. Roy. Soc., Ser. A, 106, 416 (1924).

21. See, for example, Ref. 4.

22. A. E. H. Love, Some Problems of Geodynamics (Cambridge University Press, London and New York, 1911).

23. See, for example, H. Uberall, in Physical Acoustics, Vol. 10, Edited by W. P. Mason and R. N. Thurston (Academic Press, New York and London, 1973), p. 1.

24. C. C. Tseng and R. M. White, J. Appl. Phys., 38, 4274 (1967); C. C. Tseng, ibid., 4281 (1967).

25. E. A. Kraut, Phys. Rev., 188, 1450 (1969).

26. J. L. Bleustein, Appl. Phys. Lett., 13, 412 (1968).

27. Yu. V. Gulyaev, Zh. Eksp. Teor, Fiz. Pis. v. Red., 9, 63 (1969) [Soviet Physics-JETP Letters, 9, 37 (1969)].

28. R. E. Allen, G. P. Alldredge, and F. W. de Wette, Phys. Rev. Lett., 23, 1285 (1969); 24, 301 (1970); Phys. Rev. B4, 1648 (1971); B4, 1661 (1971), B6, 632 (1972).

29. T. E. Feuchtwang, Phys. Rev., 155, 731 (1967).

30. D. Castiel, L. Dobrzynski, and D. Spanjaard, Surf. Sci., 59, 252 (1976).

31. A. Blandin, D. Castiel, and L. Dobrzynski, Solid State Commun., 13, 1175 (1973).

32. W. Ludwig, Japan. J. Appl. Phys. Suppl., 2, Part 2 (1974), p. 879.

33. S. E. Trullinger and S. L. Cunningham, Phys. Rev. Letters, 30, 913 (1973); Phys. Rev. B18, 1898 (1978).

34. G. P. Alldredge, R. E. Allen, and F. W. de Wette, J. Acoust. Soc. Am., 49, 1453 (1971).

35. G. P. Alldredge, Phys. Letters, 41A, 281 (1972).

36. B. Szigeti, Trans. Faraday Soc., 45, 155 (1949), and Proc. Roy. Soc. (London), A204, 51 (1950).

37. W. Cochran, Nature, 191, 60 (1961).

38. A. D. B. Woods, W. Cochran, and B. N. Brockhouse, Phys. Rev., 119, 980 (1960).

39. J. R. Hardy, Phil. Mag., 4, 1278 (1959); J. R. Hardy and A. M. Karo, Phil. Mag., 5, 859 (1960); J. R. Hardy, Phil. Mag., 7, 315 (1962).

40. A. A. Maradudin and L. J. Sham, in Lattice Dynamics, Edited by M. Balkanski (Flammarion Sciences, Paris, 1978), p. 296.

41. R. F. Wallis, in Progress in Surface Science, Vol. 4, Edited by Sydney G. Davison (Pergamon Press, New York, 1973), p. 3.

42. F. W. de Wette, in Lattice Dynamics, Edited by M. Balkanski (Flammarion Sciences, Paris, 1978), p. 275.

43. F. W. de Wette and G. P. Alldredge, in Methods of Computational Physics, Vol. 15, Edited by G. Gilat, B. Alder, S. Fernbach, and M. Rotenberg (Academic Press, New York, 1976), p. 163.

44. M. G. Lagally, in Surface Physics of Materials, Vol. II, Edited by J. M. Blakeley (Academic Press, New York, 1975), p. 419.

45. R. Fuchs and K. L. Kliewer, Phys. Rev., 140, A2076 (1965).

46. K. L. Kliewer and R. Fuchs, Phys. Rev., 144, 495 (1966); 150, 573 (1966).

47. A. A. Lucas, J. Chem. Phys., 48, 3156 (1968).

48. F. W. de Wette and G. E. Schacher, Phys. Rev., 137, 78 (1965).

49. S. Y. Tong and A. A. Maradudin, Phys. Rev., 181, 1318 (1969).

50. W. E. Jones and R. Fuchs, Phys. Rev., B4, 3581 (1971).

51. T. S. Chen, R. E. Allen, G. P. Alldredge, and F. W. de Wette, Solid State Commun., 8, 2105 (1970).

52. T. S. Chen, G. P. Alldredge, and F. W. de Wette, Solid State Commun., 10, 941 (1972).

53. G. Lakshmi and R. Srinivasan, in Lattice Dynamics, Edited by M. Balkanski (Flammarion Sciences, Paris, 1978), p. 305.

54. I. M. Lifshitz and L. M. Rosenzweig, Zh. Eksp. Teor. Fiz., 18, 1012 (1948); I. M. Lifshitz, Nuovo Cimento Suppl., 3, 732 (1956).

55. P. Masri and L. Dobrzynski, J. de Physique, 32, 295 (1971).

56. S. W. Musser and K. H. Rieder, Phys. Rev., B2, 3034 (1970).

57. G. Benedek, in Lattice Dynamics, Edited by M. Balkanski, (Flammarion Sciences, Paris, 1978), p. 308.

58. P. Masri and L. Dobrzynski, Surf. Science, 34, 119 (1973).

59. B. Djafari-Rouhani, P. Masri, and L. Dobrzynski, Phys. Rev., B15, 5960 (1977).

60. P. Masri, B. Djafari-Rouhani, and L. Dobrzynski, in Lattice Dynamics, Edited by M. Balkanski (Flammarion Sciences, Paris, 1978), p. 312.

61. D. C. Gazis and R. F. Wallis, Surf. Science, 5, 482 (1966).

62. See pp. 528-535, Reference 4.

63. D. C. Gazis, R. Herman, and R. F. Wallis, Phys. Rev., 119, 533 (1960).

64. G. Armand and J. B. Theeten, Phys. Rev., B9, 3969 (1974).

65. V. Bertolani, F. Nizzali, and G. Santoro, in Lattice Dynamics, Edited by M. Balkanski (Flammarion Sciences, Paris, 1978), p. 302.

66. D. Lagersie, M. Lannoo, and L. Dobrzynski, J. de Physique, 32, 963 (1971).

67. Suggested in Reference 31.

68. P. Masri, J. de Physique, 35, 433 (1974).

69. R. F. Wallis, D. L. Mills, and A. A. Maradudin, in Localized Excitations in Solids, Edited by R. F. Wallis (Plenum Press, New York, 1968), p. 403.

70. J. T. Collett, Marilyn F. Bishop, and S. E. Trullinger, Phys. Rev., B18, 2464 (1978).

71. B. Djafari-Rouhani and L. Dobrzynski, J. de Physique, 38, C4-126 (1977).

72. S. E. Trullinger and A. A. Maradudin, Phys. Rev., B10, 1350 (1974).

73. A. A. Maradudin and S. H. Vosko, Rev. Mod. Phys., 40, 1 (1968).

74. See p. 82, Reference 4.

75. G. Ya. Liubarskii, The Applications of Group Theory in Physics (Pergamon, New York, 1960), p. 95.

76. P. Rudra, J. Math. Phys., 6, 1273 (1965).

77. A. A. Maradudin and D. L. Mills, Annals of Physics, 100, 262 (1976).

78. L. Dobrzynski and A. A. Maradudin, Phys. Rev., B14, 2200 (1976) and B15, 2432 (1976).

79. See, for example, R. F. Wallis, in Dynamical Properties of Solids, Edited by G. K. Horton and A. A. Maradudin (North-Holland, 1975), p. 473.

80. See pp. 283-297, Reference 41 and pp. 582-595, Reference 4.

81. A. A. Maradudin, R. F. Wallis, and A. Eguiluz, in Statistical Mechanics and Statistical Methods in Theory and Applications, Ed. W. Landman (Plenum Press, New York, U.S.A., 1977) p. 371.

82. F. Garcia-Molinar, Annales de Physique (Paris), 2, 179 (1977).

83. R. E. Allen and F. W. de Wette, J. Chem. Phys., 51, 4820 (1969).

84. See for References, F. W. de Wette, in "Lattice Dynamics," Ed. M. Balkanski (Flammarion Sciences, Paris, 1978), p. 280.

85. L. Dobrzynski and J. Friedel, Surface Sci., 12, 469 (1968).

86. L. Dobrzynski, Ann. Phys. (Paris), 4, (1969) 637.

87. B. Djafari-Rouhani, L. Dobrzynski, and G. Allan, Surface Sci., 55, 663 (1976).

88. J. P. Biberian and M. Bienfait, Surface Sci., 30, 242 (1971).

89. See for references, Refs. 85 and 88.

90. L. Dobrzynski, Le Vide, 189, 8 (1969).

91. P. Masri, G. Allan, and L. Dobrzynski, J. de Phys. (Paris), 33, 85 (1972).

92. K. H. Rieder and E. M. Horl, Phys. Rev. Letters, 20, 209 (1968).

93. L. Dobrzynski and G. Leman, J. de Physique, 30, 116 (1969).

94. M. Born, Repts. on Prog. Phys., 9, 294 (1942).

95. C. B. Duke and G. E. Laramore, Phys. Rev., B2, 4765 (1970); G. E. Laramore and C. B. Duke, ibid., 4783 (1970).

96. D. W. Jepsen, P. M. Marcus, and F. Jona, Surface Sci., 39, 27 (1973).

97. M. G. Lagally, T. C. Ngoc, and M. B. Webb, Phys. Rev. Lett., 26, 1557 (1971).

98. A. A. Maradudin and J. Melngailis, Phys. Rev., 133, A1188 (1964).

99. P. Masri and L. Dobrzynski, J. de Physique, 32, 939 (1971).

100. See the section on Surface Atom Mean Square Displacements for Ref. and discussion, p. 226.

101. B. C. Clark, R. Herman, and R. F. Wallis, Phys. Rev., 139, A860 (1965).

102. R. E. Allen and F. W. de Wette, Phys. Rev., 188, 1320 (1969).

103. A. U. MacRae, Surface Sci., 2, 522 (1964).

104. E. R. Jones, J. T. McKinney, and M. B. Webb, Phys. Rev., 151, 476 (1966).

105. H. B. Lyon and G. A. Somorjai, J. Chem. Phys., 44, 3707 (1966); R. H. Goodman, H. H. Farrell, and G. A. Somorjai, ibid., 48, 1046 (1968).

106. R. Kaplan and G. A. Somorjai, Solid State Comm., 9, 505 (1971).

107. D. Tabor and J. Wilson, Surface Sci., 20, 203 (1970); D. Tabor, J. M. Wilson, and T. J. Bastow, ibid., 26, 471 (1971).

108. R. F. Wallis, B. C. Clark, and R. Herman, Phys. Rev., 167, 652 (1968).

109. A. Ignatiev and T. N. Rhodin, Phys. Rev., B8, 893 (1973).

110. S. Y. Tong, T. N. Rhodin, and A. Ignatiev, ibid., B8, 906 (1973).

111. See chapters 8 and 9, Reference 4.

112. E. W. Montroll and R. B. Potts, Phys. Rev., 102 72 (1956).

113. K. Yamahuzi and T. Tanaka, Prog. Theor. Phys., 20, 327 (1958).

114. M. Ashkin, Phys. Rev., 136, A821 (1964).

115. J. Hori and T. Asahi, Prog. Theor. Phys., 31, 49 (1964).

116. J. Hori and T. Asahi, Prog. Theor. Phys., 17, 523 (1957).

117. T. B. Grimley, Proc. Phys. Soc. (London), 79, 1203 (1962).

118. M. Rich, Los Alamos Sci. Lab. Rep. LA-3537, October 14, 1966.

119. R. E. Allen, G. P. Alldredge, and F. W. de Wette, J. Chem. Phys., 54, 2605 (1971).

120. J. B. Theeten, L. Dobrzynski, and J. L. Domange, Surf. Sci., 34, 145 (1973).

121. J. P. Coulomb, J. Suzanne, M. Bienfait, and P. Masri, Solid State Commun., 15, 1585 (1974).

122. H. Shechter, J. Suzanne, and J. G. Dash, Phys. Rev. Lett., 37, 706 (1976).

123. W. A. Pliskin and R. P. Eischens, Z. Phys. Chem., 24, 11 (1960).

124. See, for example, the Proceedings of the Jülich International Conference on Vibrations in Adsorbed Monolayers, June 1978, Editors S. Lehwald and H. Ibach, Zentralbibliothek der K. F. A. Jülich.

125. T. Asahi and J. Hori, in "Lattice Dynamics," Ed. R. F. Wallis (Pergamon Press, Oxford, 1965), p. 571.

126. H. Kaplan, Phys. Rev., 125, 1271 (1962).

127. L. Dobrzynski and D. L. Mills, J. Phys. and Chem. Solids, 30, 1043 (1969).

128. B. Djafari-Rouhani and L. Dobrzynski, J. Phys. Lett., 37, L. 213, (1976).

129. D. A. Degras, Molecular Processes on Solid Surfaces (McGraw-Hill, New York), 1968, p. 275.

130. G. Armand, P. Masri, and L. Dobrzynski, J. Vac. Sci. Technol., 9, 705 (1972).

131. J. Suzanne, P. Masri, and M. Bienfait, Surf. Sci., 43, 441 (1974).

132. L. Dobrzynski, Surf. Sci., 20, 99 (1970).

133. P. Masri and L. Dobrzynski, J. Phys. and Chem. Solids, 34, 847 (1973).

134. G. P. Alldredge, R. E. Allen, and F. W. de Wette, Phys. Rev., B4, 1682 (1971).

135. W. R. Lawrence and R. E. Allen, Phys. Rev., B14, 2910 (1976) and Phys. Rev., B15, 5081 (1977).

136. D. Castiel, L. Dobrzynski, and D. Spanjaard, Surf. Sci., 60, 269 (1976).

137. L. Dobrzynski and D. L. Mills, Phys. Rev., B7, 1322 (1973).

138. G. Armand and J. B. Theeten, Phys. Rev., B9, 3969 (1974).

139. G. Armand and Y. Lejay, Solid State Commun., 24, 321 (1977).

140. B. Djafari-Rouhani and L. Dobrzynski, Reference 124, p. 120.

141. J. P. Coulomb and P. Masri, Solid State Commun., 15, 1623 (1974).

142. N. S. Gillis, C. R. Fuselier, and J. C. Raich in Lattice Dynamics International Conference Paris, September 1977, Edited by M. Balkanski (Flammarion Sciences, Paris, 1978), p. 341.

143. J. Pouliquen, A. Defebvre, and F. Caudron, C. R. Acad. Sci. Paris, 272, 1033 (1971) and J. Pouliquen, to be published.

144. B. Djafari-Rouhani and L. Dobrzynski, Journal de Physique, 38, C4-126 (1977).

145. B. K. Agrawal, B. Djafari-Rouhani, and L. Dobrzynski, Phys. Rev. B, to be published.

146. L. Dobrzynski and J. Lajzerowicz, Phys. Rev., B12, 1358 (1975).

147. W. Kohn and K. H. Lau, Solid State Comm., 18, 553 (1976).

148. T. B. Grimley, Proc. Phys. Soc., 90, 751 (1967).

149. T. B. Grimley, J. Am. Chem. Soc., 90, 3016 (1968).

150. T. L. Einstein and J. R. Schrieffer, Phys. Rev., B7, 3629 (1973).

151. M. Schick and C. E. Campbell, Phys. Rev., A2, 1591 (1970).

152. S. L. Cunningham, L. Dobrzynski, and A. A. Maradudin, Phys. Rev., B7, 4643 (1973).

153. S. L. Cunningham, A. A. Maradudin, and L. Dobrzynski, Le Vide (France), 28, 171 (1973).

154. K. H. Lau and W. Kohn, Surface Science, 65, 607 (1977).

155. A. M. Stoneham, Solid State Comm., 24, 425 (1977).

156. A. A. Maradudin and R. F. Wallis, to be published.

157. See pp. 390-396, Reference 4.

158. See, for example, P. A. Carruthers, Revs. Mod. Phys., 33, 92 (1961).

159. H. B. G. Casimir, Physica, 5, 495 (1938).

160. B. A. Lippman and J. Schwinger, Phys. Rev., 79, 469 (1950).

161. A. Grimm, A. A. Maradudin, and S. Y. Tong, J. Phys. Coll., Cl, suppl. to no. 4, 31, c 1-9 (1970).

162. L. Dobrzynski and A. A. Maradudin, Phys. Rev., B7, 1207 (1973); ibid., B12, 6006 (1975).

163. D. C. Gazis and R. F. Wallis, Surface Sci., 3, 19 (1965); see also R. A. Toupin and D. C. Gazis, in Lattice Dynamics, p. 597, Ed. R. F. Wallis (Pergamon Press, Oxford, 1965).

164. B. C. Clark, R. Herman, D. C. Gazis, and R. F. Wallis, Ferro-electricity, p. 101, Ed. E. F. Weller (Elsevier Publishing Co., Amsterdam, 1967).

165. G. C. Benson, P. Balk and P. White, J. Chem. Phys., 31, 109 (1959).

166. R. Shuttleworth, Proc. Phys. Soc. (London), A, 62, 167 (1949).

167. B. J. Alder, J. R. Vaisnys, and G. Jura, J. Phys. Chem. Solids, 11, 182 (1959).

168. H. H. Schmidt and G. Jura, J. Phys. Chem. Solids, 16, 60 (1960).

169. G. C. Benson and T. A. Claxton, J. Phys. Chem. Solids, 25, 367 (1964).

170. F. Bonneton and M. Drechsler, Surface Sci., 22, 426 (1970).

171. K. S. Yun and G. C. Benson, J. Chem. Phys., 44, 2548 (1966).

172. J. Vail, Can. J. Phys., 45, 2661 (1967).

173. D. P. Jackson, Can. J. Phys., 49, 2093 (1971).

174. P. Wynblatt and N. A. Gjostein, Surface Sci., 12, 109 (1968).

175. G. C. Benson, P. I. Freeman, and E. Dempsey, J. Chem. Phys., 39, 302 (1963); G. C. Benson and T. A. Claxton, ibid., 48, 1356 (1968).

176. J. J. Burton and G. Jura, J. Phys. Chem., 71, 1937 (1967).

177. R. E. Allen and F. W. de Wette, Phys. Rev., 179, 873 (1969).

178. R. E. Allen, F. W. de Wette, and A. Rahman, Phys. Rev., 179, 887 (1969).

179. V. E. Kenner and R. E. Allen, Phys. Rev., B8, 2916 (1973).

180. J. Friedel, Annales de Physique (Paris), 1, 257 (1976).

181. G. Allan and M. Lannoo, Surf. Sci., 40, 375 (1973).

182. G. Allan, Handbook of Surfaces and Interfaces 1, Ed. L. Dobrzynski (Garland Publishing Co., New York, 1978), p. 320.

183. M. W. Finnis and V. Heine, J. Phys. F, 4, L37 (1974).

184. R. F. Wallis and A. A. Maradudin, Japan J. Appl. Phys. Suppl., 2, Pt. 2, 723 (1974).

185. S. K. Ma and G.P. Alldredge, Bull. American Phys. Soc. Series II, 20, 490 (1975)

186. S. K. Ma, F. W. de Wette, and G. P. Alldredge, Surf. Sci., 78, 598 (1978).

187. J. M. Wilson and T. J. Bastow, Surf. Sci., 26, 461 (1971).

188. C. D. Gelatt, M. G. Lagally, and M. B. Webb, Bull. Am. Phys. Soc., 14, 793 (1969).

189. A. Ignatiev and T. N. Rhodin, Phys. Rev., B8, 893 (1973).

190. D. J. Cheng, R. F. Wallis, and L. Dobrzynski, Surf. Sci., 43, 400 (1974).

191. D. J. Cheng, R. F. Wallis, and L. Dobrzynski, Surf. Sci., 49 (1975).

192. B. K. Agrawal, B. Djafari-Rouhani, and L. Dobrzynski, to be published.

193. G. Allan, Surf. Sci., to be published.

194. H. Viefhues and W. Ludwig, in Lattice Dynamics, Edited by M. Balkanski (Flammarion Sciences, Paris, 1978), p. 299.

195. F. Press and I. Healy, J. Appl. Phys., 28, 1323 (1957).

196. P. J. King and F. W. Sheard, Proc. Roy. Soc. (London), A320, 175 (1970).

197. A. A. Maradudin and D. L. Mills, Phys. Rev., 173, 881 (1968).

198. L. Landau and G. Rumer, Physik. Z. Sowjetunion, 11, 18 (1937).

199. E. Salzmann, T. Plieninger, and K. Dransfeld, Appl. Phys. Lett., 13, 14 (1968).

200. A. J. Budreau and P. H. Carr, Appl. Phys. Lett., 18, 239 (1971).

201. H. J. Maris, Phys. Rev., 188, 1308 (1969).

202. P. J. King and F. W. Sheard, J. Appl. Phys., 40, 5189 (1969).

203. R. G. Steg and P. G. Klemens, Phys. Rev. Lett., 24, 381 (1970).

204. S. G. Kalashinkov, Zh. Eksp. Teor. Fiz., 13, 295 (1943).

205. R. F. Wallis and A. A. Maradudin, Surface Sci., to be published.

206. R. F. Wallis, A. A. Maradudin, and L. Dobrzynski, Phys. Rev., B15, 5681 (1977).

207. R. L. Dennis and D. L. Huber, Phys. Rev., B5, 4717 (1972).

208. J. Lajzerowicz and L. Dobrzynski, Phys. Rev., B14, 2695 (1976).

209. R. D. Mindlin, J. Appl. Phys., 7, 195 (1936).

210. See p. 351 of Chapter 4 of this book.

211. B. Djafari-Rouhani and L. Dobrzynski, Solid State Comm., 20, 1029 (1976).

212. T. Nakayama and T. Sakuma, J. Appl. Phys., 46, 2445 (1975).

213. T. Sakuma, Phys. Rev. Lett., 29, 1394 (1972) and Phys. Rev., B8, 1433 (1973).

214. R. F. Wallis, D. L. Mills, and A. A. Maradudin, Phys. Rev. B., to be published.

215. A. A. Maradudin, in Phonons and Phonon Interactions, Edited by T. A. Bak (W. A. Benjamin, New York, 1964), pp. 462-473.

216. P. Debye, Ann. Phys., 39, 789 (1912).

217. M. J. Lighthill, Introduction to Fourier Analysis and Generalized Functions (Cambridge University Press, Cambridge, 1958), Chap. IV.

218. See p. 587 of Reference 4.

219. K. Portz and A. A. Maradudin, Phys. Rev. B., 16, 3535 (1977).

220. M. Dupuis, R. Mazo, and L. Onsager, J. Chem. Phys., 33, 1452 (1960).

221. See for references, Reference 41 and 78.

222. W. C. Overton and R. F. Swim, Phys. Rev., 84, 758 (1951).

223. T. S. Chen, G. P. Alldredge, F. W. de Wette, and R. E. Allen, J. Chem. Phys., 55, 3121 (1971).

224. J. H. Barkman, R. L. Anderson, and T. E. Brackett, J. Chem. Phys., 42, 1112 (1965).

225. V. R. Velasco and F. Garcia-Moliner, Surf. Sci., 67, 555 (1977).

226. B. Djafari-Rouhani, L. Dobrzynski, and R. F. Wallis, Phys. Rev., B16, 741 (1977).

227. B. Djafari-Rouhani and L. Dobrzynski, Phys. Rev., B14, 2296 (1976).

228. B. Djafari-Rouhani and L. Dobrzynski, Surface Sci., 61, 521 (1976).

229. K. Portz and A. A. Maradudin, to be published.

230. K. R. Subbaswamy and A. A. Maradudin, Phys. Rev., B18, 4181 (1978).

231. K. Portz and A. A. Maradudin, to be published.

Chapter 3

SURFACE POLARITONS

INTRODUCTION

We define a polariton to be an electromagnetic wave coupled to an elementary excitation of a condensed medium, i.e., a photon coupled to a plasmon, phonon, magnon, etc. A surface polariton is a polariton in which the associated electromagnetic field is localized at the surface of the medium (1,2).

In our discussion we shall restrict our attention to surface polaritons associated with a planar interface between two media. The cases of cylindrical and spherical interfaces have been studied by Ruppin and Englman (3).

THEORY OF SURFACE POLARITONS IN DIELECTRIC MEDIA

Isotropic Case

We first present an elementary theoretical treatment of surface polaritons for the case where both media are isotropic. We assume that the interface separating the two media is specified by $x_3 = 0$. The material in the half space $x_3 > 0$ is characterized by a frequency-dependent dielectric constant $\varepsilon_a(\omega)$, and the material in the half-space $x_3 < 0$ is characterized by a dielectric constant $\varepsilon_b(\omega)$. For the present, we assume that losses are sufficiently small to permit neglecting the imaginary parts of the dielectric constants. We also make the local approximation and neglect any wave vector dependence of the dielectric constants. In this section the magnetic permeability is assumed to be unity everywhere.

If the magnetic field is eliminated from Maxwell's equations, the electric field $\vec{E}$ is found to obey the equation

$$\nabla \times \nabla \times \vec{E} + \frac{1}{c^2} \frac{\partial^2 \vec{D}}{\partial t^2} = 0 \qquad (1)$$

where the displacement $\vec{D}$ is given by $\vec{D} = \varepsilon(\omega)\vec{E}$. We seek a solution to Eq. (1) corresponding to a surface polariton propagating in the z-direction. The electric field therefore varies in a wavelike manner in the x_2-direction and decreases exponentially with increasing distance from the interface. Thus, we can write

$$\vec{E}_a(\vec{r},t) = \vec{E}_a^0 \; e^{-\alpha_a x} \; e^{i(k_\parallel x_2 - \omega t)} \qquad x_3 > 0 \qquad (2a)$$

$$\vec{E}_b(\vec{r},t) = \vec{E}_b^0 \; e^{\alpha_b x} \; e^{i(k_\parallel x_2 - \omega t)} \qquad x_3 < 0 \qquad (2b)$$

where $k_\parallel$ is the wave vector describing the propagation parallel to the surface and the decay constants α_a and α_b must have positive real parts in order to have a bona fide surface polariton.

Substituting Eq. (2) into Eq. (1) and solving the resulting algebraic equations yields the following expressions for the electric fields

$$\vec{E}_a(\vec{r},t) = \left(E^o_{a1}, E^o_{a2}, \frac{ik_\parallel}{\alpha_a} E^o_{a2}\right) e^{-\alpha_a x_3 + ik_\parallel x_2 - i\omega t}, \quad x_3 > 0 \tag{3a}$$

$$\vec{E}_b(\vec{r},t) = \left(E^o_{b1}, E^o_{b2}, -\frac{ik_\parallel}{\alpha_b} E^o_{b2}\right) e^{\alpha_b x_3 + ik_\parallel x_2 - i\omega t}, \quad x_3 < 0 \tag{3b}$$

and for the decay constants

$$\alpha_a^2 = k_\parallel^2 - \varepsilon_a(\omega)\frac{\omega^2}{c^2} \tag{4a}$$

$$\alpha_b^2 = k_\parallel^2 - \varepsilon_b(\omega)\frac{\omega^2}{c^2} \tag{4b}$$

From Maxwell's equations, the magnetic fields are found to be

$$\vec{B}_a(\vec{r},t) = \frac{ic}{\omega}\left(\frac{\omega^2}{\alpha_a c^2}\,\varepsilon_a(\omega)E^o_{a2}, \alpha_a E^o_{a1}, ik_\parallel E^o_{a1}\right)$$

$$\cdot\, e^{-\alpha_a x_3 + ik_\parallel x_2 - i\omega t}, \quad x_3 > 0 \tag{5a}$$

$$\vec{B}_b(\vec{r},t) = \frac{ic}{\omega}\left(-\frac{\omega^2}{\alpha_a c^2}\,\varepsilon_b(\omega)E^o_{b2}, -\alpha_b E^o_{b1}, ik_\parallel E^o_{b1}\right)$$

$$\cdot\, e^{\alpha_b x_3 + ik_\parallel x_2 - i\omega t}, \quad x_3 < 0 \tag{5b}$$

where we have assumed that $\vec{B}(\vec{r},t) \sim e^{-i\omega t}$.

We now consider the boundary conditions. The continuity of the tangential components of $\vec{E}$ at $x_3 = 0$ yields the relations

$$E^o_{a1} = E^o_{b1}, \quad E^o_{a2} = E^o_{b2} \tag{6}$$

The continuity of the normal component of $\vec{D}$ yields

$$\frac{\varepsilon_a(\omega)}{\varepsilon_b(\omega)} = -\frac{\alpha_a}{\alpha_b} \tag{7}$$

From the continuity of the tangential components of $\vec{B}$, we obtain, in addition to Eq. (7) the relation

$$(\alpha_a + \alpha_b)E^o_{a1} = 0 \tag{8}$$

Since α_a and α_b must have positive real parts, we see that $E^o_{a1} = E^o_{b1} = 0$; hence, the electric field of the surface polariton lies in the sagittal plane i.e., the plane defined by the direction of propagation and the surface normal. The mode is therefore a TM mode.

The dispersion relation for surface polaritons is given by Eq. (7). Utilizing Eqs. (4), we can rewrite the dispersion relation in the form

$$\frac{c^2k_\parallel^2}{\omega^2} = \frac{\varepsilon_a(\omega)\varepsilon_b(\omega)}{\varepsilon_a(\omega)+\varepsilon_b(\omega)} \tag{9}$$

This may be compared with the dispersion relation for bulk polaritons $c^2k^2/\omega^2 = \varepsilon(\omega)$. In using Eq. (9) one must beware of the fact that this equation has spurious roots arising from the squaring of Eq. (7).

Let us consider a surface polariton characterized by real and positive α_a and α_b. Then Eq. (7) tells us that $\varepsilon_a(\omega)$ and $\varepsilon_b(\omega)$ must have opposite signs at the frequency ω corresponding to the surface polariton. The medium with the negative $\varepsilon(\omega)$ is called the <u>surface active</u> medium, and the medium with positive $\varepsilon(\omega)$ is called the <u>surface inactive</u> medium. The symmetry of Eq. (7) shows that either medium can be the active medium.

Let us now consider as an example the specific case of surface polaritons associated with surface plasmons in n-type GaAs against vacuum. The conduction band of GaAs to a good approximation is spherical and parabolic, so we can use the Drude form for the dielectric constant of a free electron gas

$$\varepsilon(\omega) = \varepsilon_\infty\left(1 - \frac{\omega_p^2}{\omega^2}\right) \tag{10}$$

where ε_∞ is the background (high-frequency) dielectric constant, $\omega_p^2 = 4\pi ne^2/\varepsilon_\infty m^*$ is the square of the plasma frequency, n is the carrier density, and m* is the effective mass. The dispersion relation Eq. (9) with $\varepsilon_a(\omega)$ given by Eq. (10) and $\varepsilon_b(\omega) = 1$, can be solved explicitly to give the frequency of surface plasmon polaritons

$$\omega_{sp}^2(k_\parallel) = \frac{1}{2\varepsilon_\infty}\{(1+\varepsilon_\infty)c^2k_\parallel^2 + \varepsilon_\infty\omega_p^2 - [((1+\varepsilon_\infty)c^2k_\parallel^2 + \varepsilon_\infty\omega_p^2)^2 - 4\varepsilon_\infty{}^2c^2k_\parallel^2\omega_p^2]^{\frac{1}{2}}\} \tag{11}$$

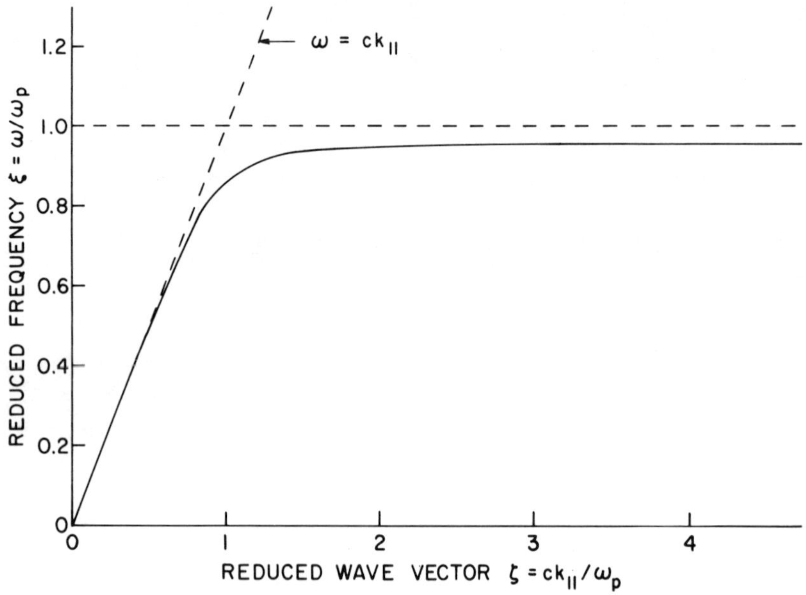

Fig. 3.1.--Dispersion curve of a surface plasmon polariton for n-type GaAs. The curve is asymptotic to the value ξ = 0.957.

A plot of the dispersion relation for n-type GaAs is given in Fig. 3.1. We note several features of the dispersion curve. First, it lies entirely to the right of the light line in vacuum, $\omega = ck_{\|}$. This means that no simple resonant interaction between the surface polariton and light in vacuum is possible. Second, at large values of $k_{\|}$ (the unretarded limit), the surface polariton frequenty approaches the asymptotic value

$$\omega_{sp}(\infty) = \frac{\omega_p}{(1+\frac{1}{\varepsilon_\infty})^{\frac{1}{2}}} \tag{12}$$

At small wave vectors, the surface polariton is primarily photon-like, whereas at large wave vectors it is primarily plasmon-like. In the latter case, the electric field is expressible in terms of an electrostatic potential $\phi(\vec{r},t)$

$$\vec{E}(\vec{r},t) = -\nabla\phi(\vec{r},t) \tag{13}$$

where

$$\phi(\vec{r},t) = \phi_o e^{-\alpha_a x_3 + ik_\parallel x_2 - i\omega t}, \quad x_3 > 0 \tag{14a}$$

$$\phi(\vec{r},t) = \phi_o e^{\alpha_b x_3 + ik_\parallel x_2 - i\omega t}, \quad x_3 < 0 \tag{14b}$$

Substitution of Eqs. (13) and (14) into the equation.

$$\nabla \cdot \vec{D} = 0 \tag{15}$$

yields

$$\alpha_a = \alpha_b = k_\parallel \tag{16}$$

which follows from Eqs. (4) in the limit $k_\parallel \to \infty$.

The second example which we wish to consider in this section is that of surface polaritons associated with surface optical phonons in a cubic ionic crystal with two atoms per unit cell. The appropriate dielectric constant is given by

$$\varepsilon(\omega) = \varepsilon_\infty\left\{1 + \frac{\omega_L^2 - \omega_T^2}{\omega_T^2 - \omega^2}\right\} \tag{17}$$

where ω_T and ω_L are the limiting long-wavelength transverse and longitudinal optical phonon frequencies, respectively. If Eq. (9) is solved with $\varepsilon_a(\omega)$ given by Eq. (17) and $\varepsilon_b(\omega) = 1$, the result is

$$\omega_{SO}^2(k_{\parallel}) = \frac{1}{2\varepsilon_\infty} \{(1 + \varepsilon_\infty)c^2k_{\parallel}^2 + \varepsilon_\infty\omega_L^2$$
$$- [((1 + \varepsilon_\infty)c^2k_{\parallel}^2 + \varepsilon_\infty\omega_L^2)^2$$
$$- 4\varepsilon_\infty c^2k_{\parallel}^2(\omega_T^2 + \varepsilon_\infty\omega_L^2)]^{\frac{1}{2}}\} \tag{18}$$

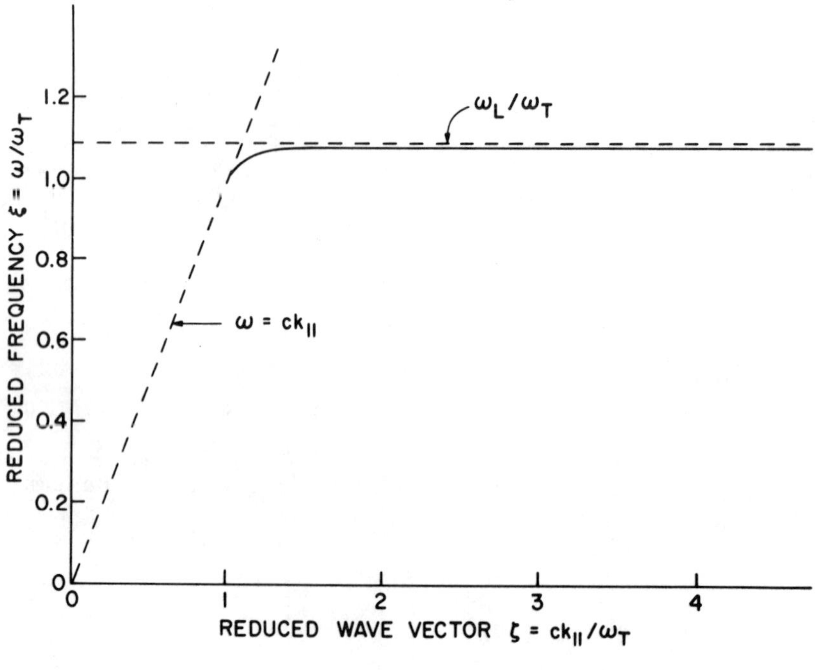

Fig. 3.2.--Dispersion curve of a surface phonon polariton for GaAs. The curve is asymptotic to the value $\xi = 0.994(\omega_L/\omega_T)$.

A plot of $\omega_{SO}(k_\parallel)$ versus $k_\parallel$ is given in Fig. 3.2 for intrinsic GaAs against vacuum. The dispersion curve starts on the light line at ω_T and, as $k_\parallel \to \infty$, approaches the asymptotic value

$$\omega_{SO}(\infty) = \left\{\omega_T^2 + \frac{\omega_L^2 - \omega_T^2}{1 + \frac{1}{\varepsilon_\infty}}\right\}^{\frac{1}{2}} \tag{19}$$

The surface optical phonon case reduces to the surface plasmon case in the limit $\omega_T \to 0$ and $\omega_L \to \omega_p$.

The final example which we present in this section involves surface polaritons associated with coupled surface plasmons and surface optical phonons. The appropriate dielectric constant now contains both plasmon and optical phonon contributions and is given by

$$\varepsilon(\omega) = \varepsilon_\infty\left\{1 - \frac{\omega_p^2}{\omega^2} + \frac{\omega_L^2 - \omega_T^2}{\omega_T^2 - \omega^2}\right\} \tag{20}$$

The dispersion curve can be calculated using Eqs. (9) and (20) and typically consists of two branches, one similar to that of Fig. 3.1 and the other similar to that of Fig. 3.2. The interaction between the surface plasmon and surface optical phonon becomes particularly strong when ω_p and ω_L are comparable. This is illustrated in the unretarded limit in Fig. 3.3 where the squares of the coupled surface mode frequencies for n-type InSb are plotted against free-carrier concentration measured by the square of the ratio of the uncoupled surface plasmon and surface optical phonon frequencies. One sees the typical repulsion of the coupled mode curves in the region where $\omega_{sp}(\infty) \sim \omega_{SO}(\infty)$. In this region the modes are mixed

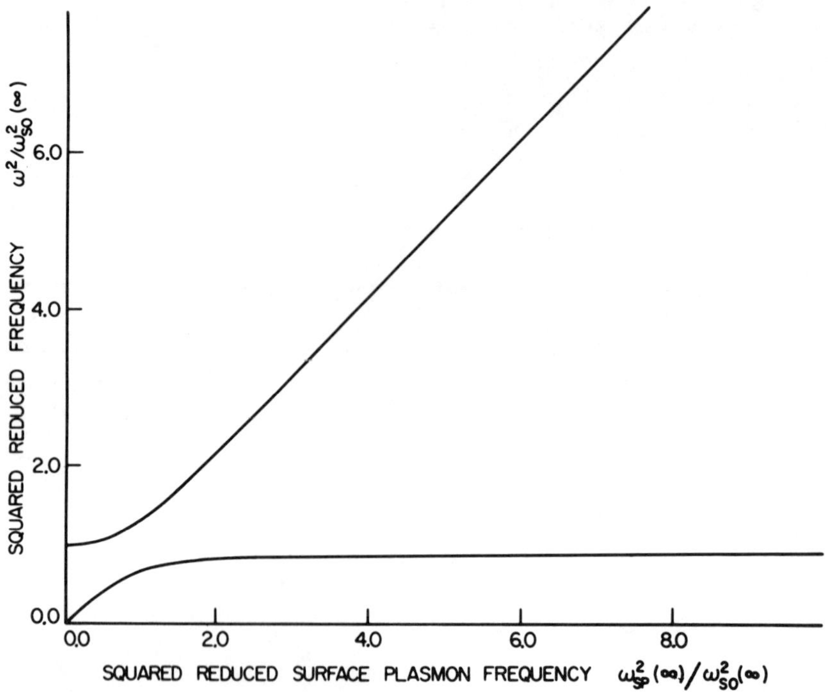

Fig.3.3.--Squared frequencies of coupled surface modes plotted against carrier concentration measured by $\omega_{sp}^2(\infty)$, for n-type InSb.

plasmon-phonon modes. At very high carrier concentrations where $\omega_{sp}(\infty) >> \omega_{SO}(\infty)$, the modes are decoupled into essentially pure plasmon and phonon modes.

Anisotropic Case

We shall restrict our attention in this section to the situation where one medium (medium b) is isotropic with dielectric constant $\varepsilon_b > 0$ and the second medium (medium a) is an anisotropic dielectric. The case for which $\varepsilon_b = 1$ has been treated by Wallis et al. (4) for the most general form of the dielectric tensor of the surface active medium. In dyadic notation, we can write for the displacement $\vec{D}_a$

$$\vec{D}_a = \overleftrightarrow{\varepsilon}_a(\omega) \cdot \vec{E}_a \tag{21}$$

If we take $\vec{E}_a$ to have the form

$$\vec{E}_a(\vec{r},t) = \vec{E}_a^o e^{-\alpha_a x_3} e^{i(k_1 x_1 + k_2 x_2 - \omega t)} \tag{22}$$

and substitute this expression into Eq. (1), we obtain a set of linear, homogeneous algebraic equations whose determinant of coefficients must vanish for a nontrivial solution. This determinantal equation can be written as

$$f(\alpha_a, \omega, \vec{k}_{\parallel}) = 0 \tag{23}$$

In general, it is a quartic equation in α_a which may possess more than one root with a positive real part. Ordinarily, the electric and magnetic vectors must consist of a superposition of solutions corresponding to the roots α_a with positive real parts in order to satisfy the boundary conditions.

A simple case which can be treated analytically and which is of experimental interest is that in which the interface is parallel to a principal axis of the anisotropic dielectric and the wave vector $\vec{k}_{\parallel}$ is parallel to a second principal axis, which we take to be the z-axis. If $\varepsilon_{\perp}(\omega)$ and $\varepsilon_{\parallel}(\omega)$ are the principal values of the dielectric tensor along axes perpendicular to the surface and parallel to $k_{\parallel}$, respectively, the dispersion relation for surface polaritons is found to be

$$\frac{\varepsilon_{\parallel}(\omega)}{\varepsilon_b} = -\frac{\alpha_a}{\alpha_b} \tag{24}$$

where

$$\alpha_a^2 = \frac{\varepsilon_{\parallel}(\omega)}{\varepsilon_{\perp}(\omega)} \left[k_{\parallel}^2 - \varepsilon_{\perp}(\omega)\frac{\omega^2}{c^2}\right] \tag{25}$$

and α_b is given by Eq. (4b). By squaring Eq. (24) one can solve for $k_{\parallel}^2$ and obtain

$$\frac{c^2k_{\parallel}^2}{\omega^2} = \frac{\varepsilon_b\varepsilon_{\perp}(\omega)[\varepsilon_b - \varepsilon_{\parallel}(\omega)]}{\varepsilon_b^2 - \varepsilon_{\parallel}(\omega)\varepsilon_{\perp}(\omega)} \tag{26}$$

Solutions to Eq. (26) correspond to surface polaritons only if either of the following two conditions are satisfied

(i) $\varepsilon_{\parallel}(\omega)$ and $\varepsilon_{\perp}(\omega)$ are both negative, $k_{\parallel} > \frac{\omega}{c}\sqrt{\varepsilon_b}$

(ii) $\varepsilon_{\parallel}(\omega) < 0$, $\varepsilon_{\perp}(\omega) > 0$, $\frac{\omega}{c}\sqrt{\varepsilon_b} < k_{\parallel} < \frac{\omega}{c}\sqrt{\varepsilon_{\perp}(\omega)}$

Case (i) describes the generalization of the surface polaritons to anisotropic materials. Such surface polaritons have been designated "type I" surface polaritons by Bryksin, Mirlin, and Reshina (5). Case (ii), on the other hand, represents a new type of surface polariton which is designated "type II" by Bryksin et al. It exists only if $\varepsilon_{\perp}(\omega) > \varepsilon_b$ and only for a limited range of values of $k_{\parallel}$. In particular, it does not exist in the unretarded limit, $c \to \infty$. The type II surface polariton is an example of what we shall call a "photon-induced" surface polariton.

Gyrodielectric Case

If an external magnetic field is applied to a semiconductor containing free charge carriers, the semiconductor exhibits a variety of magneto-optical effects including cyclotron resonance, magnetoplasma resonance, Faraday effect, the Voigt effect. If a free surface is present, the magnetic field will also modify the surface polaritons associated with surface plasmons. There are a variety of cases depending upon the types of free carriers present.

Semiconductors with a Single Type of Free Carrier

We now develop the theory of surface magnetoplasmon polaritons for a semiconductor such as n-type InSb where the energy band is to a good approximation simple and spherical. The semiconductor is assumed to be semi-infinite and to fill the half-space given by $x_3 \geq 0$. The external magnetic field $\vec{B}_0$ is taken to lie in the $x_1 - x_3$ plane and to make an angle θ with the x_3-axis. The two-dimensional wave vector $\vec{k}_{\parallel}$ describing the propagation of the surface wave parallel to the surface is taken to make an angle ϕ with the x_2-axis. For the present, we neglect effects due to damping, spatial dispersion, and the interaction of magnetoplasmons and optical phonons. The dielectric tensor can then be written in the form

$$\varepsilon_a(\omega) = \begin{bmatrix} s^2\varepsilon_3 + c^2\varepsilon_1 & ic\varepsilon_2 & sc(\varepsilon_3 - \varepsilon_1) \\ -ic\varepsilon_2 & \varepsilon_1 & is\varepsilon_2 \\ sc(\varepsilon_3 - \varepsilon_1) & -is\varepsilon_2 & s^2\varepsilon_1 + c^2\varepsilon_3 \end{bmatrix} \qquad (27)$$

where $\varepsilon_1 = \varepsilon_\infty\{1 + [\omega_p^2/(\omega_c^2 - \omega^2)]\}$, $\varepsilon_2 = \varepsilon_\infty\omega_c\omega_p^2/\omega(\omega^2 - \omega_c^2)$, $\varepsilon_3 = \varepsilon_\infty\{1 - (\omega_p^2/\omega^2)\}$, $s = \sin\theta$, $c = \cos\theta$, and ω_c is the cyclotron frequency defined by $\omega_c = eB_o/m^*c$.

We consider first a particularly simple case which can be treated analytically, namely, $\vec{B}_o$ parallel to the surface and $\vec{k}$ perpendicular to $\vec{B}_o(\theta = \pi/2,\ \phi = 0,\pi)$. If we substitute Eqs. (2) and (27) into Eq. (1), we obtain a nontrivial solution provided

$$\alpha_a^2 = k_\parallel^2 - \varepsilon_v(\omega)\frac{\omega^2}{c^2} \qquad (28)$$

with

$$\varepsilon_v(\omega) = \varepsilon_{a33}(\omega) + \frac{\varepsilon_{a32}^2(\omega)}{\varepsilon_{a33}(\omega)} \qquad (29)$$

and α_b^2 is given by Eq. (4b). In the following, we shall take $\varepsilon_b = 1$. For this configuration the x_1-component of the electric field is decoupled from the x_3- and x_2-components, just as in the isotropic case. The amplitude ratios E_{a3}/E_{a2} and E_{b3}/E_{b2} can be evaluated from Eq. (1); however, a simpler result can be obtained if the equation $\nabla \cdot \vec{D} = 0$ is used. Then one finds

$$\frac{E_{a3}}{E_{a2}} = \frac{-\alpha_a \varepsilon_{a32}(\omega) + ik_{\parallel} \varepsilon_{a33}(\omega)}{\alpha_a \varepsilon_{a33}(\omega) + ik_{\parallel} \varepsilon_{a32}(\omega)} \tag{30a}$$

$$\frac{E_{bx}}{E_{bz}} = -\frac{ik_{\parallel}}{\alpha_b} \tag{30b}$$

The boundary conditions requiring the continuity of the normal component of $\vec{D}$ and the tangential components of $\vec{E}$ at the surface can be written as

$$\varepsilon_{a33}(\omega)E_{a3} + \varepsilon_{a32}(\omega)E_{a2} = E_{b3} \tag{31a}$$

$$E_{a2} = E_{b2} \tag{31b}$$

Combining Eqs. (30) and Eqs. (31) yields the dispersion relation

$$\alpha_a + \alpha_b \varepsilon_v(\omega) + ik_{\parallel} \mathrm{sgn}\,(\vec{k}_{\parallel}) \frac{\varepsilon_{a32}(\omega)}{\varepsilon_{a33}(\omega)} = 0 \tag{32}$$

By $\mathrm{sgn}(\vec{k}_{\parallel})$ we mean +1 if the outward normal to the surface, $\vec{B}_o$, and $\vec{k}_{\parallel}$ form a right-handed triple and −1 if they form a left-handed triple. Calculated surface magnetoplasmon polariton dispersion curves for n-type InSb are shown in Fig. 3.4 for $\omega_c = 0.5\ \omega_p$. One feature, which is immediately obvious from Eq. (32), is that the dispersion curves are different for wave vectors of the same magnitude but opposite signs. Considering first the case $\mathrm{sgn}(\vec{k}_{\parallel}) = +1$,

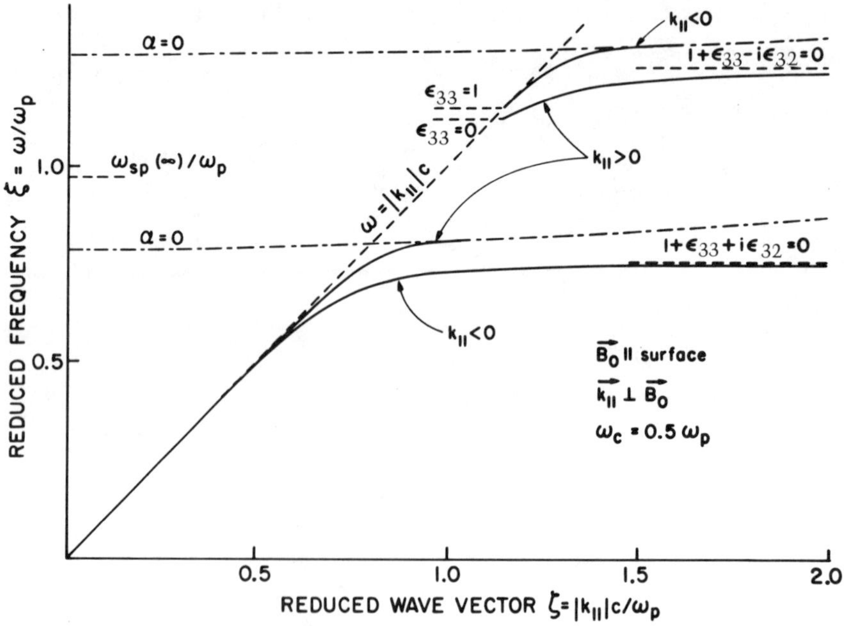

Fig. 3.4.--Surface polariton dispersion curves for n-InSb, with $\vec{B}_0$ parallel to the surface and $\vec{k}_{\parallel} \perp \vec{B}_0$. The dashed curves ($\alpha = 0$) are bulk polariton dispersion curves.

we notice a striking effect not found in the case of zero magnetic field, namely, the dispersion curve consists of two parts with a gap between them (6). Detailed analysis reveals that the gap exists only if ε_∞ and ω_c satisfy the inequality

$$\varepsilon_\infty > (\omega_c^2 + \omega_p^2)^{\frac{1}{2}}/\omega_c \tag{33}$$

It is clear that ε_∞ must exceed unity. For given ε_∞, the ratio ω_c/ω_p must exceed a critical value specified by Eq. (33). For InSb this critical value is 0.064.

Considering now the case $\mathrm{sgn}(\vec{k}_{\parallel}) = -1$, we see from the dispersion curves in Fig. 3.4 that the situation is qualitatively different from the preceding case. We now have a complete lower branch which is similar to the dispersion curve for zero magnetic field. In addition, however, we have an upper branch which starts at the light line where $\varepsilon_{a33} = 1$, rises to the right of the light line, and ends when it meets a bulk polariton dispersion curve defined by $\alpha_a = 0$. The upper branch exists for $\varepsilon_\infty > 1$ and $\omega_c > 0$. It may be remarked that for both cases $\mathrm{sgn}(\vec{k}_\parallel) = \pm 1$, there are frequencies at which both a bulk polariton and a surface polariton can propagate, in contrast to the situation for zero magnetic field.

We now turn to a more complicated case, namely, $\vec{B}_0$ parallel to the surface and $\vec{k}_\parallel$ parallel to $\vec{B}_0$($\theta = \pi/2$, $\phi = \pi/2$). For this case the electric vector no longer lies in the sagittal plane, and the x_1-component of the electric vector is coupled to the x_3- and x_2-components. Using the appropriate form of the dielectric tensor given by Eq.(27) and substituting Eq. (22) into Eq. (1), we find for a nontrivial solution that α_a must satisfy a biquadratic equation whose solutions are given by

$$\alpha_{aj}^2 = \eta_1^2[(\varepsilon_1 + \varepsilon_3)/2\varepsilon_1] + (\omega^2/c^2)(\varepsilon_2^2/2\varepsilon_1)$$
$$- (-1)^j\{[\eta_1^2((\varepsilon_1 - \varepsilon_3)/2\varepsilon_1) + (\omega^2/c^2)(\varepsilon_2^2/2\varepsilon_1)]^2$$
$$+ k_\parallel^2(\omega^2/c^2)(\varepsilon_3/\varepsilon_1)(\varepsilon_2^2/\varepsilon_1)\}^{\frac{1}{2}} \qquad (34)$$

where j = 1, 2 and where $\eta_1^2 = k_\parallel^2 - (\omega^2/c^2)\varepsilon_1$. In order to satisfy the boundary conditions, it is necessary to superpose solutions similar to those given by Eq. (22)

$$\vec{E}_a(\vec{r},t) = [K_1\vec{E}_a(\alpha_{a1})e^{-\alpha_{a1}x_3} + K_2\vec{E}_a(\alpha_{a2})e^{-\alpha_{a2}x_3}]e^{i(k_\parallel x_1 - \omega t)} \qquad (35)$$

where K_1 and K_2 are constants. Substitution of Eq. (35) into the boundary conditions leads to the following dispersion relation

$$(\alpha_o + \alpha_1 + \alpha_2)\alpha_1\alpha_2\varepsilon_1 + [\alpha_o(\alpha_1+\alpha_2) + \alpha_1^2 + \alpha_1\alpha_2 + \alpha_2^2]\alpha_o\varepsilon_1\varepsilon_3$$

$$+ \alpha_o\kappa_1^2\varepsilon_3(1 - \varepsilon_3) = 0 \qquad (36)$$

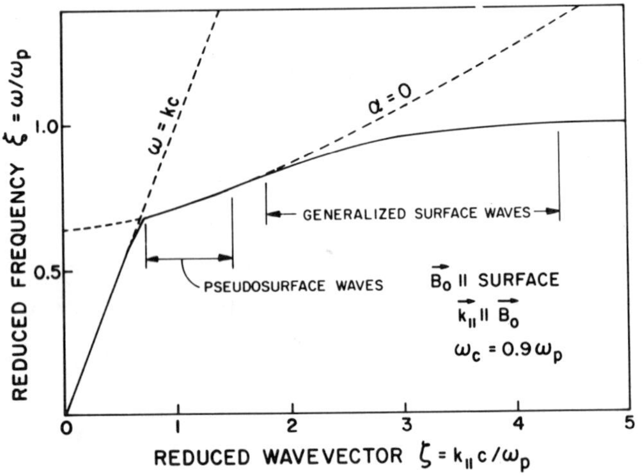

Fig. 3.5.--Surface polariton dispersion curve for n-InSb, with B_0 parallel to the surface and $\vec{k}_\parallel \parallel \vec{B}_0$. The dashed curve ($\alpha = 0$) is a bulk polariton dispersion curve.

The dispersion curve is plotted in Fig. 3.5 for n-type InSb with $\omega_c = 0.9\omega_p$. Two new features are exhibited by Fig. 3.5. In the region labeled "pseudosurface waves," one decay constant in the surface-active medium is positive real but the other is pure imaginary. The polariton therefore consists of a superposition of a localized surface part and a nonlocalized propagating part. In the pseudosurface wave region, the dispersion curve lies very slightly above the bulk polariton curve indicated by the dashed line.

The second feature of interest in Fig. 3.5 is the region labeled "generalized surface waves." Here the decay constants in the surface-active medium are complex conjugates of one another. The polariton is truly localized at the surface, but the electric and magnetic fields decay in an oscillatory rather than a monotonic fashion into the medium.

In the nonretarded limit, only one decay constant is required. The surface polariton frequency is determined by the equation

$$\varepsilon_1(\varepsilon_3/\varepsilon_1)^{\frac{1}{2}} = -1 \qquad (37)$$

and the decay constant is given by

$$\alpha_a = k_\parallel(\varepsilon_3/\varepsilon_1)^{\frac{1}{2}} \qquad (38)$$

We see from Eqs. (37) and (38) that both ε_1 and ε_3 must be negative in order to have a bona fide surface polariton in the unretarded limit, i.e., $\omega_c < \omega_p$.

The configuration $\vec{B}_o$ perpendicular to the surface gives results which are qualitatively similar to those for the configuration just discussed. Both psudosurface waves and generalized surface waves are exhibited. This geometry possesses cylindrical symmetry about the external magnetic field, so the results are independent of the direction of propagation in the surface.

Semiconductors with Multiple Carriers

Many semiconductors are characterized by conduction or valence bands which have several equivalent extrema rather than a single extremum, and which have ellipsoidal rather than spherical constant energy surfaces. In general, free carriers associated with different extrema will have different effective masses and hence different cyclotron frequencies. This leads to additional surface polariton modes.

As an example, let us consider p-type PbTe which has valence band maxima in the [111] directions in wave vector space. We restrict our attention to the geometry where the surface normal is in the [001] direction and the external magnetic field is parallel to the surface and in the [100] direction. The dielectric tensor then has the form (See Ref. 7)

$$\varepsilon_a(\omega) = \begin{bmatrix} \varepsilon_B & 0 & 0 \\ 0 & \varepsilon_A & \varepsilon_c \\ 0 & -\varepsilon_c & \varepsilon_A \end{bmatrix} \tag{39}$$

where

$$\varepsilon_A = \varepsilon_\infty\{1 - [\omega_{p11}^2/(\omega^2 - \omega_{c11}^2)]\} \tag{40a}$$

$$\varepsilon_B = \varepsilon_\infty\{1 - [\omega_{p11}^2(\omega^2 - \omega_{c33}^2)/\omega^2(\omega^2 - \omega_{c11}^2)]\} \tag{40b}$$

$$\varepsilon_C = i\varepsilon_\infty \, \omega_{c11}\omega^2{}_{p12}/\omega(\omega^2 - \omega_{c11}^2) \tag{40c}$$

and where

$$\omega_{c11}^2 = \omega_{ct}^2(K + 2)/3K \quad , \quad \omega_{p12}^2 = \omega_{pt}^2[(K + 2)/3K]^{\frac{1}{2}}$$

$$\omega_{c33}^2 = 3\omega_{ct}^2/(2K + 1) \quad , \quad \omega_{pt}^2 = 4\pi e^2 n/m_T^* \varepsilon_\infty$$

$$\omega_{p11}^2 = \omega_{pt}^2(2K + 1)/3K \quad , \quad \omega_{ct} = eB_o/m_T^* c$$

Also, $K = m_L^*/m_T^*$, m_L^* and m_T^* are the longitudinal and transverse effective masses, n is the free carrier concentration, and ε_∞ is the high-frequency dielectric constant. The principal difference between the dielectric tensor elements of n-type InSb and p-type PbTe is in the elements ε_3(InSb) and ε_B(PbTe). The former is independent of $\vec{B}_o$ whereas the latter depends on $\vec{B}_o$ and, in fact, involves two cyclotron frequencies, ω_{c11} and ω_{c33}.

The surface polariton dispersion curves can be calculated using the procedures already described. We shall illustrate the situation by considering the case $\vec{k}_\parallel \perp \vec{B}_o$ where, we recall, $\vec{B}_o$ is parallel to the surface. The dispersion curves are presented in Fig. 3.6. Compared to n-type InSb (Fig. 3.4) we see that p-type PbTe has two additional branches, one for $\mathrm{sgn}(\vec{k}_\parallel) = +1$ and one for $\mathrm{sgn}(\vec{k}_\parallel) = -1$. This behavior is typical of the multiple carrier case and becomes even more pronounced when the external magnetic field does not make equal angles with the major axes of all the constant energy ellipsoids. Thus, when $\vec{B}_o$ is in the [101] direction, p-type PbTe exhibits additional surface polariton branches beyond those shown in Fig. 3.6.

Another example of multiple carriers is the electron-hole plasma which arises if a semiconductor is maintained at a sufficiently high temperature. Here again one finds additional surface polariton modes (8).

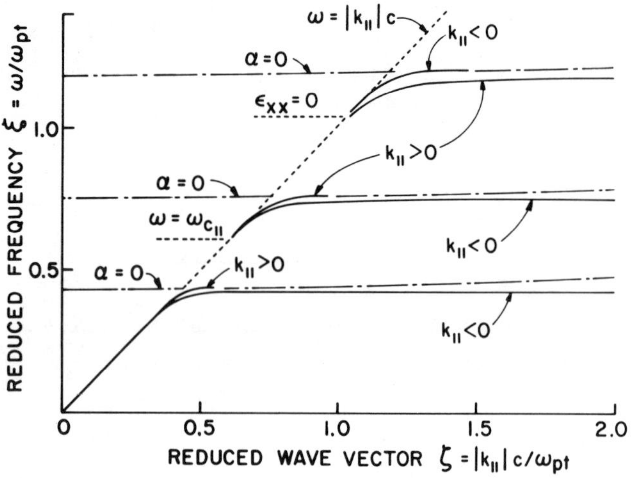

Fig. 3.6.--Surface polariton dispersion curves for p-PbTe with $\vec{B}_0 \parallel [010] \parallel$ surface, $\vec{k}_\parallel \perp \vec{B}_0$, and $\omega_{ct} = \omega_{pt}$. The dashed lines ($\alpha = 0$) are bulk polariton dispersion curves.

Coupled Magnetoplasmon-Optical Phonon Surface Polaritons

The coupling of surface magnetoplasmon polaritons to surface optical phonon polaritons can be investigated by adding the term

$$\varepsilon_\infty \left(\frac{\omega_L^2 - \omega_T^2}{\omega_T^2 - \omega^2} \right)$$

to the diagonal elements of the dielectric tensor given by Eq. (27) or Eq. (39). The coupled mode frequencies can then be calculated as described previously.

We now present some results for n-type InSb (9) when retardation is neglected and the external magnetic field $\vec{B}_0$ is parallel to both the surface and the wave vector $\vec{k}_\parallel$. The dispersion relation is given by Eq. (37). In Fig. 3.7

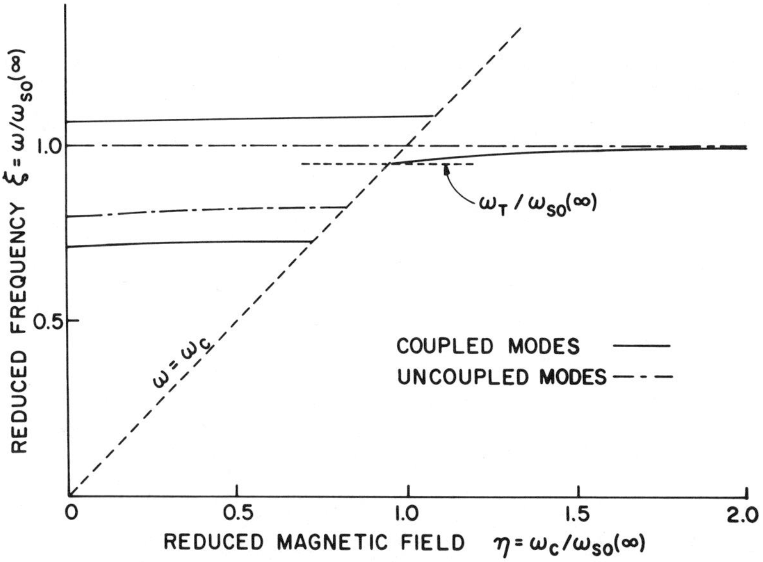

Fig. 3.7.--Coupled surface mode frequencies (unretarded) for n-InSb plotted against magnetic field for $\vec{B}_0$ parallel to the surface and $\omega_{sp}(\infty)/\omega_{SO}(\infty) = 0.8$.

the coupled mode frequencies are plotted against magnetic field for the case $\omega_{sp}(\infty) = 0.8\omega_{SO}(\infty)$. We see that the interaction forces the plasmon and phonon branches apart in the low-field region. We also note that the plasmon-like branch terminates at the cyclotron line $\omega = \omega_c$ and the phonon-like line undergoes a discontinuity at this line. The phonon-like branch approaches the value $\omega_{SO}(\infty)$ at large magnetic fields.

When retardation is included the discussion becomes complicated except for the case in which $\vec{B}_O$ is parallel to the surface and perpendicular to $\vec{k}_{\parallel}$. The dispersion curves for n-type InSb are shown in Fig. 3.8 for $\omega_c = 0.5\ \omega_p$, $\omega_{SO}(\infty) = \omega_{sp}(\infty)$, and $\text{sgn}(\vec{k}_{\parallel}) = +1$. Both polariton branches

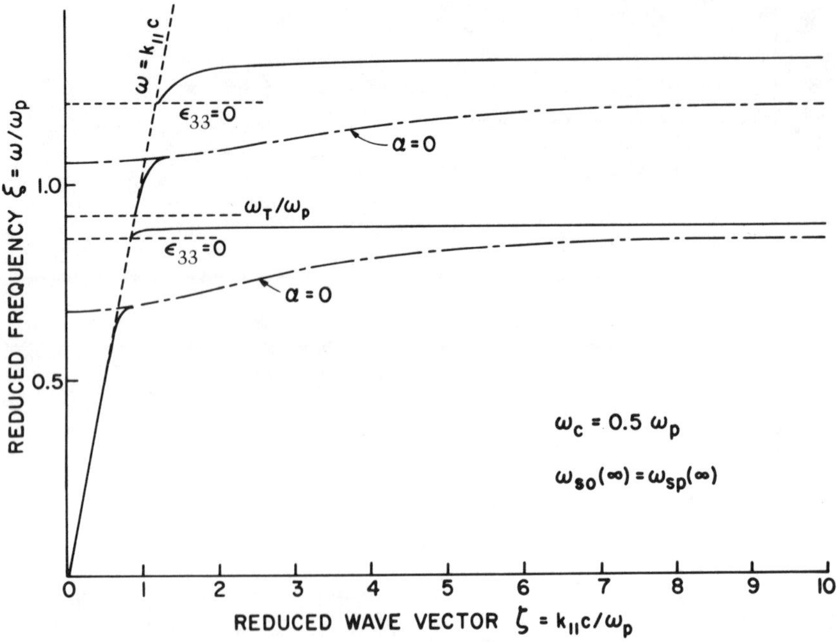

Fig. 3.8.--Coupled surface polariton dispersion curves for n-InSb, with $\vec{B}_0$ parallel to the surface, $\vec{k}_{\parallel} \perp \vec{B}_0$, and $\mathrm{sgn}(\vec{k}_{\parallel}) = +1$.

have gaps of the type found in Fig. 3.4. In a similar fashion one can discuss materials with multiple carriers such as p-type PbTe.

Surface Polaritons with Damping

In our previous discussion we have neglected dissipative processes and assumed that the dielectric tensors in both media are real. We now relax this assumption and let medium a have a complex dielectric tensor. We shall restrict our attention to the isotropic case and write

$$\varepsilon_a(\omega) = \varepsilon_a'(\omega) + i\varepsilon_a''(\omega) \tag{41}$$

The dispersion relation is still given by Eq. (7), but we must now recognize that α_a, α_b, and either $\vec{k}_\parallel$ or ω are complex. We shall take ω real and $\vec{k}_\parallel$ complex. Then the dispersion relation is

$$\frac{\varepsilon'_a(\omega) + i\varepsilon''_a(\omega)}{\varepsilon_b(\omega)} = -\frac{\alpha'_a + i\alpha''_a}{\alpha'_b + i\alpha''_b} \tag{42}$$

where $\varepsilon_b(\omega)$ is real. The existence of a nontrivial solution of Maxwell's equations requires that α_a and α_b satisfy Eqs. (4). Combining the latter equations with Eq. (42) yields the analogue of Eq. (9)

$$(k'_\parallel + ik''_\parallel)^2 = \left(\frac{\omega^2}{c^2}\right) \frac{\varepsilon_b(\omega)[\varepsilon'_a(\omega) + i\varepsilon''_a(\omega)]}{\varepsilon'_a(\omega) + i\varepsilon''_a(\omega) + \varepsilon_b(\omega)} \tag{43}$$

By equating the real and imaginary parts, respectively, of both sides of each of the equations just discussed, one can solve for α'_a, α''_a, α'_b, α''_b, $k'_\parallel$ and $k''_\parallel$ individually. The results in general are cumbersome and will not be given here. However, the situation becomes simpler for a very lossy material where $\varepsilon''_a(\omega) \gg \varepsilon'_a(\omega)$, 1. Then Eq. (43) can be solved approximately to give

$$k'_\parallel \approx \frac{\omega}{c} \left[1 - \frac{\varepsilon'_a(\omega) + \frac{3}{4}}{2(\varepsilon''_a(\omega))^2}\right] \tag{44a}$$

$$k''_\parallel \approx \frac{\omega}{2c} \cdot \frac{1}{\varepsilon''_a(\omega)} \tag{44b}$$

where we have taken $\varepsilon_b(\omega) = 1$ for simplicity. The attenuation length L is the distance required for the energy density of the wave to decrease to 1/e of its initial value and is given by

$$L = \frac{1}{2k''_{\parallel}} \qquad (45a)$$

$$\approx \frac{c}{\omega} \cdot \varepsilon''_a(\omega) \qquad (45b)$$

which is very long compared to a wavelength.

We note from Eq. (44a) that if $\varepsilon'_a(\omega) < -3/4$, $k'_{\parallel}$ lies to the right of the light line, and we have a damped surface polariton which attenuates as it propagates with the attenuation length L. On the other hand, if $\varepsilon'_a(\omega) > -3/4$, $k'_{\parallel}$ lies to the left of the light line, we have a new type of surface polariton known as a Zenneck mode (1). It should be noted that Zenneck modes can exist even when the real part of $\varepsilon_a(\omega)$ is positive. Since α_a and α_b are complex quantities, the Zenneck modes are "generalized" surface waves. The real and imaginary parts of α_a and α_b are given approximately by

$$\alpha'_a = \frac{\omega}{c}\left(\frac{\varepsilon''_a}{2}\right)^{\frac{1}{2}} = -\alpha''_a \qquad (46a)$$

$$\alpha'_b = \frac{\omega}{c}(2\varepsilon''_a)^{-\frac{1}{2}} = \alpha''_b \qquad (46b)$$

Effect of Spatial Dispersion on Surface Polaritons

Our discussion so far has ignored the effect of spatial dispersion on the properties of surface polaritons.

Spatial dispersion is a consequence of a nonlocal relationship between the electric displacement $\vec{D}$ and the macroscopic electric field $\vec{E}$ as expressed by

$$D_\mu(\vec{x},t) = \sum_\nu \int d^3x' \int_{-\infty}^{t} dt' \varepsilon_{\mu\nu}(\vec{x},\vec{x}';t,t')E_\nu(\vec{x}',t') \tag{47}$$

where the integral on $\vec{x}'$ extends over the volume of the medium. If the medium is spatially homogeneous and the Hamiltonian is time-independent, the nonlocal dielectric tensor becomes a function of $\vec{x} - \vec{x}'$ only and $t - t'$ only, i.e., $\varepsilon_{\mu\nu}(\vec{x} - \vec{x}';\ t-t')$. In such a case, a Fourier analysis of Eq. (47) yields

$$D_\mu(\vec{q},\omega) = \sum_\nu \varepsilon_{\mu\nu}(\vec{q},\omega)E_\nu(\vec{q},\omega) \tag{48}$$

The wave vector dependence of the dielectric tensor $\varepsilon_{\mu\nu}(\vec{q},\omega)$ is a manifestation of nonlocality or spatial dispersion.

An analysis of spatial dispersion when a free surface is present has been carried out by Maradudin and Mills (10). If the surface is defined by $x_3 = 0$, with the medium in the half-space $x_3 > 0$, and the displacement and macroscopic field are specified by

$$D_\mu(\vec{x},t) = D_\mu(x_3)e^{ik_\parallel x_1}e^{-i\omega t} \tag{49a}$$

$$E_\mu(\vec{x},t) = E_\mu(x_3)e^{ik_\parallel x_1}e^{-i\omega t} \tag{49b}$$

then the constitutive relation takes the form

$$D_\mu(x_3) = \sum_\nu \int_0^\infty dx_3' \varepsilon_{\mu\nu}(\vec{q}_\parallel, \omega; x_3 - x_3') E_\nu(x_3') \qquad (50)$$

where

$$\varepsilon_{\mu\nu}(\vec{q}, \omega; x_3 - x_3') = \int_{-\infty}^{\infty} \frac{dq_3}{2\pi} \varepsilon_{\mu\nu}(\vec{q}, \omega) e^{iq_3(x_3 - x_3')} \qquad (51)$$

The integral relationship embodied in Eq. (50) has the consequence that Maxwell's equations are now integrodifferential equations whose general solution is difficult to obtain. Progress can be made, however, if one adopts a particular model. For example, excitons and infrared-active T phonons in cubic crystals can be described by the dielectric tensor

$$\varepsilon_{\mu\nu}(q, \omega) = \delta_{\mu\nu}\varepsilon(q, \omega) = \delta_{\mu\nu}\{\varepsilon_o + \frac{\Omega_p^2}{\omega_T^2 + Dq^2 - \omega^2 - i\omega\gamma}\} \qquad (52)$$

where ω_T is the frequency of the excitation at $\vec{q} = 0$, Ω_p is an effective "plasma" frequency (equal to $\varepsilon_o(\omega_L^2 - \omega_T^2)^{\frac{1}{2}}$ in the phonon case), D is the coefficient of the term responsible for spatial dispersion effects, and γ is the damping constant. If we define $\Gamma(q_\parallel)$ by

$$\Gamma(q_\parallel) = \left(\frac{\omega^2 - \omega_T^2 - Dq_\parallel^2 + i\omega\gamma}{D}\right)^{\frac{1}{2}} \qquad (53)$$

then Eq. (50) becomes

$$D_\mu(x_3) = \varepsilon_o E_\mu(x_3) + \frac{i\Omega_p^2}{2D\Gamma(k_\parallel)} \int_o^\infty dx_3' e^{i\Gamma(k_\parallel)|x_3 - x_3'|} E_\mu(x_3') \tag{54}$$

If Eq. (54) is substituted into Maxwell's equations and if $E_\mu(x_3)$ has the form

$$E_\mu(x_3) = E_\mu^c \, e^{-\alpha x_3}, \quad x_3 \geq 0 \tag{55a}$$

$$E_\mu(x_3) = E_\mu^v \, e^{\alpha_o x_3}, \quad x_3 > 0 \tag{55b}$$

then a nontrivial solution results only if

$$\varepsilon(\vec{k}_\parallel, i\alpha;\omega) \left[\frac{c^2(k_\parallel^2 - \alpha^2)}{\omega^2} - \varepsilon(\vec{k}_\parallel, i\alpha;\omega)\right] = 0 \tag{56}$$

and

$$\alpha_o^2 = k_\parallel^2 - (\omega^2/c^2) \tag{57}$$

where

$$\varepsilon(k_\parallel, i\alpha;\omega) = \varepsilon_o + \frac{\Omega_p^2}{\omega_T^2 - \omega^2 + D(k_\parallel^2 - \alpha^2) - i\omega\gamma} \tag{58}$$

For given values of ω and $k_\parallel$, there are three values of α which satisfy Eq. (56) and which have positive real parts. Maradudin and Mills show that there is an additional boundary condition when spatial dispersion is present. This means that solutions corresponding to all three values of α, with $\mathrm{Re}(\alpha) > 0$, must be superposed in order to satisfy the boundary conditions. When this is done, the dispersion relation for surface polaritons is found to be

$$\alpha_1(i\Gamma+\alpha_1)(k_\parallel^2 - \alpha_2\alpha_3) - \alpha_2(i\Gamma+\alpha_2)(k_\parallel^2 - \alpha_3\alpha_1)$$
$$+ k_\parallel^2(i\Gamma+\alpha_3)(\alpha_2 - \alpha_1)$$
$$+ \alpha_o[\varepsilon_1(i\Gamma+\alpha_1)(k_\parallel^2 - \alpha_2\alpha_3)$$
$$- \varepsilon_2(i\Gamma+\alpha_2)(k_\parallel^2 - \alpha_1\alpha_3)] = 0 \qquad (59)$$

where $\varepsilon_i = \varepsilon(k_\parallel, i\alpha_i; \omega)$ for i = 1,2. The quantities α_1 and α_2 are the roots obtained by setting the factor in square brackets on the left-hand side of Eq. (56) equal to zero.

The surface polariton frequency obtained by solving Eq. (59) is found to have an imaginary part which produces a decay of the surface polariton and which arises from the absence of a gap in the bulk excitation spectrum between ω_T and $\omega_L = [\omega_T^2 + (\Omega_p^2/\varepsilon_o)]^{\frac{1}{2}}$ when spatial dispersion is present. The surface polariton frequency therefore occurs in a region where the excitation energy density of states for bulk excitations is nonzero, and energy can leak from the surface into the bulk of the crystal. Another consequence of spatial dispersion is that as $k_\parallel \to \infty$, the frequency $\omega_s(k_\parallel)$ of the surface polariton does not approach a constant asymptotic value corresponding to $\varepsilon(\omega) = -1$, but instead increases linearly with $k_\parallel$ when $ck_\parallel \gg \omega$. The shift in ω_s which is linear in $k_\parallel$ can be shown to be

$$\Delta\omega_s(k_{\parallel}) = D^{\frac{1}{2}} k_{\parallel} \left[\frac{(\varepsilon_s - \varepsilon_o)^{\frac{1}{2}}}{2(1+\varepsilon_o)^{3/2}} \quad \frac{\omega_T}{\omega_s(o)} \right] \left[\varepsilon_o^{\frac{1}{2}} - i \left(\frac{\varepsilon_o + 3}{2} \right) \right] \tag{60}$$

where $\varepsilon_s = \varepsilon_o + (\Omega_p^2/\omega_T^2)$ and $\omega_s(o)$ is the solution to $\varepsilon(\omega) = -1$. For ZnSe, one finds that

$$\frac{\mathrm{Re}\Delta\omega_s(k_{\parallel})}{\omega_T} \quad 0.007 \quad \left(\frac{ck_{\parallel}}{\omega_T} \right)$$

so spatial dispersion effects are not entirely negligible in this material.

THEORY OF SURFACE POLARITONS IN MAGNETIC MEDIA

We now consider magnetic media and relax our previous assumption that the magnetic permeability $\mu(\omega)$ is unity. In the absence of free charges Maxwell's equations are invariant under the interchange of $\vec{E}$ and $\vec{H}$, of $\varepsilon(\omega)$ and $\mu(\omega)$, and the interchange of ω and $-\omega$. Consequently, surface polaritons should exist at the interface between two magnetic media or between a magnetic and a dielectric medium, provided the magnetic medium is surface active in some frequency range.

The theoretical development of surface polaritons propagating along the interface between two magnetic media can be carried out in complete analogy to the case of two dielectrics (11). As an example, we shall consider a semi-infinite ferromagnetic insulator with vacuum beyond a planar boundary defined by $x_3 = 0$. We assume a single domain magnetized in the $+x_1$-direction and propagation in the

$\pm x_2$-directions. For this geometry the permeability tensor takes the form

$$\mu_a(\omega) = \begin{bmatrix} \mu_{11} & 0 & 0 \\ 0 & \mu_{32} & -\mu_{31} \\ 0 & \mu_{31} & \mu_{32} \end{bmatrix} \tag{61}$$

Proceeding as in the corresponding gyrodielectric case, we find from Maxwell's equations that the decay constants are given by

$$\alpha_a^2 = k_\parallel^2 - \left(\frac{\omega^2}{c^2}\right) \mu_v \tag{62a}$$

$$\alpha_b^2 = k_\parallel^2 - \frac{\omega^2}{c^2} \tag{62b}$$

where

$$\mu_v = \mu_{22} + \frac{\mu_{32}^2}{\mu_{22}} \tag{63}$$

and $\varepsilon_a = 1$. The boundary conditions lead to the dispersion relation

$$\alpha_a + \alpha_b \mu_v + i k_\parallel \, \mathrm{sgn}(\vec{k}_\parallel) \mu_{32}/\mu_{22} = 0 \tag{64}$$

Just as in the gyrodielectric case, the dispersion relation exhibits the nonreciprocity associated with nonequivalent propagation in the $+\vec{k}_\parallel$ and $-\vec{k}_\parallel$ directions. The magnetic vector of the surface polariton lies in the sagittal plane, and the mode is a TE mode.

To further illustrate the nature of the surface polaritons we consider the permeability associated with magnons in a single-domain ferromagnetic insulator. The required elements of the permeability tensor are

$$\mu_{22} = \mu_\infty \left(1 + \frac{4\pi\omega_s\omega_o}{\omega_o^2 - \omega^2}\right) \tag{65a}$$

$$\mu_{32} = i\mu_\infty \frac{4\pi\omega_s\omega}{\omega_o^2 - \omega^2} \tag{65b}$$

where $\omega_o = \gamma H_o$, $\omega_s = \gamma M_o$, H_o is the magnitude of the static magnetic field, M_o is the static magnetization, γ is the gyromagnetic ratio, and μ_∞ is the permeability due to other magnetic dipole excitations such as optical magnons. Using Eqs. (62)-(65) one can calculate dispersion curves explicitly.

The results for the case $\mu_\infty = 10.0$ and $\omega_s/\omega_o = 1$ are shown in Fig. 3.9. There is considerable similarity to the corresponding gyrodielectric case of Fig. 3.4, except that the lower branches of both the $+\vec{k}_\parallel$ and $-\vec{k}_\parallel$ curves are missing. This is a consequence of the different frequency dependences of the off-diagonal elements of the dielectric and permeability tensors. For $\mathrm{sgn}(\vec{k}_\parallel) = +1$, the dispersion curve approaches an asymptotic limit as $k_\parallel \to \infty$. This is given by $\omega_{sm} = (\omega_o + \mu_\infty\omega_s)/(1 + \mu_\infty)$, which corresponds to the Damon-Eshbach mode (12).

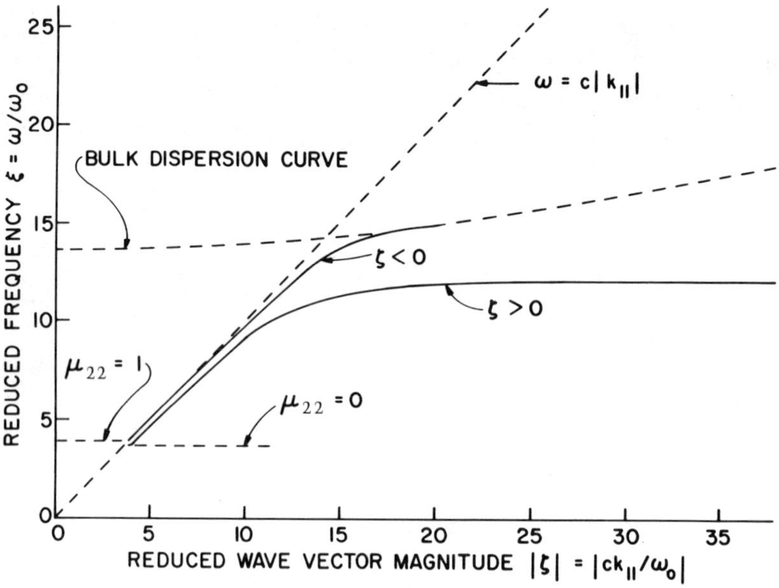

Fig. 3.9.--Dispersion curves for surface polaritons propagating on a gyromagnetic medium, with $\vec{H}_0$ parallel to the surface and $\vec{k}_\parallel \perp H_0$.

ROLE OF SURFACE POLARITONS IN FORMING THE IMAGE CHARGE

The classical interaction between an external static point charge and a semi-infinite dielectric medium can be understood in terms of the image charge. Recently, Mahan (13) has shown that surface polarization modes play the central role in determining the image charge when the static charge is outside the dielectric medium. The presence of the static charge induces a static polarization field which can be expanded as a linear combination of the polarization normal modes of the dielectric. In a cubic dielectric, the long-wavelength polarization modes can be classified as bulk longitudinal, bulk transverse, and surface modes. Mahan demonstrates by a microscopic treatment that when the source charge is outside the medium, the surface polarization modes are the only ones induced, and in the long-wavelength limit they produce a polarization field equivalent to that of an image charge. The contributions of the shorter wavelength polarization modes become important only when the source charge is within a few interatomic spacings of the surface.

Let a charge q be located at $\vec{R}_1$ and let the dielectric medium be composed of ions of polarizability $\alpha(\omega)$ located at sites $\vec{R}_j$. The potential $\Phi(\vec{R}_2,\vec{R}_1)$ at a point $\vec{R}_2$ can be written as (See Ref. 14)

$$\Phi(\vec{R}_2,\vec{R}_1) = \frac{q}{R_{12}} - \sum_{\mu j} E_\mu(\vec{R}_{2j})P_\mu^{(1)}(\vec{R}_j) \tag{66}$$

where $\vec{R}_{ij} = \vec{R}_i - \vec{R}_j$. The polarization component $P_\mu^{(1)}(\vec{R}_j)$ satisfies the equation

$$P_\mu^{(1)}(\vec{R}_i) = q\alpha(o)E_\mu(\vec{R}_{1i}) - \alpha(o) \sum_{j,\nu} \Phi_{\mu\nu}(\vec{R}_{ij})P_\nu^{(1)}(\vec{R}_j) \tag{67}$$

and

$$E_\mu(\vec{R}) = R_\mu/R^3 \tag{68}$$

$$\Phi_{\mu\nu}(\vec{R}) = \left(\delta_{\mu\nu} - \frac{3R_\mu R_\nu}{R^2}\right)\frac{1}{R^3} \tag{69}$$

Equation (67) can be solved with the aid of the Green's function $G_{\mu\nu}(\vec{R}_i,\vec{R}_j)$, which obeys the equation

$$\sum_{j,\nu}[\delta_{\mu\nu}\delta_{ij} + \alpha(o)\ \Phi_{\mu\nu}(\vec{R}_{ij})]G_{\mu\lambda}(R_j,R_k) = \delta_{\mu\lambda}\delta_{ik} \tag{70}$$

Then

$$P_{\mu}^{(1)}(R_i) = q\alpha(o) \sum_{j,\nu} G_{\mu\nu}(\vec{R}_i,\vec{R}_j)E_{\nu}(\vec{R}_{j1}) \tag{71}$$

and

$$\Phi(\vec{R}_2,\vec{R}_1) = q[\frac{1}{R_{12}} - \alpha(o) \sum_{i,j}\sum_{\mu,\nu} E_{\mu}(\vec{R}_{2j})G_{\mu\nu}(\vec{R}_j,\vec{R}_i) \cdot E_{\nu}(\vec{R}_{i1})] \tag{72}$$

The Green's function can be constructed from the eigenfrequencies ω_n and eigenfunctions $\psi_{\nu}(\vec{R}_j,n)$ of the homogeneous equation

$$\sum_{j,\nu}[\delta_{\mu\nu}\delta_{ij} + \alpha(\omega_n)\Phi_{\mu\nu}(\vec{R}_{ij})]\Psi_{\nu}(\vec{R}_j,n) = 0 \tag{73}$$

We obtain

$$G_{\mu\nu}(\vec{R}_i,\vec{R}_j) = \sum_n \frac{\Psi_{\mu}(\vec{R}_i,n)\Psi_{\nu}^{*}(\vec{R}_j,n)}{[1 - \alpha(o)/\alpha(\omega_n)]} \tag{74}$$

When this result is substituted into Eq. (72), the potential $\Phi(\vec{R}_2,\vec{R}_1)$ is found to be given by

$$\Phi(\vec{R}_2,\vec{R}_1) = q\left[\frac{1}{R_{12}} - \alpha(o) \sum_n \frac{\Lambda^*(\vec{R}_1,n)\Lambda(\vec{R}_2,n)}{[1 - \alpha(o)/\alpha(\omega_n)]}\right] \tag{75}$$

where

$$\Lambda(\vec{R}_1,n) = \sum_{\nu j} E_\nu(\vec{R}_{1j})\Psi^*_\nu(\vec{R}_j,n) \tag{76}$$

The eigenfunctions obtained by solving Eq. (73) can be classified in the long-wavelength limit as longitudinal, transverse, and surface modes. Only the longitudinal and surface modes can contribute to $\Phi(\vec{R}_2,\vec{R}_1)$, since the polarization induced by the source charge does not contain any transverse mode components. Considering the case where $\vec{R}_1$ and $\vec{R}_2$ are outside the medium, we find after calculating the long-wavelength eigenvalues and eigenfunctions of Eq. (73) that the longitudinal modes contribute nothing and that the surface modes yield

$$\Phi^{(s)}(\vec{R}_2,\vec{R}_1) = -q \left(\frac{\varepsilon-1}{\varepsilon+1}\right) \frac{1}{R'_{12}} \tag{77}$$

Here R'_{12} is the distance from the image charge and is given by

$$R'_{12} = [(x_{11} - x_{21})^2 + (x_{12} - x_{22})^2 + (x_{13} + x_{23})^2]^{\frac{1}{2}} \tag{78}$$

where

$$\vec{R}_i \equiv (x_{i1},x_{i2},x_{i3})$$

and ε is the dielectric constant of the medium. The total potential is then given by

$$\Phi(\vec{R}_2,\vec{R}_1) = \frac{q}{\varepsilon}\{\frac{1}{R_{12}} - (\frac{\varepsilon -1}{\varepsilon +1}) \frac{1}{R'_{12}}\} \tag{79}$$

which is the standard macroscopic result.

ROLE OF SURFACE POLARITONS IN THE INTERACTION BETWEEN MACROSCOPIC BODIES

Let us consider two semi-infinite dielectric media, which are characterized by an isotropic dielectric constant $\varepsilon(\omega)$, filling the regions $x_3 < -d/2$ and $x_3 > d/2$. They are separated by a slab of thickness d whose dielectric constant is taken to be unity. In the absence of retardation, the dispersion relation giving the frequencies of the electromagnetic modes localized at each of the two surfaces is given by (see Ref. 14)

$$g(\omega,k,d) \equiv \left(\frac{\varepsilon(\omega) + 1}{\varepsilon(\omega) - 1}\right)^2 e^{2kd} - 1 = 0 \tag{80}$$

where k is the magnitude of the wave vector parallel to the surface. Equation (80) is equivalent to the pair of equations

$$\varepsilon(\omega) = - \coth \frac{kd}{2} \tag{81a}$$

$$\varepsilon(\omega) = - \tanh \frac{kd}{2} \tag{81b}$$

If we assume periodic boundary conditions in the two directions parallel to the surfaces, the zero-point energy of the electromagnetic field associated with the surface excitations can be written in the form

$$U(d) = \left(\frac{L}{2\pi}\right)^2 \int_0^\infty 2\pi k dk \, \frac{1}{2\pi i} \int_C^\infty \frac{\hbar\omega}{2} \, d\ell n \, g(\omega, k, d) \tag{82}$$

where L is the normalization length associated with the periodic boundary conditions and the contour C encircles the positive half of the real ω-axis in the counterclockwise sense. The quantity U is divergent in general, but its derivative with respect to d, which is the negative of the force between the two dielectric media, is finite. If L^2 is set equal to unity to give the force per unit area, and if the contour C is deformed into a semicircle of infinite radius in the right-hand half of the ω-plane together with the imaginary ω-axis, only the latter contributes to the force, which becomes

$$F(d) = \frac{\hbar}{16\pi^2 d^3} \int_0^\infty x^2 dx \int_0^\infty \frac{d\xi}{\left(\frac{f(\xi) + 1}{f(\xi) - 1}\right)^2 e^x - 1} \tag{83}$$

where $f(\xi) = \varepsilon(i\xi)$.

The force specified by Eq. (83) is determined entirely by surface excitations. It is only these excitations which have electric fields that extend into the vacuum outside the media. When retardation is neglected, the bulk transverse excitations in the media have no macroscopic electric field associated with them. The longitudinal excitations do possess a macroscopic electric field, but this field is confined within the medium just as the electric field in a parallel plate condenser is confined to the volume of the condenser.

When retardation is included, the dispersion relation analogous to Eq. (80) becomes (see Ref. 15):

$$g_1(\omega,k,d) \equiv \left(\frac{\alpha_a + \alpha_b\varepsilon_a(\omega)}{\alpha_a - \alpha_b\varepsilon_a(\omega)}\right)^2 e^{2\alpha_b d} - 1 = 0 \tag{84}$$

where the subscripts a and b refer to the medium and vaccum, respectively, and are α_a and α_b given by Eqs. (4). In the present configuration, magnetic waves (TE modes) are possible. Their dispersion relation is given by

$$g_2(\omega,k,d) \equiv \left(\frac{\alpha_a + \alpha_b}{\alpha_a - \alpha_b}\right)^2 e^{2\alpha_b d} - 1 = 0 \tag{85}$$

The zero-point energy can be obtained by a straightforward generalization of Eq. (82). Deforming the contour and making the change of variables $\omega = i\xi$ and $ck = \xi(p^2-1)^{\frac{1}{2}}$, we obtain for the force

$$F = -\frac{\hbar^2}{2\pi^2c^3}\int_1^\infty p^2dp\int_o^\infty \xi^3\left[\frac{1}{f_1(\xi,p)} + \frac{1}{f_2(\xi,p)}\right]d\xi \tag{86}$$

where $f_i(\xi,p) = g_i\left(i\xi,\xi(p^2-1)^{\frac{1}{2}}/c,d\right)$, for $i = 1,2$. The force given by Eq. (86) is identical to that derived by Lifshitz (16).

EXPERIMENTAL STUDIES OF SURFACE POLARITONS

It is not possible to observe surface polaritons directly by conventional optical absorption measurements. Conservation of energy requires that the frequency of the incident light (assumed incident from the vacuum) must equal the frequency of the surface polariton

$$c(k_{\parallel}^2 + k_3^2)^{\frac{1}{2}} = \omega_s(k_{\parallel}) \tag{87}$$

On the other hand, the requirement that the electromagnetic fields of the polariton decay with increasing distance from the crystal-vacuum interface imposes the condition

$$ck_{\parallel} > \omega_s(k_{\parallel}) \tag{88}$$

These two conditions are compatible only if $k_3^2 < 0$, so the incident light must be attenuated in the x_3-direction in the vacuum above the crystal surface. One can achieve such an attentuated field by the method of attenuated total reflection (ATR) developed by Otto (17). In this method a prism of dielectric constant ε_p is placed above the crystal and is separated from it by a gap of thickness d and dielectric constant ε_g. Light with its electric vector in the x_3x_2-plane is incident on the interface between the prism and gap and makes an angle θ with the x_3-axis, where $\theta > \sin^{-1}(\varepsilon_g/\varepsilon_p)$. The light undergoes total internal reflection within the prism, and at the same time gives rise to an electric field in the gap which varies exponentially with x, i.e., a field for which $k_{x_3}^2 < 0$. One can now have the frequency of the incident light equal to the frequency of the surface polariton for some value of $k_{\parallel}$, and when this occurs there is a dip in the reflectivity of the system as a function of the frequency of the incident light.

The ATR method has provided the most detailed experimental results available concerning the dispersion relations of surface polaritons. The first observation of surface polaritons associated with surface optical phonons was reported by Bryksin, Gerbstein, and Mirlin (18) for NaCl films. These authors subsequently investigated this type of surface polariton in films of KBr, NaF, LiF, CdF_2, and CaF_2 (19). The ATR method was used to determine the dispersion curves of surface optical phonon polaritons on GaP by Marschall and Fischer (20), on $CaMoO_4$ by Barker (21), and on LiF by Fischer, Tyler, and Bell (22). By studying a series of n-type InSb samples with free carrier concentrations in the range $1.5 \times 10^{16} cm^{-3}$ to $1.1 \times 10^{18} cm^{-3}$, Bryksin, Mirlin, and Reshina (23) have been able to observe the interaction between surface plasmons and surface optical phonons. The results are in good agreement with the theoretical predictions, although the measured frequencies of the plasmon-like surface polaritons are somewhat lower than the calculated ones.

The first observations of surface polaritons at the interface between an anisotropic dielectric medium and the vacuum have been made by Bryksin, Mirlin, and Reshina (5) on MgF_2 and TiO_2, both of which have the rutile structure. Surface polaritons of both types I and II were observed and were found to obey the dispersion relation given by Eq. (26). The same procedure has been used by Falge and Otto (24) on α-quartz. They studied both the ordinary surface polaritons which occur when the optical, or c, axis of the crystal is parallel to the surface and $\vec{k}_{\parallel}$ is perpendicular to c, and extraordinary surface polaritons, which occur when the c-axis is perpendicular to the surface with $\vec{k}_{\parallel}$ in an arbitrary direction or when the c-axis is parallel to the surface and to $\vec{k}_{\parallel}$. Both type I and type II surface polaritons were observed and their dispersion curves determined. The agreement between the experimental and theoretical dispersion curves is excellent.

Several experimental investigations have been carried out on surface polaritons associated with surface magnetoplasmons in n-type InSb. Palik et al. (25) have used the configuration in which $\vec{B}_o$ is parallel to both the surface and the wave vector $\vec{k}_{\parallel}$ and have observed the coupled surface magnetoplasmon-surface optical phonon modes at fields up to 80 kG. The configuration in which $\vec{B}_o$ is parallel to the surface and perpendicular to $\vec{k}_{\parallel}$ has been investigated by Hartstein and Burstein (26). They observed the nonreciprocity of the dispersion curves and the gap in the dispersion curve predicted by the theory (which was discussed earlier in this chapter).

An alternative to the ATR method as a means of studying surface polaritons is the use of a laser beam as an energy source and prism couplers for launching and detecting the surface polaritons. This procedure has been used by Schoenwald, Burstein, and Elson (27) to investigate surface polaritons propagating along a surface of copper. Their exciting radiation was the 10.6 μm line of a CO_2 laser. The attenuation length L defined in Eq. (45a) can be measured by this method and was found to be 1.6 cm, which can be compared to the calculated value of 1.9 cm.

Another experimental method which is somewhat related to the ATR method is the observation of the reflectivity from a surface upon which a grating has been ruled. The presence of the grating relaxes the condition of wavevector conservation for the component parallel to the surface, $\vec{k}_{\parallel}$. The effective values of $k_{\parallel}$ are given by

$$k_{\parallel}^{(n)} = \frac{\omega}{c} \sin \theta + \frac{2\pi n}{d} \qquad (89)$$

where ω is the incident frequency, θ is the angle of incidence, d is the grating spacing, and n is an integer. For d sufficiently small, one can have $ck_{\parallel}^{(n)} > \omega$ even when n is a small integer. The reflectivity exhibits a dip whenever the surface polariton frequency, $\omega_s(k_{\parallel}^{(n)})$, is equal to ω for some value of n, corresponding to the excitation of the surface polariton by the incident field. This procedure has been used by Marschall, Fischer, and Queisser (28) to obtain the dispersion curve for surface plasmon polaritons in n-type InSb. However, the method has the drawback that the surface is perturbed in a rather complicated way, and the effect of this perturbation on the surface polariton is difficult to determine precisely.

Another technique which is applicable to the investigation of surface polaritons, particularly in the unretarded regime, is the observation of energy loss by electrons transmitted through or reflected by the sample. A peak in the energy loss spectrum can occur as a result of the excitation of surface polaritons. Surface optical phonons have been studied experimentally, for example, by energy loss experiments in which either high-energy ($\sim$50 KeV) electrons were transmitted through a thin film of LiF (29) or low-energy ($\sim$10 eV) electrons were specularly reflected from a $(1\bar{1}00)$ surface of ZnO (30). The penetration of the electric field of the surface optical phonons into the vacuum outside the crystal provides a mechanism for strong coupling of these phonons to the incident electrons.

Recently, the techniques of Raman scattering have been applied to the study of surface polaritons. Evans, Ushioda, and McMullen (31) observed Raman scattering from surface polaritons in a thin film (∿2500 Å thick) of GaAs deposited on a sapphire substrate using the 4880 Å line of an argon ion laser and near forward scattering at normal incidence. The thinness of the sample made possible the study of the scattered radiation in transmission. The surface polariton dispersion curve has two branches, the higher frequency one corresponding to surface polaritons localized at the vacuum-film interface and the lower frequency one corresponding to surface polaritons at the film-substrate interface. Only the latter is observed, presumably because throughout the range of $k_{\parallel}$ explored by Evans et al., the frequency of the former lies within the line width of the L phonon line. It is this fact which seems to be responsible for the failure to observe surface polaritons in earlier Raman scattering experiments in which only a single interface, between the vacuum and crystal, existed.

REFERENCES

1. J. Zenneck, Ann. Phys. (Leipz.), 23, 846 (1907).

2. A. Sommerfeld, Ann. Phys. (Leipz.), 28, 665 (1909).

3. R. Ruppin and R. Englman, J. Phys., C 1, 630 (1968); R. Englman and R. Ruppin, J. Phys., C 1, 1515 (1968).

4. R. F. Wallis, J. J. Brion, E. Burstein, and A. Hartstein, Phys. Rev., B9, 3424 (1974).

5. V. V. Bryksin, D. N. Mirlin, and I. I. Reshina, Zh. Eksp. Teor. Fiz. Pis. v Red., 16, 445 (1972) [Soviet Physics JETP Lett., 16, 315 (1972)].

6. J. J. Brion, R. F. Wallis, A. Hartstein, and E. Burstein, Phys. Rev. Lett., 28, 1455 (1972).

7. R. Nii, J. Phys. Soc. Japan, 19, 58 (1964).

8. P. G. Flahive and J. J. Quinn, Phys. Rev. Lett., 31, 586 (1973).

9. J. J. Brion, R. F. Wallis, A. Hartstein, and E. Burstein, Surf. Sci., 34, 73 (1973).

10. A. A. Maradudin and D. L. Mills, Phys. Rev., B7, 2787 (1973).

11. A. Hartstein, E. Burstein, A. A. Maradudin, R. Brewer, and R. F. Wallis, J. Phys., C 6, 1266 (1973).

12. R. W. Damon and J. R. Eshbach, J. Phys. Chem. Solids, 19, 308 (1960).

13. G. D. Mahan, Phys. Rev., B5, 739 (1972).

14. N. G. van Kampen, B. R. A. Nijboer, and K. Schram, Phys. Lett., 26A, 307 (1968).

15. E. Gerlach, Phys. Rev., B4, 393 (1971).

16. E. M. Lifshitz, Zh. Eksp. Teor. Fiz., 29, 94 (1955) [Sov. Phys. JETP, 2, 73 (1956)].

17. A. Otto, Z. Phys., 216, 398 (1968).

18. V. V. Bryksin, Yu. M. Gerbshtein, and D. N. Mirlin, Fiz. Tverd. Tela, 13, 2125 (1972) [Sov. Phys. Solid State, 13, 1779 (1972)].

19. V. V. Bryksin, Yu. M. Gerbshtein, and D. N. Mirlin, Fiz. Tverd. Tela, 14, 543, 3368 (1972) [Sov. Phys. Solid State, 14, 453, 2849 (1972)].

20. N. Marschall and B. Fischer, Phys. Rev. Lett., 28, 511 (1972).

21. A. S. Barker, Phys. Rev. Lett., 28, 892 (1972).

22. B. Fischer, I. L. Tyler, and R. J. Bell, in Polaritons, Edited by E. Burstein and F. De Martini (Pergamon, Oxford) 123 (1974).

23. V. V. Bryksin, D. N. Mirlin, and I. I. Reshina, Solid State Comm., 11, 695 (1972); see also I. I. Reshina, Yu. M. Gerbshtein, and D. N. Mirlin, Fiz. Tverd. Tela, 14, 1280 (1972) [Sov. Phys. Solid State, 14, 1104 (1972)].

24. H. J. Falge and A. Otto, Phys. Stat. Sol., (b) 56, 523 (1973).

25. E. D. Palik, R. Kaplan, R. W. Gammon, H. Kaplan, J. J. Quinn, and R. F. Wallis, Phys. Lett., 45A, 143 (1973).

26. A. Hartstein and E. Burstein, Solid State Comm., 14, 1223 (1974).

27. J. Schoenwald, E. Burstein, and J. M. Elson, Solid State Comm., 12, 185 (1973).

28. N. Marschall, B. Fischer, and H. J. Queisser, Phys. Rev. Lett., 27, 95 (1971).

29. H. Boersch, J. Geiger, and W. Stickel, Phys. Rev. Lett., 17, 379 (1966).

30. H. Ibach, Phys. Rev. Lett., 24, 1416 (1970).

31. D. J. Evans, S. Ushioda, and J. D. McMullen, Phys. Rev. Lett., 31, 369 (1973).

Chapter 4

EXAMPLES

INTRODUCTION

In this section, we illustrate several physical effects due to surface phonons on the basis of a very simple and pedagogical model. The calculations are analytical and simply reproducible. So these examples can be considered as problems for graduate students, as well as qualitative and semiquantitative descriptions of surface phonon effects.

The crystal will be taken to be a semi-infinite simple-cubic crystal with a (100) surface. The surface lies in the x_1-x_2 plane. The atoms in the bulk have mass M. The lattice vibrations of the bulk crystal are described by a simple model introduced by Rosenstock and Newell (1) and popularized by Montroll and Potts (2).

Let $u_\alpha(\ell)$ denote the α component ($\alpha = x_1, x_2$, or x_3 of the displacement from equilibrium of the atom at lattice site $\vec{x}(\ell) = a_o(\ell_1\hat{x}_1 + \ell_2 x_2 + \ell_3 x_3)$, where a_o is the lattice parameter. The potential energy Φ associated with the lattice vibrations has the simple form

$$\Phi = \frac{1}{2}\gamma \sum_{\ell}\sum_{\delta}\sum_{\alpha} [u_\alpha(\ell) - u_\alpha(\ell + \delta)]^2 \qquad (1)$$

where ℓ ranges over all sites of the crystal, and δ over the six nearest neighbor sites of the atom ℓ. For an atom in the surface layer, one neighbor is missing, so the sum over δ is confined to the five nearest neighbors in this case.

This model is not rotationally invariant (see p. 24 and Ref. 3). However, the model is quite useful for estimating the effect of the surface and alterations of the surface region on the thermodynamic properties of the crystal (4-7) and on the local-mode frequencies associated with adsorbed atoms (7-9). We feel that semiquantitative estimates of surface effects on lattice vibrations can be obtained from the model, so long as one does not deal with long-wavelength low-frequency surface waves where a model with proper rotational invariance is required.

In this chapter, we will discuss the following properties of a clean surface, such as specific heat, entropy, and mean square displacements of atoms; surface waves in the presence of an adsorbed layer; the phonon contribution to the free energy of interacting adatom pairs; and the scattering of phonons by a crystal surface.

CLEAN SURFACES

In the Born-Oppenheimer approximation, the thermodynamic properties of a solid are the sum of an electronic and a vibrational contribution. Depending on the experimental conditions, one or the other contribution is often dominant. The phonon part may be dominant in particular in the specific heat, the entropy, the mean square displacements. Before calculating these quantities, we develop the model used here.

Formalism

From the above form of the potential energy and by assuming a sinusoidal time dependence for the displacements, we obtain in the bulk the three uncoupled equations of motion

$$M\omega^2 u_\alpha(\ell) = \gamma \sum_\delta [u_\alpha(\ell) - u_\alpha(\ell + \delta)] \qquad (2)$$

In matrix notation, these equations can be written as $\overleftrightarrow{L}\vec{u} = 0$, where

$$L_{\alpha\beta}(\ell\ell',\omega^2) = [(M\omega^2 - 6\gamma)\delta_{\ell\ell'} + \sum_\delta \gamma\delta_{\ell,\ell'+\delta}]\delta_{\alpha\beta} \qquad (3)$$

By standard diagonalization of $\overleftrightarrow{L}$, one finds that in the infinitely extended crystal there is one threefold degenerate phonon branch for each wave vector $\vec{k}$. The frequency of a phonon of wave vector $\vec{k}$ is given by

$$\omega^2(\vec{k}) = \frac{2\gamma}{M}(3 - \cos K_1 a_0 - \cos K_2 a_0 \cos K_3 a_0) \qquad (4)$$

A convenient choice of normalized eigenvectors (for $\hat{n}_1 = \hat{x}_1$, $\hat{n}_2 = \hat{x}_2$, and $\hat{n}_3 = \hat{x}_3$) is the set

$$u_i(\ell) = \hat{n}_i \exp\{i\vec{k} \cdot \vec{x}(\ell)\} \qquad (5)$$

Now, we create two adjacent (001) free surfaces by setting to zero all atomic constants that describe interactions which cross a fictitious plane that lies between the layers with $\ell_3 = 0$ and $\ell_3 = 1$. We denote the dynamical matrix of the crystal in the presence of the two free surfaces by $\overleftrightarrow{L}'$, and write $\overleftrightarrow{L}' = \overleftrightarrow{L} - \delta\overleftrightarrow{L}$, where $\delta\overleftrightarrow{L}$ is the perturbation produced by the "bond breaking" procedure. If $\delta L_{\alpha\beta}(\ell\ell')$ is an element of the perturbation matrix $\delta\overleftrightarrow{L}$, it is convenient to introduce the Fourier transformed quantity

$$\delta L_{\alpha\beta}(\vec{k}_{\parallel};\ell_3\ell'_3) = \sum_{\ell'_1\ell'_2} \delta L_{\alpha\beta}(\ell\ell') \cdot e^{i\vec{k}_{\parallel} \cdot (\vec{x}_{\parallel}(e) - \vec{x}_{\parallel}(e'))} \tag{6}$$

since introduction of the free surfaces does not destroy the periodicity of the crystal in the x_1 and x_2 directions. For the model used here, one has the simple form

$$\delta L_{\alpha\beta}(\vec{k}_{\parallel};\ell_3\ell'_3) = \delta_{\alpha\beta}\Big(-\delta_{\ell_3 0}\delta_{\ell'_3 1} - \delta_{\ell_3 1}\delta_{\ell'_3 0} + \delta_{\ell_3 0}\delta_{\ell'_3 0} + \delta_{\ell_3 1}\delta_{\ell'_3 1}\Big)\gamma \tag{7}$$

where $\delta_{\ell\ell'}$ is the usual Kronecker symbol. There is no explicit dependence on $\vec{k}_{\parallel}$ as all the removed interactions are parallel to the x_3 axis.

It is convenient to introduce the following Green's functions

$$\overleftrightarrow{G} \equiv \overleftrightarrow{L}^{-1} \tag{8a}$$

$$\overleftrightarrow{U} \equiv \overleftrightarrow{L}'^{-1} \tag{8b}$$

which are related by

$$\overleftrightarrow{U} = \overleftrightarrow{G} + \overleftrightarrow{G}\delta\overleftrightarrow{L}\overleftrightarrow{U} \tag{9}$$

Due to translational symmetry parallel to the surface, it is of interest to introduce the following Fourier transform

$$G(\ell,\ell';\omega^2) = \frac{1}{N_s} \sum_{\ell_1'\ell_2'} G(\vec{k}_\parallel;\ell_3 - \ell_3';\omega^2)$$

$$\exp i\vec{k}_\parallel \cdot \left(\vec{x}_\parallel(\ell) - \vec{x}_\parallel(\ell')\right) \tag{10}$$

where $N_s = L^2$ is the number of atoms in a (001) layer. Let us introduce also

$$H(\vec{k}_\parallel;\omega^2) = |\overleftrightarrow{I} - \overleftrightarrow{G}(\vec{k}_\parallel;\omega^2)\delta\overleftrightarrow{L}(\vec{k}_\parallel)| \tag{11}$$

As on p. 80 and Ref. 4, we may calculate the frequencies of the localized modes and resonance modes that arise from the perturbation $\delta\overleftrightarrow{L}$ from the condition

$$\mathrm{Re}\ H(\vec{k}_\parallel;\omega^2 + i\varepsilon) = 0 \tag{12}$$

where ε is an infinitesimally small number.

Suppose we consider a quantity F which is a function of the normal mode frequencies $\omega_j(\vec{k})$ of the perfect crystal (j is the band index). Let F have the form

$$F = \sum_{\vec{k}j} f\left(\omega_j(\vec{k})\right) \tag{13}$$

The change in the quantity F that results from the perturbation $\overleftrightarrow{\delta L}$ is given by (see Refs. 4 and 7)

$$F = -\frac{1}{\pi}\int_0^\infty d\omega \, \frac{df(\omega)}{d\omega}\, \eta(\omega) \tag{14}$$

where the phase angle $\eta(\omega)$ is given by

$$\eta(\omega) = \sum_{\vec{k}_\parallel} \eta(\vec{k}_\parallel, \omega) \tag{15a}$$

with

$$\eta(\vec{k}_\parallel, \omega) = -\operatorname{Arg} H(\vec{k}_\parallel; \omega^2 + i\varepsilon) \tag{15b}$$

and is related to the variation of the density of states by

$$\Delta\rho(\omega) = \frac{1}{\pi}\frac{d\eta}{d\omega} \tag{16}$$

The result in Eq. (14) takes due account of localized states, when Eq. (15b) is understood to be a generalized phase shift (10,11). When a localized state ω_L appears below the bulk band $\eta(\vec{k}_\parallel, \omega)$ jumps from 0 to π at this frequency, and remains at this value till the bottom of the bulk band. When the bound state ω_L appears above the bulk band, $\eta(\vec{k}_\parallel, \omega)$ is equal to $-\pi$ at the top of the bulk band, and jumps back to zero at ω_L.

Let us now calculate the matrix elements of the bulk Green's function $\overleftrightarrow{G}$ defined by Eq. (8). The virtue of the model used here is that this Green's function is of a rather simple form. We have, due to degeneracy

$$G_{\alpha\beta}(\ell,\ell';\omega^2) = \delta_{\alpha\beta}G(\ell,\ell';\omega^2) \tag{17}$$

and to translation symmetry

$$G(\ell,\ell';\omega^2) = G(\ell-\ell';\omega^2) \tag{18}$$

Using Eqs. (3)-(5) one finds easily

$$G(\ell;\omega^2) = \frac{1}{N}\sum_{\vec{k}_\parallel} \frac{\exp\left(i\vec{k}_\parallel \cdot \vec{x}_\parallel(\ell)\right)}{M\omega^2 - 2\gamma(3 - \cos k_1a_0 - \cos k_2a_0 - \cos k_3a_0)} \tag{19}$$

where $N = L^3$ is the total number of atoms in the periodicity volume of the infinitely extended crystal. Using Eqs. (10) and (19) one has

$$G(\vec{k}_\parallel;\ell_3;\omega^2) = \frac{N_s}{2\gamma N}\sum_{k_3} \frac{\exp(ik_3\ell_3a_0)}{\cos k_3a_0 - \xi}, \tag{20}$$

where we have defined

$$\xi = X - E \tag{21}$$

and where

$$X = 3 - \cos k_1a_0 - \cos k_2a_0, \tag{22}$$

$$E = \frac{M}{2\gamma}(\omega^2 + i\varepsilon) \tag{23}$$

For an infinite crystal, we can transform the discrete summation in Eq. (20) into an integration

$$G(\vec{k}_\parallel ; \ell_3 ; \omega^2) = \frac{a_0}{2\pi\gamma} \int_0^{\pi/a_0} dk_3 \frac{\cos(\ell_3 k_3 a_0)}{\cos k_3 a_0 - \xi} \qquad (24)$$

Let us note at this stage that when using $\overleftrightarrow{G}$ in the above form to calculate surface properties, we assume $N \gg N_s$. This is rigorous for the study of semi-infinite crystals, but not for thin slabs. This supposes that the surface-to-bulk ratio is small and we can expect the surface contribution to the thermodynamic functions to be proportional to this ratio. By standard integration, one finds that

$$G(\vec{k}_\parallel ; \ell_3 ; \omega^2) = \frac{1}{\gamma} \frac{t^{1 + |\ell_3|}}{t^2 - 1} \qquad (25)$$

where

$$t = \begin{cases} \xi - \sqrt{\xi^2 - 1} & \xi > 1 \\ \xi + i\sqrt{1 - \xi^2} & -1 < \xi < 1 \\ \xi + \sqrt{\xi^2 - 1} & \xi < -1 \end{cases} \qquad (26)$$

These three cases correspond to ω^2 below ($\xi > 1$), inside, and above ($\xi < -1$) the bulk band. Inside the band, we calculated $\overleftrightarrow{G}(\omega^2 + i\varepsilon)$.

We can now calculate, from Eqs. (7), (11), and (25)

$$H(\vec{k}_\parallel ; \omega^2 + i\varepsilon) = \frac{1 - t}{1 + t} \qquad (27)$$

Using the value of t below and above the bulk band, one sees from Eq. (12) that this model has no localized phonons on a clean surface. For a more sophisticated model in which the force constants satisfy the conditions imposed by rotational invariance (12), one would necessarily find a Rayleigh wave mode (13). However, we will see that our model gives a good order of magnitude for the surface thermodynamic functions.

Surface Thermodynamic Functions by the Phase-Shift Method (4)

The variation due to the surface of the thermodynamic functions can be calculated from Eqs. (14) and (15). Inside the bulk band ($-1 < \xi < 1$), Eq. (27) gives for a given ($\vec{k}_{\parallel}$)

$$H\left(\vec{k}_{\parallel};\omega^2 + i\varepsilon\right) = -\,i\left(\frac{1-\xi}{1+\xi}\right)^{\frac{1}{2}} \tag{28}$$

The total phase shift is then, from Eq. (15)

$$\eta(E) = \frac{3\pi}{2}\sum_{\vec{k}_{\parallel}}\left[\theta(E - X + 1) - \theta(E - X - 1)\right] \tag{29}$$

where $\theta(E)$ is the Heaviside unit step function

$$\begin{aligned}\theta(E) &= 1 \quad E > 0\\ &= 0 \quad E < 0\end{aligned} \tag{30}$$

The factor 3 in Eq. (29) is due to degeneracy.

On defining

$$a = \frac{h\omega_j(\vec{k})}{k_B T} \tag{31}$$

the quantity f introduced in Eq. (13) is given by Eqs. (32)-(35), for the free energy F, the specific heat C_V, the entropy S, the energy E, respectively

$$f_F = k_B T\left[\frac{a}{2} + \mathrm{Ln}\left(1 - e^{-a}\right)\right] \tag{32}$$

$$f_{C_V} = k_B \frac{a^2 e^a}{\left(e^a - 1\right)^2} \tag{33}$$

$$f_S = k_B\left[-\mathrm{Ln}\left(1 - e^{-a}\right) + \frac{a}{e^a - 1}\right] \tag{34}$$

$$f_E = k_B T\left[\frac{a}{2} + \frac{a}{e^a - 1}\right] \tag{35}$$

Due to the simple form of the phase shift, Eq. (29), we obtain from Eq. (14)

$$\Delta F = -\frac{3}{2}\frac{N_s^2 a_0^2}{\pi^2}\int_0^\infty dE\,\frac{df(E)}{dE}\int_0^{\pi/a_0} dk_1\int_0^{\pi/a_0} dk_2[\theta(E - X + 1) - \theta(E - X - 1)] \tag{36}$$

or, by interchanging the orders of integration

$$\Delta F = \frac{3N_s^2 a_0^2}{2\pi^2}\int_0^{\pi/a_0} dk_1\int_0^{\pi} dk_2[f(X - 1) - f(X + 1)] \tag{37}$$

Low-Temperature Results

At low temperatures, the functions $f\left(\hbar\omega/k_B T\right)$ have non-negligible values only for small frequencies. We can then use the long-wavelength approximation (see also

p. 247) in which Eq. (37) can be replaced by

$$\Delta F = \frac{3N_s^{\,2} a_0^{\,2}}{\pi} \int_0^\infty k_\parallel dk_\parallel f\,|(\gamma/M)^{\frac{1}{2}}\hbar\phi/k_B T| \tag{38}$$

After integration one easily finds

$$\Delta C_v = KT^2 \tag{39}$$

$$\Delta S = \frac{K}{2}\,T^2 \tag{40}$$

$$\Delta F = -\frac{K}{6}\,T^3 \tag{41}$$

where

$$K = \frac{9S_0}{4\pi c_\ell^{\,2}}\,\xi(3)\,\frac{k_B^{\,3}}{\hbar^2} \tag{42}$$

and where S_0 is the surface area, $\xi(3)$ is the Riemann zeta function, and $c_\ell = c_t = (\gamma/M)^{\frac{1}{2}}a_0$ is the velocity of sound in this model. In an isotropic model satisfying the conditions required by rotation invariance, the coefficient K was found (14) to be bigger by a factor (10/9) than the one derived here (14). Inasmuch as the Montroll-Potts model does not give rise to surface vibration modes, the surface contribution to the thermodynamic functions given by Eqs. (39)-(42) arises only from the perturbation of the bulk modes by the introduction of a crystal surface. This illustrates the point that the existence of surface waves is not necessary for the existence of a surface contribution to thermodynamic functions.

High-Temperature Results

Due to the simple form of Eq. (37) one can also find analytically the surface specific heat at temperatures T of the order of and higher than the Debye temperature T_D. When substituting into Eq. (37) the expression for f_{C_v} [Eq. (33)] and expanding in powers of $\hbar\omega/k_BT$ for $\hbar\omega/k_BT \ll 1$, we obtain

$$\Delta C_v = -3N_s k_B \sum_{n=1}^{\infty} (-)^n B_{2n} \frac{(1-2n)}{2n!} \left(\frac{\hbar}{k_BT}\right)^{2n} K_{2n} \tag{43}$$

where B_{2n} are the Bernoulli numbers and

$$K_{2n} = -\frac{a_0^2}{8\pi^2} \int_{-\pi/a_0}^{+\pi/a_0} dk_1 \int_{-\pi/a_0}^{+\pi/a_0} dk_2 \left(\frac{2\gamma}{M}\right)^n [(X+1)^n - (X-1)^n] \tag{44}$$

Let us define the Debye temperature by

$$k_B T_D = \hbar\omega_L \tag{45}$$

where ω_L is the maximum frequency of the bulk phonons

$$\omega_L^2 = 12\gamma/M \tag{46}$$

Then, we find that the surface contribution to the specific heat is given by

$$\Delta C_v = \frac{3}{2} \frac{k_B S_0}{a_0^2} \left[\frac{1}{72}\left(\frac{T_D}{T}\right)^2 - \frac{1}{1440}\left(\frac{T_D}{T}\right)^4 + \dots\right] \tag{47}$$

This quantity is negligible compared to the bulk specific heat $(C_v = 3Nk_B)$, and goes to zero as T increases. This is obvious since at high temperatures the total specific heat is proportional to the number of degrees of freedom, which was not changed by the "bond breaking" procedure.

Variation with Temperature

By a simple numerical integration of Eq. (36) we obtain the variation with temperature of the surface thermodynamic functions. In Figs. 4.1 and 4.2 we give these curves for the specific heat and the entropy.

Surface Thermodynamic Functions from the Moment Method

The phase shift method used above for calculating the change in the thermodynamic functions is an exact one. However, for a more realistic model where one would have to calculate the Green's functions numerically for each set (k_1, k_2) this method is long. One can choose, as did Allen et al. (15), to perform a direct slab calculation. This method is straightforward and limited only by the dimensions of the matrices to diagonalize on the computer. Another approach is to consider the present precision of the experimental values and to use an approximative method yielding the same order of precision.

Let us write $\overleftrightarrow{L}$ in the general form

$$\overleftrightarrow{L} = \overleftrightarrow{M}^{\frac{1}{2}}(\omega^2\overleftrightarrow{I} - \overleftrightarrow{D})\overleftrightarrow{M}^{\frac{1}{2}} \qquad (48)$$

Since we are dealing with a Bravais lattice and are not

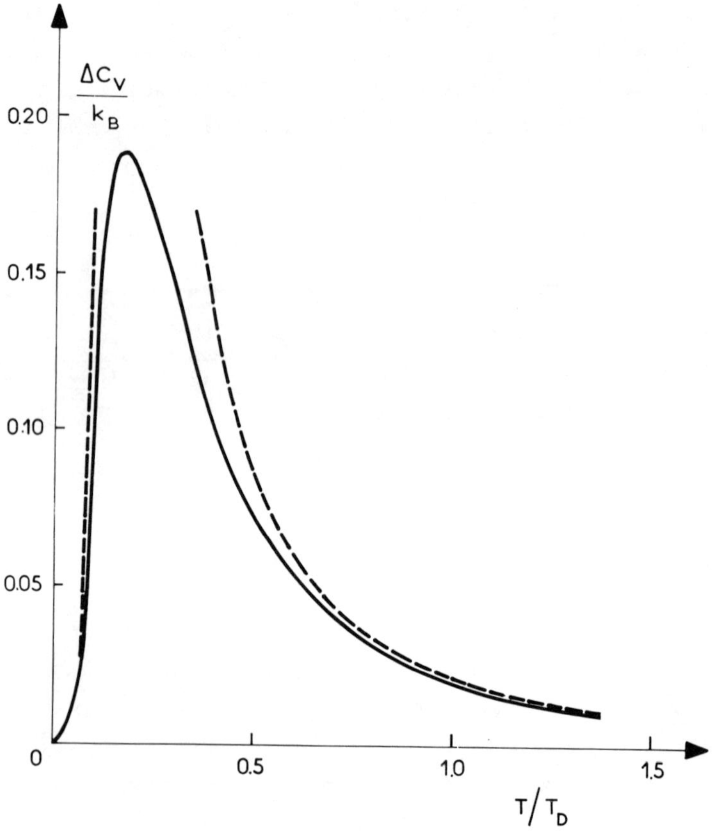

Fig. 4.1.--Variation of specific heat per surface atom as a function of T/T_D, where T_D is the bulk Debye temperature. The dotted lines show the low- and high-temperature expansions.

perturbing the mass matrix $\overleftrightarrow{M}$, we can treat M as a scalar, here, instead of as a matrix.

It is well known (16,17) that the knowledge of the first moments of the density of states, that is

$$\mu_{2n} = \frac{1}{N} \operatorname{Tr}(\overleftrightarrow{D})^n \tag{49}$$

enables one to obtain good approximate values of the

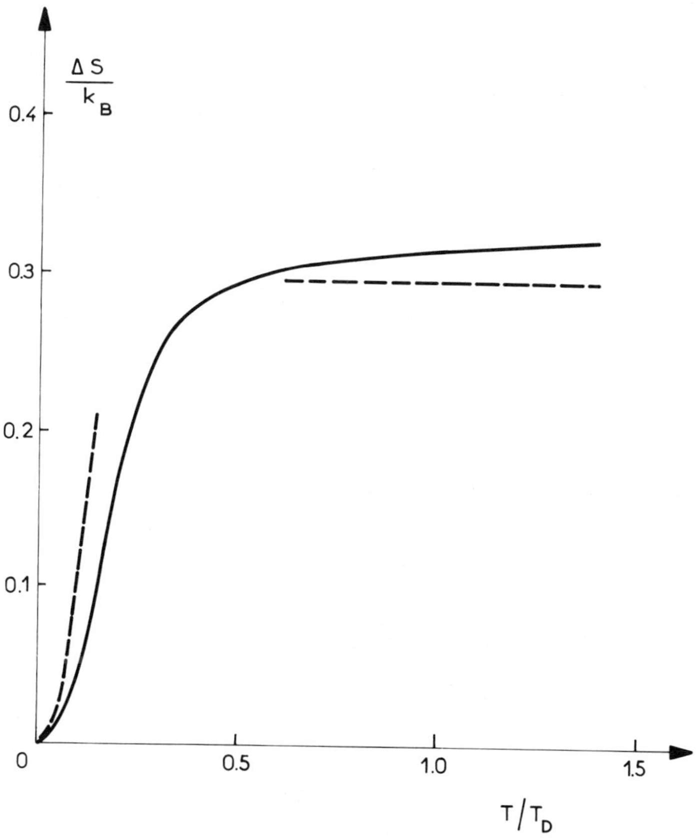

Fig. 4.2.--Variation of entropy per surface atom as a function of T/T_D, where T_D is the bulk Debye temperature. The dotted lines show the low- and high-temperature expansions.

thermodynamic functions for temperatures $T \gtrsim T_D/(2\pi)$. This method may be simplified (18,19) by writing the dynamical matrix $\overleftrightarrow{D}$ as the sum of its diagonal part $\overleftrightarrow{d}$ and its nondiagonal part $\overleftrightarrow{R}$ (20), as follows

$$\overleftrightarrow{D} = \overleftrightarrow{d} + \overleftrightarrow{R} \tag{50}$$

In the Einstein approximation, one neglects $\overleftrightarrow{R}$ in comparison with $\overleftrightarrow{d}$. Then all atoms in a monoatomic crystal have one and the same Einstein frequency ω_E. Noting that $\overleftrightarrow{d}$ is the preponderant part of $\overleftrightarrow{D}$, one develops (17, 21) the thermodynamic function F in a Schafroth (22) expansion around $\overleftrightarrow{d}$ (generalization of the Taylor expansion to the case where $\overleftrightarrow{d}$ and $\overleftrightarrow{R}$ do not commute). The first term in this expansion is exactly the one given by Einstein's approximation.

The thermodynamic function of Eq. (13) can be written in the two following ways, where $n(\omega)$ is the density of states

$$F(T) = \int d\omega\, f(\omega, T) n(\omega) \tag{51}$$

$$F(T) = \text{Tr } f(\overleftrightarrow{D}, T) \tag{52}$$

Before using the second of these expressions, let us discuss the Einstein approximation to the first one.

This approximation replaces the exact density of states $n(\omega)$ by a delta function situated at the Einstein frequency ω_E. To see how this affects the integral in Eq. (51), it is instructive (23) to look at the curves of f_{C_V}, f_S, and f_F of Eqs. (32)-(34) (Figs. 4.3-4.5) plotted as functions of ω/ω_L at temperatures low and high compared to T_D [see Eq. (45)]. From these figures it becomes obvious that the Einstein approximation will be good at high temperatures for the specific heat and the entropy and better at low than at high temperatures for the free energy.

High-Temperature Surface Entropy

Let us illustrate the method described above for the calculation of the high-temperature surface entropy.

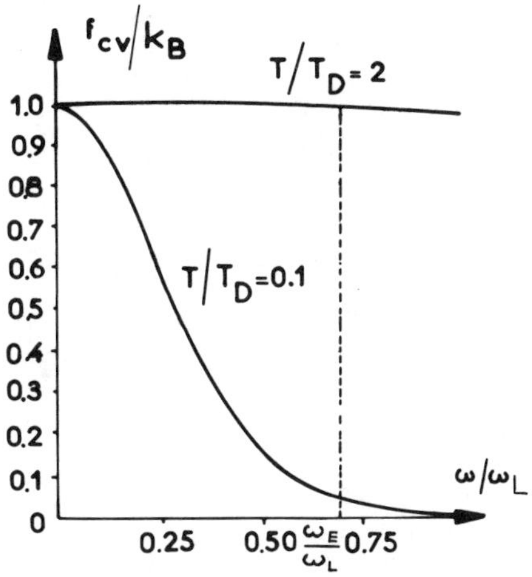

Fig. 4.3.--Planck factor for the specific heat as a function of ω/ω_L, where ω_L is the bulk maximum frequency, at low- and high-temperatures as compared to the Debye temperature $T_D\left(\hbar\omega_L = k_B T_D\right)$

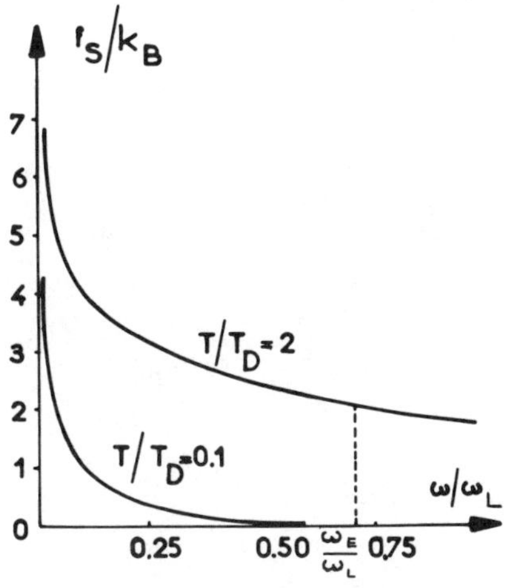

Fig. 4.4.--Planck factor for the entropy as a function of ω/ω_L, where ω_L is the bulk maximum frequency, at low- and high-temperatures as compared to the Debye temperature $T_D\left(\hbar\omega_L = k_B T_D\right)$.

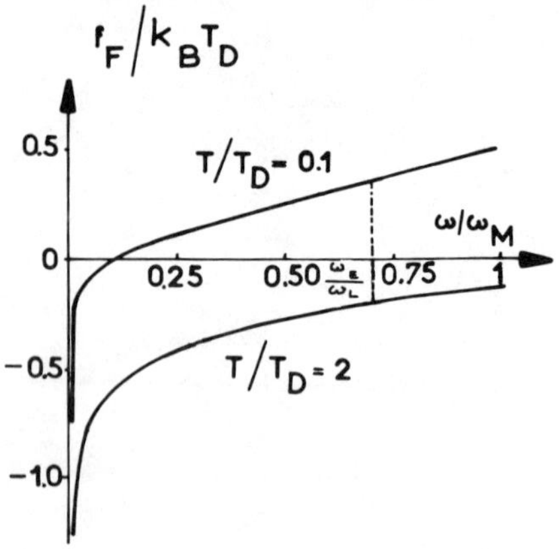

Fig. 4.5.--Planck factor for the free energy as a function of ω/ω_L, where ω_L is the bulk maximum frequency, at low- and high-temperatures as compared to the Debye temperature $T_D\left(\hbar\omega_L = k_B T_D\right)$.

In the limit $\hbar\omega/\left(k_B T\right) < 1$, Eqs. (52) and (34) can be expanded in the following form

$$S/k_B = \mathrm{Tr}\left[\overleftrightarrow{I} - \ell n \frac{\hbar\overleftrightarrow{D}^{\frac{1}{2}}}{k_B T}\right] + \sum_{n=1}^{\infty} 0\left(\frac{1}{T^{2n}}\right) \tag{53}$$

where B_{2n} are the Bernoulli numbers and where

$$0\left(\frac{1}{T^{2n}}\right) = (-)^{n-1} \frac{2n - 1}{2n(2n)!} B_{2n}\left(\frac{\hbar}{k_B T}\right)^{2n} \mu_{2n} \tag{54}$$

We write $\ell n\overleftrightarrow{D}$ in the form:

$$\ell n(\overleftrightarrow{d} + \overleftrightarrow{R}) = \ell n\ \overleftrightarrow{d} + \sum_{n=1}^{\infty} \frac{(-)^{n-1}}{n}\left(\overleftrightarrow{d}^{-1}\overleftrightarrow{R}\right)^n \tag{55}$$

When $\overleftrightarrow{d}$ and $\overleftrightarrow{R}$ commute, this is a straightforward Taylor expansion; when they do not commute, one obtains Eq. (55) from a Schafroth (22) expansion or by noting that

$$\mathrm{Tr}\ \ell n(\overleftrightarrow{A}\overleftrightarrow{B}) = \mathrm{Tr}\ \ell n\ \overleftrightarrow{A} + \mathrm{Tr}\ \ell n\ \overleftrightarrow{B} \tag{56}$$

This is easily deduced from the general property of a diagonalizable matrix $\overleftrightarrow{D}$

$$\det \overleftrightarrow{D} = \exp(\mathrm{Tr}\ \ell n\ \overleftrightarrow{D}) \tag{57}$$

Finally, we find that

$$S/k_B = \mathrm{Tr}\left[\overleftrightarrow{I} - \ell n\left(\frac{\hbar \overleftrightarrow{d}^{\frac{1}{2}}}{k_B T}\right)\right] + \sum_{n=1}^{\infty}\left[\frac{(-)^{n-1}}{2n}\ \mathrm{Tr}\left(\overleftrightarrow{d}^{-1}\overleftrightarrow{R}\right)^n + 0\left(\frac{1}{T^{2n}}\right)\right] \tag{58}$$

The change in entropy ΔS due to the free surface is obtained by the difference between Eq. (58) evaluated for a semi-infinite crystal and an infinite one. Salter (20) remarked that this expansion was rapidly convergent. Let us, however, specify the rapidity of the convergence by calculating the high-temperature bulk entropy for the Montroll-Potts model. The calculation of $\mathrm{Tr}\left(\overleftrightarrow{d}^{-1}\overleftrightarrow{R}\right)^n$ can be made in real space (24) by counting the number of closed circuits of interatomic jumps between successive nearest neighbors

$$\mathrm{Tr}\left(\overleftrightarrow{d}^{-1}\overleftrightarrow{R}\right)^n = \sum_{i_1 \ldots i_p}\left(\overleftrightarrow{d}^{-1}\overleftrightarrow{R}\right)_{i_1 i_2} \cdots \left(\overleftrightarrow{d}^{-1}\overleftrightarrow{R}\right)_{i_p i_1} \tag{59}$$

From this expression one sees easily that in the simple-cubic Montroll-Potts model

$$\mathrm{Tr}\left(\overleftrightarrow{d}^{-1}\overleftrightarrow{R}\right)^{2n+1} = 0 \tag{60}$$

$$\frac{1}{3N}\,\mathrm{Tr}\left(\overleftrightarrow{d}^{-1}\overleftrightarrow{R}\right)^{2n} = \frac{1}{6^{2n}}\,P_{2n} \tag{61}$$

where P_{2n} is the number of closed circuits of nearest neighbor interatomic jumps starting from one atom and coming back to the same atom after passing through (2n - 1) atoms. P_{2n} is calculated easily in the case of a simple cubic lattice. On a one-dimensional chain, the number of closed circuits of 2n interatomic jumps on first nearest neighbors (starting from a given atom) is the same as in a random walk C_{2n}. In a simple cubic lattice, the calculation of P_{2n} can be separated into three random walks along the three directions $\hat{x}$, $\hat{y}$, $\hat{z}$. Let $2n = 2\ell + 2m + 2p$. One will make 2ℓ jumps along $\hat{x}$, 2m jumps along $\hat{y}$, and 2p jumps along $\hat{z}$. Their total number will be $C_{2\ell}^{\ell}C_{2m}^{m}C_{2p}^{p}$. But one has to count the number of ways one can choose the 2ℓ jumps along $\hat{x}$, the 2m jumps along $\hat{y}$, and the 2p jumps along $\hat{z}$ among the total 2n jumps. Finally

$$P_{2n} = \sum_{\substack{\ell,m,p\\ \ell+m+p=n}} C_{2n}^{2\ell}C_{2(n-1)}^{2m}C_{2\ell}^{\ell}C_{2m}^{m}C_{2p}^{p} \tag{62}$$

or

$$P_{2n} = \sum_{\substack{\ell,m,p\\ \ell+m+p=n}} \frac{(2n)!}{(\ell!)^2(m!)^2(p!)^2} \tag{63}$$

Thibaudier et al. (24,25) compared this expansion with those of the Bessel function $J_0(2x)$

$$J_0(2x) = \sum_{r=0}^{\infty} \frac{(ix)^{2r}}{(r!)^2} \tag{64}$$

Then they obtained the characteristic function

$$f(x) = \sum_{n=0}^{\infty} \frac{(-ix)^{2n}}{(2n)!} \gamma^{2n} P_{2n} = J_0{}^3(2\gamma x) \tag{65}$$

and the bulk density of states

$$n(\omega^2) = \frac{1}{2\pi} \int_{-\infty}^{+\infty} e^{ix\omega^2} f(x)dx = \frac{1}{2\pi} \int_{-\infty}^{+\infty} e^{ix\omega^2} J_0{}^3(2\gamma x)dx \tag{66}$$

providing a proof of the rapid convergence of this expansion.

Finally, putting Eqs. (45), (46), (61), and (63) into Eq. (58) one has

$$\frac{S}{3Nk_B} = 1 + \ln \frac{2^{\frac{1}{2}}T}{T_D} + \sum_{n=1}^{\infty} \left[v_{2n} + 0\left(\frac{1}{T^{2n}}\right) \right] \tag{67}$$

where we have defined

$$v_{2n} = - \frac{(2n - 1)!}{2\left(6^{2n}\right)} \sum_{\substack{\ell,m,p \\ \ell+m+p=n}} \frac{1}{(\ell!)^2(m!)^2(p!)^2} \tag{68}$$

In order to specify the rapidity of the convergence of this expansion we present the partial sums

$$v_2 = - \frac{1}{24} \tag{69a}$$

$$\sum_{n=1}^{2} v_{2n} = - \frac{1.20}{24} \tag{69b}$$

$$\sum_{n=1}^{5} v_{2n} = - \frac{1.35}{24} \tag{69c}$$

In the case of Ag, where $T_D \simeq 226°K$, we obtain the following values at the lowest temperature (26) for which a surface entropy was measured ($T \simeq 350°K$)

$$1 + \ell n\, 2^{\frac{1}{2}} \frac{T}{T_D} \simeq 1.783$$

$$\sum_{n=1}^{5} v_{2n} = - 0.056$$

When calculating the surface excess of entropy ΔS one has to keep in mind that $\overleftrightarrow{d}$ and $\overleftrightarrow{R}$ do not commute. One can nevertheless calculate easily the first corrections to the Einstein value of ΔS using Eq. (59). Per atom of a (001) surface, one obtains

$$\Delta S/k_B = (0.273 + 0.024 + 0.012 + \ldots) - \left[0.010\left(\frac{T_D}{T}\right)^2 + 0.0002\left(\frac{T_D}{T}\right)^4 + 0\left(\frac{1}{T^6}\right)\right] \quad (70)$$

The phase-shift method gave the exact result (Fig. 4.2)

$$\Delta S/k_B \xrightarrow[T\to\infty]{} 0.329 \quad (71)$$

One sees then that the Einstein approximation $(\Delta S/k_B = 0.273)$ gives an error of 17%. With the first and second correction terms, the error is 10% and 6%, respectively.

These calculations are easily done for realistic models (18,19,26,27), and the comparison of the theoretical and experimental order of magnitude is satisfactory. Expansion (58) can also be used for other cases as, for example, the calculation of the entropy of lattice vacancies (28) or the study of the stability of the bcc phase at high temperatures in metals (29).

Surface Mean Square Displacements

When interpreting the low-energy electron diffraction (LEED) data in a kinematic theory (30) one obtains a measure of the surface mean square displacements. This measure is of some imprecision (31,32) due to experimental difficulties and to the multiple diffraction aspect of LEED. Here also the expansion around the Einstein approximation will give results of a precision comparable to the experimental one.

Let us write the mean square displacement of atom ℓ in the direction σ in the form (see p. 162)

$$\langle u_\sigma^2(\ell)\rangle = \frac{\hbar}{2M}\left[\overleftrightarrow{D}^{-\frac{1}{2}}\coth\frac{h\overleftrightarrow{D}^{\frac{1}{2}}}{2k_BT}\right]_{\ell\sigma,\ell\sigma} \tag{72}$$

At high temperatures, this expression reduces to

$$\langle u_\sigma^2(\ell)\rangle \cong \frac{k_BT}{M}[\overleftrightarrow{D}^{-1}]_{\ell\sigma,\ell\sigma} \tag{73}$$

One can do a Schafroth expansion (21,33) of these expressions around $\overleftrightarrow{d}$. This was done for surface and bulk atoms of several systems (21,33-40) and at all temperatures. Let us give an example here on the basis of the Montroll-Potts model for a (001) surface atom. The high-temperature expansion of Eq. (72) may be used in the following form

$$\langle u_\sigma^2(\ell)\rangle \simeq \frac{k_BT}{M}[\overleftrightarrow{d}^{-1}]_{\ell\sigma,\ell\sigma}\sum_{n=0}^{\infty}(\overleftrightarrow{d}^{-1}\overleftrightarrow{R})^{2n}_{\ell\sigma,\ell\sigma} \tag{74}$$

It is easy to calculate the firm terms of this development by the method described above for ΔS. For an atom of a (001) surface, one obtains

$$\langle u_\sigma^2(1)\rangle = \frac{k_BT}{5\gamma}(1 + 0.193 + 0.088 + \ldots) \simeq 0.256\,\frac{k_BT}{\gamma} \tag{75}$$

where $\sigma = x_1, x_2, x_3$.

In order to find the exact value of this surface mean square displacement, we can note from Eq. (48) that

$$(M\overleftrightarrow{D})^{-1} = - \overleftrightarrow{L}^{-1}(\omega^2 = 0) = - \overleftrightarrow{G}(\omega^2 = 0) \qquad (76)$$

Using Eqs. (7) and (9), it is straightforward (7) to find a relation between the Green's functions of the infinite crystal and those of the semi-infinite one bounded by a (001) surface

$$U_{\alpha\beta}(\ell,\ell';\omega^2) = \delta_{\alpha\beta}U(\ell,\ell';\omega^2) \qquad (77a)$$

$$U(\ell,\ell';\omega^2) = G(\ell_1 - \ell_1',\ell_2 - \ell_2',\ell_3 - \ell_3';\omega^2)$$

$$+ G(\ell_1 - \ell_1',\ell_2 - \ell_2',\ell_3 + \ell_3' - 1;\omega^2) \quad (77b)$$

for ℓ_3 and $\ell_3' \geq 1$. The surface plane is at $\ell_3 = 1$. Using this relation, the tabulated values of the bulk Green's functions (41,42) and the expression for the bulk Green's functions for $\ell = \left(\ell_1{}^2 + \ell_2{}^2 + \ell_3{}^2\right)^{\frac{1}{2}} >> 1$, given by

$$G\left(\ell;-\omega_L{}^2f^2\right) = - \frac{\exp|-(12)^{\frac{1}{2}}\ell f|}{4\pi\gamma\ell} \qquad (78)$$

one obtains, Ref. (43)

$$<u_\sigma^2(\ell_3)> = - k_BT[G(0,0,0;0) + G(0,0,2\ell_3 - 1;0)] \quad (79)$$

$$<u_\sigma^2(1)> = 0.338 \frac{k_BT}{\gamma} \qquad (80a)$$

$$<u_\sigma^2(\text{bulk})> = 0.252 \frac{k_BT}{\gamma} \qquad (80b)$$

$$\frac{<u_\sigma^2(\ell_3)> - <u_\sigma^2(\text{bulk})>}{<u^2(\text{bulk})>} \simeq \frac{1}{2\pi\ell_3} \quad \text{for } \ell_3 >> 1 \qquad (81)$$

We now see by comparing the exact result of Eq. (80a) with the expansion of Eq. (75) that the error introduced by the Einstein approximation is $\sim$ 40%, and drops to 30% and 24% when taking into account the first and second corrections, respectively.

At low temperatures, the mean square displacements of Eq. (72) can be calculated from

$$<u_\sigma^2(\ell)> \cong \frac{\hbar}{2M}[\overleftrightarrow{D}^{-\frac{1}{2}}]_{\ell\sigma,\ell\sigma} \tag{82}$$

When doing the expansion around $\overleftrightarrow{d}$ of this expression, we expect a better precision than for the high-temperature expression of Eq. (73), since $\overleftrightarrow{D}^{-\frac{1}{2}}$ converges faster than $\overleftrightarrow{D}^{-1}$. That is indeed what one finds (33) from a comparison with an exact calculation. For $T/T_D = 0.2$, where T_D is the Debye temperature, the error is $\sim$ 27% and reaches 18 and 14% when the first and second corrections are included.

As a general remark about the precision of these expansions around the Einstein contribution, one can say that the precision will be better when the ratio between the diagonal elements and the off-diagonal ones of the dynamical matrix is larger. In the case of an atom of a (001) free surface, this ratio is equal to 5 in the Montroll-Potts model.

The precision of this expansion can even be improved by taking into account the translational symmetry parallel to the surface [see Eq. (6)]. One defines $\overleftrightarrow{d}(\phi_1,\phi_2)$ and $\overleftrightarrow{R}(\phi_1,\phi_2)$ to be the diagonal and nondiagonal part of $\overleftrightarrow{D}(\phi_1,\phi_2) = \overleftrightarrow{d}(\phi_1,\phi_2) + \overleftrightarrow{R}(\phi_1,\phi_2)$, respectively.

Expansions (58) and (74) are carried out as before but around $\overleftrightarrow{d}(\phi_1,\phi_2)$. We are then left with a one-dimensional problem for the evaluation of the $(\overleftrightarrow{d}^{-1}\overleftrightarrow{R})^n$ and a summation on k_1,k_2 which can easily be done numerically.

This has the advantage of taking into account all the interactions parallel to the surface and to simplify the calculation of the correction terms $(\overleftrightarrow{d}^{-1}\overleftrightarrow{R})^n$. For more details, see Ref. 44. However, using the expansion described here, one can obtain very easily the order of magnitude of the entropy or the mean square displacements; this is done by a simple count of the number of nearest neighbor interactions.

As we noticed before, there are no surface phonons in the Montroll-Potts model. To obtain them, one must use central potentials between first and second nearest neighbors for the simple cubic lattice. In such a model, one can still do analytical calculations (16,17,45) but they are more cumbersome and we will not give them in this section. When taking into account the first derivatives of the potentials (46) or more sophisticated effects like a surface dipole layer (47,48) it is possible to obtain "soft surface phonons" by this simple cubic model and to get some insight into why real surfaces present superstructures. We do not detail here these important problems because we feel that more work has to be done in this area before any conclusion for real surfaces can be drawn.

Let us continue with the Montroll-Potts model, and present simple investigations of the properties of adsorbed monolayers.

ADSORBED MONOLAYERS

We will examine here the influence of an adsorbed layer of atoms on the dynamical properties of the semi-infinite crystal described in the preceding section. The values of the atomic force constants within the adsorbed layer differ in general from their values in the bulk crystal, while the force constants that couple the adsorbed layer to the remainder of the crystal will also be different. It is straightforward to include these effects in the formalism and we will take them into account when studying the mean square displacements of adsorbed atoms. But for a qualitative study of the other physical properties, we shall only examine in detail the case where the mass of the atoms in the layer on the surface differs from the mass of the atoms in the bulk. Since it is well known that a stiffening of force constants will increase vibrational frequencies, while the converse is true if they are decreased, the effect of force constant softening near the surface may be expected to be qualitatively similar to an increase in mass in this region. The converse will hold if the force constants are stiffened.

We will first study the modifications of the phonon dispersion curves due to this change. Then we will study the modification of the optical phonons when a superstructure appears at the surface. Finally we will take into account the changes in surface force constants to study the mean square displacements of adsorbed atoms.

<u>Surface Phonons at an Adsorbed Monolayer</u> (7)

We introduce an adsorbed layer in our model by changing the masses of the atoms in the surface layer from M to M_s. Let us define

$$\alpha = \frac{M_s - M}{M} \tag{83}$$

The modification of the dynamical matrix is due to this change [see Eq. (3)]

$$\delta L'_{\alpha\beta}(\ell_3\ell_3') = \delta_{\alpha\beta}(M - M_s)\omega^2\delta_{\ell_3 0}\delta_{\ell_3' 0} \tag{84}$$

All the information we require is contained in the quantity $\Delta(\omega^2 + i\varepsilon)$. From the preceding results, Eqs. (25) and (77), it is straightforward to show that

$$H(k_{\parallel};\omega^2 + i\varepsilon) = 1 + 2\alpha \frac{Et}{t - 1} \qquad (85)$$

where t and E are defined by Eqs. (26) and (21)-(23). We can use this result to study the occurrence of surface modes and in band resonance modes on the basis of Eq. (11).

Localized Surface Modes Outside the Bulk Band

(1) Exact results. For a given value of k_1 and k_2, one obtains the frequency of the localized modes by examining the frequency at which the determinant, Eq. (85), vanishes. By rearranging this equation, and employing the definition of the variable t, one finds the condition for the occurrence of a localized mode to be

$$(1 + 2\alpha E)\left[X - E \pm \sqrt{((X - E)^2 - 1)}\right] = 1 \qquad (86)$$

where the upper sign is employed for frequencies above the band, and the lower sign for frequencies below the band. For a surface layer of light masses ($\alpha < 0$), we find a band of localized states with frequency given by

$$\frac{M\omega^2}{2\gamma} = \frac{2\alpha X - 1 - 2\alpha - \sqrt{4\alpha^2(X^2 - 1) + 4\alpha(X - 1) + 1}}{4\alpha(\alpha + 1)} \qquad (87)$$

These localized surface modes exist only for $\frac{M\omega^2}{2\gamma} > -\frac{1}{\alpha}$. When $\alpha > 0$ (a heavy adsorbed layer), one obtains a localized mode below the bulk band, with a frequency given by

$$\frac{M\omega^2}{2\gamma} = \frac{2\alpha X - 1 - 2\alpha + \sqrt{4\alpha^2(X^2 - 1) + 4\alpha(X - 1) + 1}}{4\alpha(\alpha + 1)} \tag{88}$$

In Fig. 4.6 we show the position of the localized bands and their widths as a function of the mass change parameter α. Figure 4.7 illustrates the dispersion curves

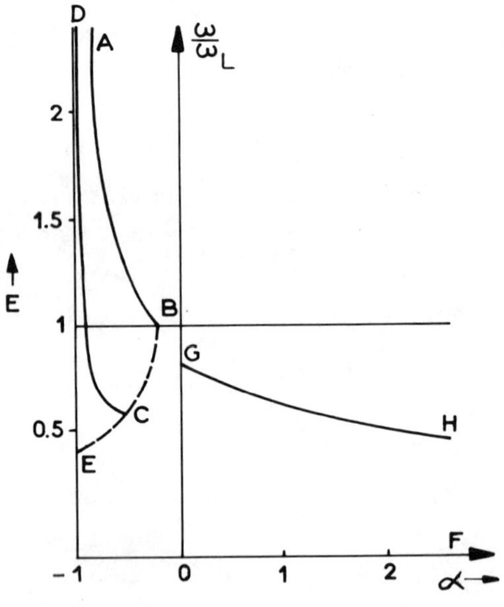

Fig. 4.6.--Frequencies of localized states associated with an adsorbed layer of mass defect atoms. The modes in the region ABCD lie above the bulk band, while those in the region FOGH lie below. The dashed line represents the resonant states at E = - (1/α).

for special directions for the case $M_s = M/2$. Figure 4.8 gives the directions of propagation employed in Fig. 4.7.

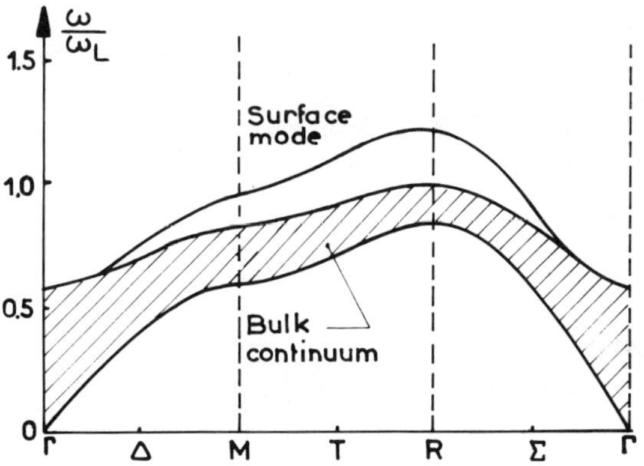

Fig. 4.7.--Excitation spectrum for a (100) adsobred layer of atoms of mass $M_S = (1/2)M$ on the crystal surface. Figure 4.8 gives the directions of propagation employed in this figure.

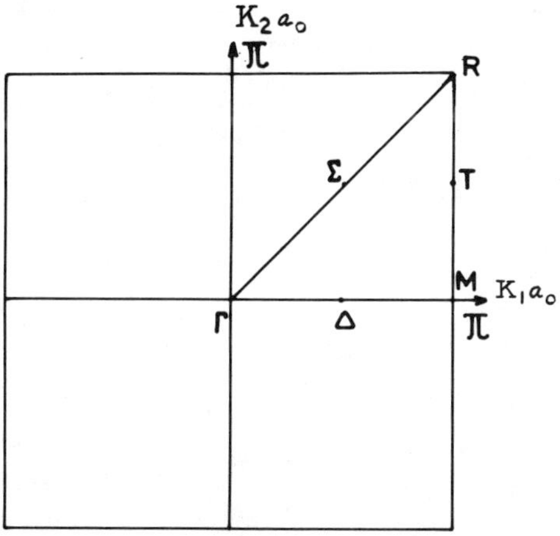

Fig. 4.8.--Brillouin zone for surface problems associated with a (100) face of a simple cubic crystal.

(2) Simple approximate derivations. The results given above are exact within the model used. The Green's solutions become cumbersome when dealing with a more sophisticated model where analytical answers cannot be obtained. It is then interesting to note that when one is

interested in well localized surface phonons, molecular type approaches will yield good approximate values for the frequencies of these modes. One can simply use a "frozen substrate model" (46) allowing only a few planes to vibrate, or a little more elaborate method based on expansions of the Green's functions (49,50) or on the moments of the density of states (51,52). The first two methods are straightforward. We will therefore describe only the third one here. As a general remark about these approximations we can say that they will give good results when the modes are closely localized near the surface: i.e., when they have short exponential penetration into the bulk and localized frequency not too close to the bulk bands. Of course, these methods cannot be applied for long-wavelength acoustic surface phonons which penetrate too far into the bulk.

We introduced the moments of the density of states by Eq. (49). Let us introduce the change in these moments due to the introduction of a perturbation as, for example, the adsorbed layer described in this section

$$\delta\mu_{2n} = \mathrm{Tr}(\overleftrightarrow{D}_A)^n - \mathrm{Tr}(\overleftrightarrow{D}')^n \tag{89}$$

where $\overleftrightarrow{D}_A$ and $\overleftrightarrow{D}'$ were introduced by Eq. (48) and describe, respectively, the crystal with the adsorbed layer and the one with the free surface. Changing M to M_S introduces a perturbation only on the first two layers

$$\overleftrightarrow{D}_A - \overleftrightarrow{D}' = \gamma\left(\frac{1}{M_S} - \frac{1}{M}\right)\begin{pmatrix} 2X-1 & -1 \\ 0 & 0 \end{pmatrix} \tag{90}$$

and the changes in the first two moments are easily found to be

$$\delta\mu_2 = -\frac{\gamma}{M}\frac{\alpha}{1+\alpha}(2X - 1) \tag{91}$$

$$\delta\mu_4 = -\left(\frac{\gamma}{M}\right)^2 \frac{\alpha}{1+\alpha}\left[\frac{2+\alpha}{1+\alpha}(2X - 1)^2 + 2\right] \tag{92}$$

One can imagine (52) a great number of models where this partial information about the variation in the density of states can be used for calculating the position of a localized state. Let us just give one here where one approximates the change in the bulk density of states by two delta functions of equal weight (- 1/2) situated at frequencies ω_1 and ω_2. The localized state whose position we want to calculate is assumed to have the frequency $\omega(k_\parallel)$. In this model

$$\delta\mu_2 = \omega^2(k_\parallel) - \frac{1}{2}(\omega_1^2 + \omega_2^2) \tag{93}$$

$$\delta\mu_4 = \omega^4(k_\parallel) - \frac{1}{2}(\omega_1^4 + \omega_2^4) \tag{94}$$

and from these relations

$$\omega^2(k_\parallel) = \frac{1}{2}\left[\delta\mu_2 + \frac{\delta\mu_4}{\delta\mu_2} + \frac{1}{\delta\mu_2}\left(\frac{\omega_2^2 - \omega_1^2}{2}\right)^2\right] \tag{95}$$

In this model $\omega_2^2 - \omega_1^2$ can be determined by using for example the exact value of ω_L directly calculated at the high symmetry point $k_1 = k_2 = 0$. In the case of light adsorbed atoms ($M_S < M$), one easily obtains (52)

$$\left(\frac{\omega(\vec{k}_\parallel)}{\omega_M}\right)^2 = \frac{1}{12}\left[\frac{1}{1+\alpha}\,(2X - 1) - \frac{1}{\alpha}\,\frac{1}{2X-1}\right] \tag{96}$$

where

$$\omega_L^2 = 12(\gamma/M) \tag{97}$$

This simple result approximates the exact one (Eq. (87)) with a precision of better than 1%. This approach may also give (52) good results for the branch of acoustic surface phonons outside the long-wavelength limit.

For a more detailed discussion of surface phonons due to adsorption in a model displaying Rayleigh waves and still remaining analytical, the reader may study Ref. 53.

<u>The Occurrence of In-Band Resonance Modes</u>.

For values of $E = M\omega^2/(2\gamma)$ inside the bulk spectrum, one finds, from Eq. (85)

$$H(\vec{k}_\parallel;\omega^2 + i\varepsilon) = 1 + \alpha E - i\alpha E\left(\frac{1 + X - E}{1 - X + E}\right)^{1/2} \tag{98}$$

Recall that in terms of the variable E, the bulk spectrum lies in the range

$$X - 1 \le E \le X + 1 \tag{99}$$

X is a function of k_1 and k_2 and ranges from 1 to 5. The corresponding phase shift, Eq. (15), is

$$\eta'(E) = \tan^{-1}\left\{\frac{\alpha E}{1+\alpha E}\left(\frac{1 + X - E}{1 - X + E}\right)^{1/2}\right\} \tag{100}$$

At the bottom of the band, where E = X - 1

$$\eta'(E) = \frac{\pi}{2}\,\mathrm{sgn}\,(\alpha) \tag{101}$$

At the top of the band, where E = X + 1

$$\eta'(E) = \begin{cases} 0 \text{ for } \alpha > (-1/E) \\ -\pi \text{ for } -1 < \alpha < (-1/E) \end{cases} \tag{102}$$

When the real part of Eq. (98) vanishes, an in-band resonance occurs. The corresponding value of E is given by

$$E = E_R = -(1/\alpha) \tag{103}$$

The conditions under which this resonance level is a sharp, well-defined feature in the density of states may be explored by noting that the change in density of phonon modes associated with the presence of the adsorbed level is

$$\Delta n(E) = \frac{3}{\pi}\frac{d\eta'}{dE} = \frac{3}{\pi}\frac{\partial}{\partial E}\tan^{-1}\left(\frac{J(E)}{R(E)}\right) \tag{104}$$

where

$$R(E) = \text{Re } H(\vec{k}_{\parallel};E + i\varepsilon) \tag{105}$$

and

$$J(E) = \text{Im } K(\vec{k}_{\parallel};E + i\varepsilon) \tag{106}$$

For energies near the resonance energy E_R, $\Delta n(E)$ is given by the well-known relation

$$\Delta n(E) \cong \frac{1}{\pi} \frac{\Gamma}{(E-E_R)^2 + \Gamma^2} \tag{107}$$

where

$$\Gamma = J(E)/(\partial R/\partial E) \tag{108}$$

The resonance is well-defined when $(\Gamma/E_R) << 1$. After some algebra, one finds

$$|\Gamma/E_R| = \left(\frac{1 + X - E_R}{1 - X + E_R}\right)^{\frac{1}{2}} \tag{109}$$

The resonance level is sharp and well-defined if $E_R = -(1/\alpha)$ is near the top of the band ($E_R \cong 1 + X$).

Change in the Low-Temperature Specific Heat

From the knowledge of the localized states and of the phase shift it is possible to obtain the change of the thermodynamic functions due to the adsorbed layer (7). We will just derive here the change in the low-temperature specific heat, since this can be done analytically and exactly and provides an instructive exercise.

This property is dominated by the long-wavelength vibrations. One may expand the various quantities in a power series

$$k_{\parallel}^2 = k_1^2 + k_2^2 \tag{110}$$

For $M_S > M$, we found a localized surface mode below the bulk band. The frequency of this mode is given by Eq. (88) as

$$E = \frac{M\omega^2}{2\gamma} \cong \frac{1}{2}k_{\parallel}^2 a_0^2 - \frac{1}{2}\alpha k_{\parallel}^4 a_0^4 \tag{111}$$

The adsorbed layer does not modify the speed of the long-wavelength vibrations. This is quite general and remains true also for Rayleigh waves (53). It is easily understood when one notices that the penetration of the acoustic surface phonon is large compared to the width of one adsorbed layer.

Then for $M_S > M$ this localized mode gives for a given $\vec{k}_{\parallel}$ a contribution to the change in the density of states

$$3\delta\left(E - \frac{1}{2}k_{\parallel}^2 a_0^2\right) \tag{112}$$

As we mentioned above [Eq. (101)], because of the jump in η' at the bottom of the bulk, one has another contribution for a given $\vec{k}_{\parallel}$ to the change in the density of states

$$-\frac{3}{2}\,\delta\left(E - \frac{1}{2}k_{\parallel}^2 a_0^2\right) \tag{113}$$

In the bulk region $E > X - 1$, another contribution comes from Eq. (100)

$$\frac{1}{\pi}\frac{\partial}{\partial E}\tan^{-1}\left\{\frac{2^{\frac{1}{2}}\alpha E}{\sqrt{E - \frac{1}{2}k_{\parallel}{}^2 a_0{}^2}}\right\}$$

The change in the low-temperature specific heat is easily obtained when these three contributions to $\Delta n(E)$ are summed on $\vec{k}$ and substituted into Eq. (13)

$$\Delta F = \int_0^{\infty} dE\ f_{C_v}(E)\Delta n(E) \tag{114}$$

The total change in C_v that occurs because the surface layer has mass M_S rather than M is

$$\delta C_v = \frac{6\pi^2}{5}\ \alpha\left(\frac{M}{\delta}\right)^{\frac{3}{2}}\frac{k_B{}^4 T^3}{\hbar^3} \tag{115}$$

This result is to be compared with the surface specific heat given by Eqs. (39) and (42). An adsorbed monolayer does not change the T^2 behavior of the surface specific heat.

Influence of a Surface Superstructure on Optical Phonons

We assumed till now that the symmetry in the adsorbed layer was the same as in the substrate. However, in many instances, when a crystalline surface is formed (perhaps by cleaving the crystal), the atoms in and near the surface shift their positions, and assume a configuration with symmetry lower than that associated with the bulk (54). In the particular case of Si, for example, the (111) surface exhibits a (2 x 1) arrangement, at room temperatures, where the unit cell of the surface layer is twice as long in one direction as the unit cell appropriate to the bulk (111) atomic layer.

One finds (9) that a superstructure has two important qualitative effects on the properties of the surface optical phonons. These effects are illustrated in a schematic fashion in Fig. 4.9. Consider first the case where the surface has no superstructure, and examine the vibrational spectrum of the semi-infinite crystal, for modes with wave vector $\vec{k}_{\parallel}$ directed along the $\hat{x}_2$ direction. For $k_1 = 0$, the vibrational spectrum is sketched in Fig. 4.9a. For each value of k_2 one has a range of frequencies associated with bulk vibrations (shaded area), and we suppose that a surface optical mode branch exists, as shown in Fig. 4.9a.

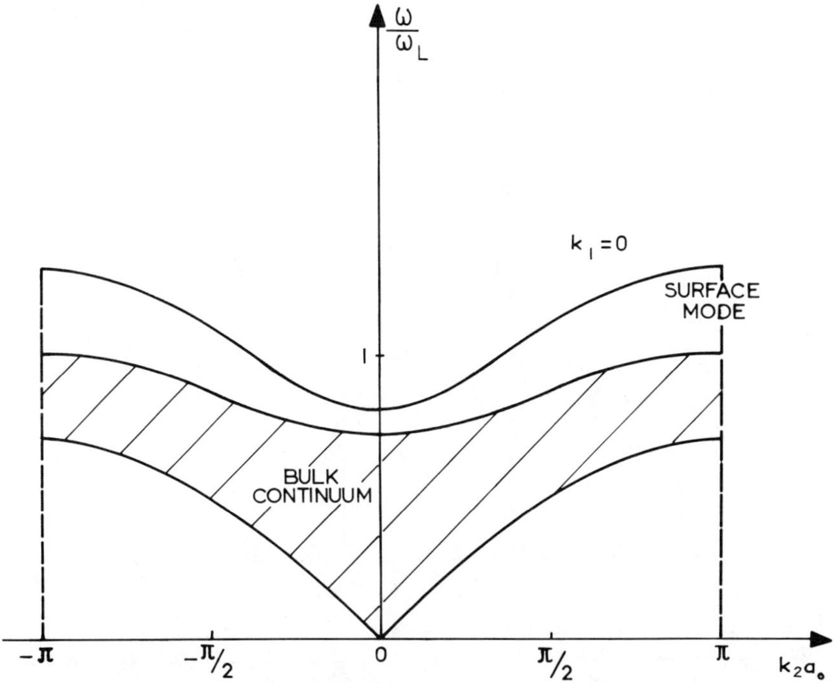

Fig. 4.9a.--Optical surface modes on an unreconstructed (001) surface.

Now suppose the surface has a superstructure, in such a manner that the new repeat distance parallel to the $\hat{x}_2$ axis assumes a value twice that appropriate to the bulk atomic plane. We can plot the normal mode spectrum of this surface as indicated in Fig. 4.9b, which could be called the extended-zone scheme for this crystal. The effect of the superstructure is to introduce new zone boundaries at $k_2 = \pm \pi/2a_0$, respectively. The region $-\pi/2a_0 < k_2 < \pi/2a_0$ is now the first Brillouin zone and the two regions $-\pi/2a_0 < k_2 < \pi/a_0$, and $-\pi/a_0 < k_2 < -\pi/2a_0$ are the second zone. A gap in the surface optical mode dispersion relation opens up at the zone boundary, as indicated in Fig. 4.9b. If the surface optical phonon dispersion relation is plotted in the <u>reduced</u> zone, there are now two distinct branches, and two

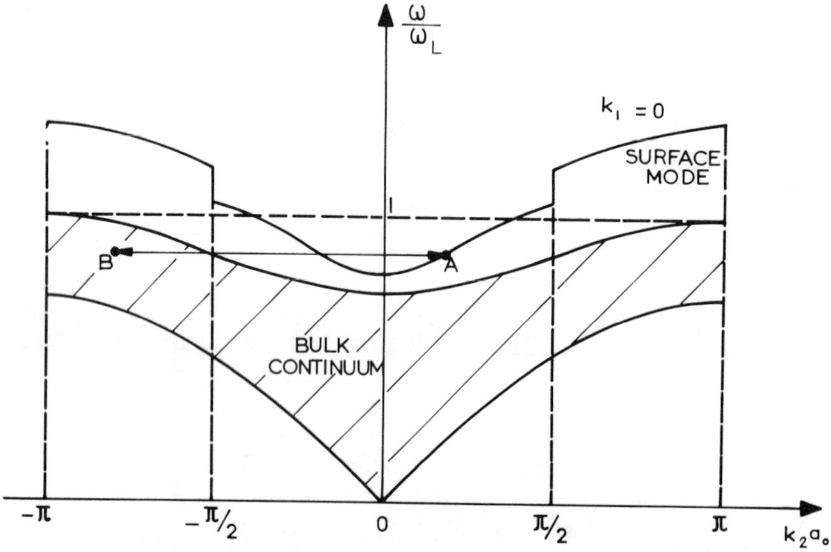

Fig. 4.9b.--Optical surface modes on a (2 x 1) reconstructed (001) surface. Note the opening of gaps at the new zone boundaries and the mixing of the surface phonon A with the bulk one B, transforming the localized mode A into a virtual state.

surface optical modes at $\vec{k}_{\parallel} = 0$. In Fig. 4.9b the two $\vec{k}_{\parallel} = 0$ modes are at $k_2 = 0$ and $k_2 = \pi/a_0$.

There is one other effect that is quite striking. Consider the surface optical mode at point A in Fig. 4.9b. This mode has a frequency ω_S lower than the maximum vibrational frequency ω_L associated with bulk vibrations with $\vec{k}$ directed along the $\hat{x}_2$ axis. When the surface has a superstructure, the resulting perturbation term in the Hamiltonian mixes the surface optical phonon A with the <u>bulk</u> phonon B displaced in $\vec{k}$ from A by the reciprocal lattice vector $\vec{G} = -\hat{x}_2\pi/a_0$. As a result of this admixture, the mode A is no longer a true surface mode, but becomes a virtual surface state, since the displacement field is no longer localized at the surface. For the crystal with a frequency spectrum similar to that sketched in Fig. 4.9b all surface optical modes (with $\vec{k}_{\parallel}$ directed along $\hat{x}_2$) having a frequency lower than the bulk maximum frequency ω_M may become virtual states when the superstructure appears. In general, for a surface optical phonon to be converted to a virtual state by this process, its frequency must be lower than ω_L and also the frequency of the surface optical mode must be below the maximum frequency of bulk vibrations with the wave vector $k_2 - \pi\hat{x}_2/a_0$.

For a surface layer of light masses ($\alpha < 0$), we found a threefold degenerate band of surface phonons given by Eq. (87).

Now suppose that in the surface layer the new repeat distance parallel to the $\hat{x}_2$ axis assumes a value twice that appropriate to the bulk atomic plane.

Let us now denote the coordinates of a site by $(\ell_1\ell_2\ell_3\kappa)$ where $\kappa = 1$ or 2 is a site index in the new unit cell containing two atoms, with the length of the new unit cell being $2a_0$ in the $\hat{x}_2$ direction.

In the bulk crystal, the unit cell will now contain two atoms, and the corresponding reduced Brillouin zone extends over the region $-\pi < k_1 a_0 < +\pi$, $-\pi < k_3 a_0 < +\pi$, and $-\pi/2 < k_2 a_0 < +\pi/2$. In this reduced Brillouin zone the dynamical matrix is

$$D(k;\kappa\kappa') = \sum_{\ell'} e^{-i\vec{k}\left(\vec{x}(\ell) - \vec{x}(\ell')\right)} D(\ell\kappa,\ell'\kappa') \qquad (116)$$

where $D(\ell\kappa,\ell''\kappa'')$ is the dynamical matrix written in real space. In the Montroll-Potts model

$$D(\vec{k};11) = D(\vec{k};22) = \frac{2\gamma}{M}\left(3 - \cos k_1 a_0 - \cos k_3 a_0\right) \qquad (117a)$$

$$D(\vec{k};12) = D(\vec{k};21)^{*} = -\frac{\gamma}{M}\left(1 + e^{2ik_2 a_0}\right) \qquad (117b)$$

The bulk Green's functions are now given in the reciprocal space by

$$G(\vec{k};11) = G(\vec{k};22) = \frac{1}{2M}\left\{\frac{1}{\omega^2 - \omega^2(k_1,k_2,k_3)} + \frac{1}{\omega^2 - \omega^2\left(k_1,k_2 + \frac{\pi}{a_0},k_3\right)}\right\} \qquad (118a)$$

$$G(k;12) = G(k;21)^{*} = \frac{1}{2M}\left\{\frac{e^{ik_2 a_0}}{\omega^2 - \omega^2(k_1,k_2,k_3)}\right.$$

$$+ \frac{e^{i\left(k_2 a_0 + \pi\right)}}{\omega^2 - \omega^2\left(k_1, k_2 + \frac{\pi}{a_0}, k_3\right)}\Bigg\} \qquad (118b)$$

These Green's functions are clearly equivalent to bulk ones used before. It is evident therefore that the Green's function of the semi-infinite crystal we are studying here can be obtained by relation (77) in the new Brillouin zone scheme just as in the old cubic one. Their form in real space will be obtained from the following relation

$$G(\ell\kappa, \ell'\kappa') = \frac{1}{N_c} \sum_{\vec{k}} e^{i\vec{k}\cdot\left(\vec{x}(\ell) - \vec{x}(\ell')\right)} G(\vec{k}; \kappa\kappa') \qquad (119)$$

where N_c is the number of unit cells in the crystal.

Suppose now that the mass of the surface layer $\ell_3 = 1$ is changed to M_S and let us assume that this adsorbed layer exhibits a (2 x 1) superstructure. We assume the force constants are alternatively softened and stiffened by the same amount Δ along the x_2 direction. We also assume for simplicity that the force constants remain equal to γ along the x_1 and x_3 directions (Fig. 4.10).

After taking advantage of the translational symmetries in the x_1 and x_2 direction, it is a straightforward matter to show that the total perturbation described above on the semi-infinite crystal may be represented by the matrix

$$\overleftrightarrow{V} = \begin{pmatrix} (M - M_S)\omega^2 & -2i\Delta e^{ik_2 a_0} \sin k_2 a_0 \\ 2i\Delta e^{-ik_2 a_0} \sin k_2 a_0 & (M - M_S)\omega^2 \end{pmatrix} \qquad (120)$$

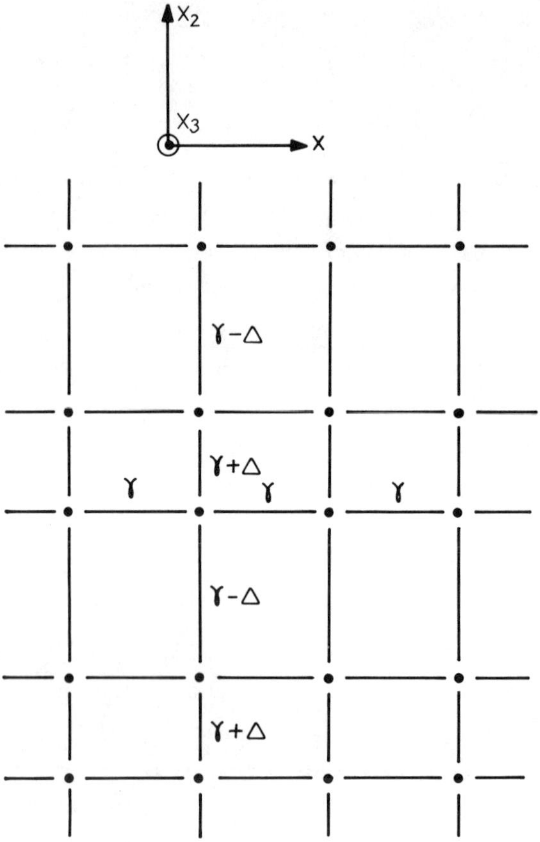

Fig. 4.10.--Model for a (2 x 1) superstructure on a (001) surface of a simple cubic lattice. The quantity γ is the force constant between first neighbors.

The elements of the matrix $\overleftrightarrow{V}$ are taken between (ℓ_3 = 1; κ = 1,2) and (ℓ_3 = 1;κ' = 1,2) and $\overleftrightarrow{V}$ was defined as being the difference between the dynamical matrix of the perturbed crystal and that of the unperturbed semi-infinite one. The corresponding Green's function matrix is

$$\overleftrightarrow{U} = \begin{pmatrix} U(11;11) & U(11;12) \\ U(12;11) & U(12;12) \end{pmatrix} \tag{121}$$

In the matrix elements the dependence on k_1, k_2, and ω^2 is not explicitly written for simplicity, and the indices are for $(\ell_3\kappa;\ell_3'\kappa')$.

Using Eqs. (25), (77), and (118), one easily sees that

$$\overleftrightarrow{U} = \begin{pmatrix} a + b & e^{ik_2a_0}(a - b) \\ e^{-k_2a_0}(a - b) & a + b \end{pmatrix} \tag{122}$$

where

$$a = \frac{1}{2\gamma}\,\frac{t}{t - 1} \tag{123a}$$

$$b = \frac{1}{2\gamma}\,\frac{t'}{t' - 1} \tag{123b}$$

The t and t' are given by relations (26) and (21)-(23). For t one has

$$X = 3 - \cos k_1a_0 - \cos k_2a_0 \tag{124a}$$

and for t', X is replaced by

$$X' = 3 - \cos k_1a_0 - \cos (k_2a_0 + \pi) \tag{124b}$$

All the physical information we need is contained in the determinant $H = |\overleftrightarrow{I} - \overleftrightarrow{U}\overleftrightarrow{V}|$, which is readily found to be

$$H(E) = H_1(E)H_2(E) - \Gamma(E) \tag{125}$$

where

$$H_1(E) = 1 + \frac{2\alpha Et}{t - 1} \tag{126}$$

$$H_2(E) = 1 + \frac{2\alpha Et'}{t' - 1} \tag{127}$$

$$\Gamma(E) = 4\left(\frac{\Delta}{\beta}\right)^2 \frac{tt'}{(t - 1)(t' - 1)} \sin^2 k_2 a_0 \tag{128}$$

Let us now consider only the case of a light adsorbed layer $M_S < M$. Since the light masses vibrate with high frequency, for calculating the frequency of the localized optical modes we note that $|\zeta| >> 1$ [Eq. (21)] and that

$$\frac{1}{t} = -2(E - X) + \frac{1}{2(E - X)} + \ldots$$

The same property holds for t'. This enables us to obtain from Eqs. (125), (12), and (21) in a first approximation the following expression for the localized optical mode dispersion relation

$$\omega^2 = \frac{\beta}{M_S}\left[5 - 2\cos k_1 a_0 \pm 2\left(\cos^2 k_2 a_0 + \left(\frac{\Delta}{\gamma}\right)^2 \sin^2 k_2 a_0\right)^{\frac{1}{2}}\right] + \ldots \tag{129}$$

Since M_S is the only mass that appears in this first approximation expression, this result describes the light mass adsorbate beating against the underlying lattice structure. All the distinct roots are found by restricting $k_2 a_0$ to the interval $-\pi/2 < k_2 a_0 < \pi/2$ and $k_1 a_0$ to the interval $-\pi < k_1 a_0 < \pi$.

One sees clearly in expression (129) that without superstructure ($\Delta = 0$) at $k_2a_0 = \pi/2$, there is one two-fold degenerate surface mode frequency and that with a superstructure there are two nondegenerate modes. This is the first effect described in the introduction. A gap in the surface optical mode dispersion relation opens up at the new zone boundary $k_2a_0 = \pi/2$.

The second effect described in the introduction, that is, the possible transformation of a localized mode into a resonant one, is also rather obvious when one looks at Figs. 4.9. However, one needs to make a careful study (9) of these resonances using the method described above to find that these resonances remain well-defined features in the density of states.

The case of a (2 x 2) superstructure due to adsorption of a monolayer of atoms on the (001) surface of Ni was studied in detail by Armand and Theeten (55), who found the same qualitative effects described above. But since they studied a realistic case where measurements were made (56), their results may show direct quantitative comparison with experiment.

Analogous effects will also appear for other surface excitations (electrons, magnons, . . .). It was possible to show (57,58) that there may be an appreciable gain in the electronic energy when one allows a surface to have a superstructure. This may be one possible driving mechanism causing superstructures to occur.

Let us now give another illustration of a possible change in the stability of the adsorbed layer.

Stability of a Physisorbed Layer from a Study of the Atomic Mean Square Displacements

It is well known that a two-dimensional solid is unstable against long-wavelength acoustic vibrations:

the mean square displacements of its atoms are infinite (59). We will therefore study the physisorbed monolayer especially for these long-wavelength phonons. Other experimental and theoretical studies of the MSD of adsorbed atoms have been carried out, but the stability of the physisorbed monolayer has not been discussed (60-62).

Let us call γ, γ', and γ'' the force constants bonding, respectively, the bulk atoms among themselves, the physisorbed atoms to the substrate atoms, and the physisorbed atoms among themselves (see Fig. 4.11). When γ' tends to zero, the mean square displacements of the physisorbed atoms are found to diverge as $\ln(\gamma''/\gamma')$. After taking due account of the translational symmetry in the directions $\hat{x}_1$ and $\hat{x}_2$ parallel to the surface, the perturbation in the equation of motion of the free surface crystal may be written as

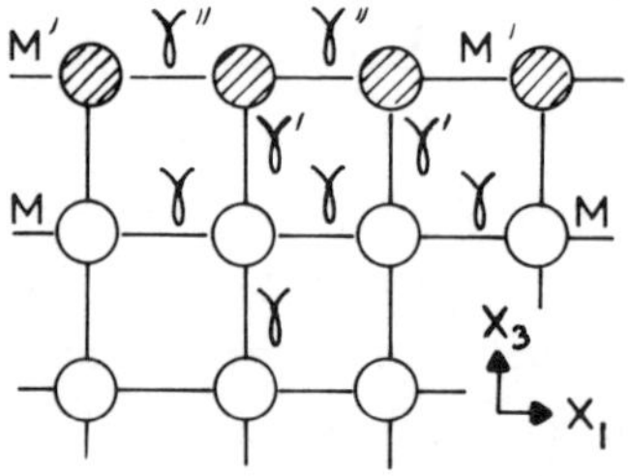

Fig. 4.11.--Force constants at an adsorbed monolayer.

$$\delta L(\vec{k}_{\parallel};\ell_3\ell_3') = \Big[(M_S - M)\omega^2 + (\gamma - \gamma') + 2(\gamma - \gamma'')$$

$$\cdot\ (2 - \cos k_1a_0 - \cos k_2a_0)\Big]\delta_{\ell_3 1}\delta_{\ell_3' 1}$$

$$+ (\gamma' - \gamma)\left(\delta_{\ell_3 1}\delta_{\ell_3' 2} + \delta_{\ell_3 2}\delta_{\ell_3' 1} - \delta_{\ell_3 2}\delta_{\ell_3' 2}\right) \qquad (130)$$

Let us now use a "frozen substrate" model (46). The adsorbed monolayer is allowed to vibrate and the substrate atoms are frozen. The frequencies of vibration of the adsorbed atoms are

$$M_S\omega^2_\alpha \simeq \gamma' + 2\gamma''\left(2 - \cos k_1a_0 - \cos k_2a_0\right), \quad \alpha = x_1, x_2, x_3 \tag{131}$$

At high temperatures, the mean square displacements of these atoms in the direction α [Eq. (73)] are

$$\langle u^2_\alpha(\ell)\rangle \simeq k_BT \sum_{\vec{k}_\parallel} \frac{1}{M_S\omega^2_\alpha} \tag{132}$$

The elastic limit contribution is

$$\langle u^2_\alpha(\ell)\rangle \simeq \frac{k_BT}{2\pi} \int_0^{k_D} \frac{a_0^2 k_\parallel \, dk_\parallel}{\gamma' + \gamma'' k_\parallel^2 a_0^2} \tag{133}$$

where $k_\parallel^2 = k_1^2 + k_2^2$. Finally, we obtain

$$\langle u^2_\alpha(1)\rangle \simeq \frac{k_BT}{4\pi\gamma''} \ell n\left(1 + \frac{\gamma''}{\gamma'} k_D^2 a_0^2\right) \tag{134}$$

where k_D is the limit value of $k_\parallel$ for which the elastic approximation remains valid. For $\gamma' << \gamma''$ we have then a very simple derivation of the logarithmic divergence of the MSD of physisorbed atoms. When a two-dimensional solid is physisorbed on a crystal surface, the substrate tends to stabilize it even for a weak coupling γ' between adsorbate and substrate, since the divergence in the mean square displacement for γ' going to zero is of a logarithmic type (43).

THE INTERACTION OF POINT DEFECTS WITH CRYSTAL SURFACES

We will illustrate here by simple examples the physical effects discussed on pp. 166-195. We will use the general theories described there in the frame of the Montroll-Potts model in order to derive as simply as possible the expressions whose physical meaning was discussed before. We will therefore not discuss these results here.

Let us first calculate the energy of interaction of a point defect with a crystal surface. Then we will come to the interaction of two adatoms on a surface. Finally, we will give the frequencies of a localized mode due to a substitutional isotopic defect as a function of its distance from a free surface.

The Energy of Interaction of a Defect with a Crystal Surface

We will confine this study to the zero-point energy of interaction. We can calculate the change in the zero-point energy when the point defect is near a free surface or in an infinite crystal, from the following expression (see p. 168)

$$\Delta E_0 = \frac{-3\hbar\omega_L}{2\pi} \int_0^\infty f\Omega(f)df \tag{135}$$

$$= \frac{-3\hbar\omega_L}{2\pi} \left\{ \left[f \ln D\left(-\omega_L^2 f^2\right) \right]_0^\infty - \int_0^\infty \ln D\left(-\omega_L^2 f^2\right) df \right\} \tag{136}$$

where

$$D(\omega^2) = \frac{|\overleftrightarrow{I} - \overleftrightarrow{U}(\omega^2)\overleftrightarrow{\delta L}(\omega^2)|}{|\overleftrightarrow{I} - \overleftrightarrow{G}(\omega^2)\overleftrightarrow{\delta L}(\omega^2)|} \tag{137}$$

$$\Omega(f) = \frac{d}{df} \ln D\left(-\omega_L^2 f^2\right) \tag{138}$$

As a simple example, we calculate the energy of interaction of an isotopic impurity at the ℓ lattice site of our simple cubic lattice with a free surface at $\ell_3 = 1$. The matrix $\delta L(n,m)$ due to the substitution of an isotopic impurity is given by

$$\delta L(nm;\omega^2) = \varepsilon M\omega^2 \delta_{n\ell}\delta_{m\ell} \tag{139}$$

where

$$\varepsilon = 1 - (M'/M) \tag{140}$$

and M' is the mass of the impurity.

Using Eq. (77) for $\overleftrightarrow{U}$, we obtain

$$D(\omega^2) = \frac{1 - \varepsilon M\omega^2\left[G(0,0,0;\omega^2) + G\left(0,0,2\ell_3 - 1;\omega^2\right)\right]}{1 - \varepsilon M\omega^2 G(0,0,0;\omega^2)} \tag{141}$$

In the weak defect limit, we can replace this expression by

$$D(\omega^2) = 1 - \varepsilon M\omega^2 G\left(0,0,2\ell_3 - 1;\omega^2\right) \tag{142}$$

For $\ell_3 \gg 1$, let us use Eq. (78). Finally, in this weak defect limit

$$\Delta E_0 = \frac{-3\hbar\omega_L}{2\pi} \frac{\varepsilon M\omega_L^2}{4\pi\gamma(2\ell_3 - 1)} \int_0^\infty df\ f^2 \exp\left[-(12)^{\frac{1}{2}}(2\ell_3 - 1)f\right] \tag{143}$$

After a simple integration, and recalling that $M\omega_L^2 = 12\gamma$, one has

$$\Delta E_0 = -\frac{3^{\frac{1}{2}}}{8\pi^2} \frac{\hbar\omega_L \varepsilon}{(2\ell_3 - 1)^4} \tag{144}$$

Phonon-Mediated Indirect Interaction of Adatoms on a Surface (63)

We will derive here, with the use of the Montroll-Potts model, the simple expressions discussed on pp. 194-195 for the energy of interaction of two adatoms through the crystal phonon field.

Equations of Motion

The model geometry showing two adatoms of mass M' bound in the "on-site" configuration to a (001) surface atom of mass M by a force constant γ' is described in Fig. 4.12. Let $\xi_\alpha(m_1 m_2 0)$ denote the α component of the displacement from equilibrium of the adatom adsorbed at the surface site $(m_1 m_2 0)$. Its equation of motion is

$$M'\omega^2 \xi_\alpha(m_1 m_2 0) = \gamma'\left[\xi_\alpha(m_1 m_2 0) - u_\alpha(m_1 m_2 1)\right] \tag{145}$$

This can be rearranged into the form

$$\xi_\alpha(m_1 m_2 0) = \frac{\omega_0^2}{\omega_0^2 - \omega^2} u_\alpha(m_1 m_2 1) \tag{146}$$

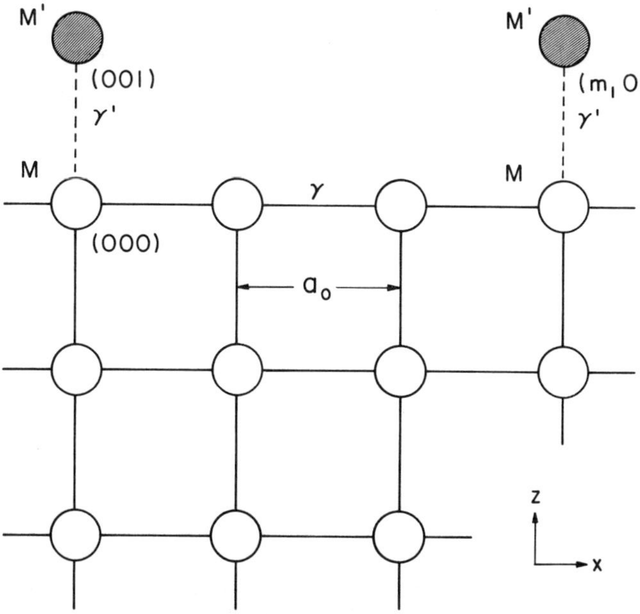

Fig. 4.12.--Model geometry showing two adatoms of mass M' bound in the "on-site" configuration to a surface atom of mass M by a force constant γ'. The force constant in the simple cubic monatomic crystal is γ.

where

$$\omega_0^2 = \gamma'/M' \tag{147}$$

The equation of motion for the system which consists of the semi-infinite crystal with two adsorbed atoms located at the lattice sites (001) and $(m_1 m_2 1)$ is in matrix form

$$(\overleftrightarrow{L}' - \delta\overleftrightarrow{L}')\vec{u} = 0 \tag{148}$$

where $\overleftrightarrow{L}'$ is the dynamical matrix of the semi-infinite crystal, and

$$\delta\overleftrightarrow{L}'\vec{u} = \gamma'\left[u_\alpha(\ell_1\ell_2 1) - \xi_\alpha(\ell_1\ell_2 0)\right]\delta_{\alpha\beta}\delta_{\ell\ell'}$$

$$\times\left[\delta_{\ell_1 0}\delta_{\ell_2 0} + \delta_{\ell_1 m_1}\delta_{\ell_2 m_2}\right] \qquad (149)$$

By substituting the adatom displacements of Eq. (146) into the above equation we obtain

$$\delta L'_{\alpha\beta}(\ell\ell') = \frac{\gamma'\omega^2}{\omega^2 - \omega_0^2}\,\delta_{\alpha\beta}\delta_{\ell\ell'}\delta_{\ell_3 1}\left[\delta_{\ell_1 0}\delta_{\ell_2 0} + \delta_{\ell_1 m_1}\delta_{\ell_2 m_2}\right] \qquad (150)$$

For the case of a single adatom on the surface at site (001) the perturbation matrix $\delta\overleftrightarrow{L}'$ becomes

$$\delta L'_{\alpha\beta}(\ell\ell') = \frac{\gamma'\omega^2}{\omega^2 - \omega_0^2}\,\delta_{\alpha\beta}\delta_{\ell\ell'}\delta_{\ell_3 1}\delta_{\ell_1 0}\delta_{\ell_2 0} \qquad (151)$$

For the case where the perturbation is a single adsorbed atom at site 1 or site 2, the above equation combined with Eq. (343) of Chapter 2 gives

$$\Delta_1(\omega^2) = \Delta_2(\omega^2) = 1 - \frac{\gamma'\omega^2}{\omega^2 - \omega_0^2}\,U(001,001;\omega^2) \qquad (152)$$

For the case of two adsorbed atoms located at (001) and $(m_1 m_2 1)$, we have

$$\Delta_{12}(\omega^2) = \begin{vmatrix} \Delta_1(\omega^2) & \chi(\omega^2) \\ \chi(\omega^2) & \Delta_2(\omega^2) \end{vmatrix} \tag{153}$$

where

$$\chi(\omega^2) = \frac{-\gamma'\omega^2}{\omega^2 - \omega_0^2} U\left(001, m_1 m_2 1; \omega^2\right) \tag{154}$$

To obtain the interaction free energy of the adatom pair, we need

$$\Delta(\omega^2) = \frac{\Delta_{12}(\omega^2)}{\Delta_1(\omega^2)\Delta_2(\omega^2)}$$

$$= 1 - \frac{(\gamma'\omega^2)^2 U^2\left(001; m_1 m_2 1; \omega^2\right)}{\left[\omega^2 - \omega_0^2 - \gamma'\omega^2 \; U(001;001;\omega^2)\right]^2} \tag{155}$$

where the Green's function $\overleftrightarrow{U}$ of the semi-infinite solid can be calculated with the help of Eq. (77) as a function of the bulk Green's function, $\overleftrightarrow{G}$.

Zero-Point Energy of Interaction

The interaction free energy at 0°K can be found from Eq. (136) and from

$$\Delta\left(-\omega^2_L f^2\right) = 1 - \frac{(\gamma' f^2)^2 U^2\left(001, m_1 m_2 1; -\omega^2_L f^2\right)}{\left\{f_0^{\,2} + f^2\left[1 - \gamma' U\left(001, 001, -\omega^2_L f^2\right)\right]\right\}^2} \tag{156}$$

where we have introduced

$$f_0^{\,2} = \frac{\omega^2_0}{\omega^2_L} = \frac{\gamma' M}{12\gamma M'} \tag{157}$$

In the limit of weak binding and large separation distance

$$\left|\gamma' U\left(001, 001; -\omega^2_L f^2\right)\right| << 1 \tag{158}$$

$\Delta\left(-\omega_L^{\,2} f^2\right)$ can be written as

$$\Delta\left(-\omega^2_L f^2\right) = 1 - \frac{(\gamma' f^2)^2 U^2\left(001, m_1 m_2 1; -\omega^2_L f^2\right)}{\left(f^2_0 + f^2\right)^2} \tag{159}$$

The magnitude of the Green's function in Eq. (158) is maximum when f = 0 and is very nearly equal to $-(3\gamma)^{-1}$. Thus, this condition translates to

$$\gamma'/\gamma << 3 \tag{160}$$

In the region of large separation distances, where $m = \left(m_1^{\,2} + m_2^{\,2}\right)^{\frac{1}{2}} >> 1$, using Eqs. (77) and (78) one obtains the leading term for the surface Green's function

$$U(001, m_1 m_2 1; \omega^2_L f^2) = -\frac{\exp(-12^{\frac{1}{2}} mf)}{2\pi\gamma m} \qquad (161)$$

By substituting this result into Eq. (159), we obtain

$$\Delta(-\omega^2_L f^2) = 1 - \frac{1}{4\pi^2}\left(\frac{\gamma'}{\gamma}\right)^2 \frac{f^4}{(f^2_0 + f^2)^2} \frac{\exp(-48^{\frac{1}{2}} mf)}{m^2} \qquad (162)$$

The second term in this expression is small compared to the first in the limits we are considering. Therefore the logarithm in Eq. (136) can be expanded

$$E_0 = \frac{-3\hbar\omega_L}{8\pi^3 m^2}\left(\frac{\gamma'}{\gamma}\right)^2 \int_0^\infty df \, \frac{f^4 \exp(-48^{\frac{1}{2}} mf)}{(f^2_0 + f^2)^2} \qquad (163)$$

If we confine our attention to the region where m is large, such that

$$(48)^{\frac{1}{2}} m f_0 >> 1 \qquad (164)$$

then the important contribution to the integral in Eq. (163) comes from values of f smaller than f_0. Hence, the denominator of the integral can be expanded about f_0 and the lowest-order term in the zero-point energy expression becomes

$$E_0 = -\frac{3\hbar\omega_L}{8\pi^3 m^2 f^4_0}\left(\frac{\gamma'}{\gamma}\right)^2 \int_0^\infty f^4 \exp(-48^{\frac{1}{2}} mf) df \qquad (165)$$

Evaluating the integral and substituting for f_0 gives, finally

$$\Delta E_0 = \frac{-(27)^{\frac{1}{2}}}{64\pi^3}\left(\frac{M'}{M}\right)^2 \frac{1}{m^7} \hbar\omega_L^2 \tag{166}$$

The condition in Eq. (164) can be rewritten as

$$m >> \tfrac{1}{2}(M'\gamma/M\gamma')^{\frac{1}{2}} \tag{167}$$

In practice, the result derived above for ΔE_0 is only good if the two adatoms are separated by more than about eight lattice sites (i.e., $m \geq 8$). For separation distances smaller than this, numerical methods were used (63) in evaluating Eqs. (155) and (135). ΔE_0 was studied as a function of the separation distance between the adatoms, and the ratios γ'/γ and M'/M. The results are given in Ref. 63.

Temperature Dependence of Free Energy

In order to find the leading temperature-dependent term in both the high- and low-temperature limits, we need the Green's function $G(m_1m_2m_3;-\omega^2{}_L f^2)$ in the limit as $f \to 0$ and $f \to \infty$. It can be easily shown from previous results (41,42) that

$$\lim_{f\to 0} G(m;-\omega^2{}_L f^2) = \gamma^{-1}\left[A_0(m) + A_1(m)|f| + \ldots\right] \tag{168}$$

and

$$\lim_{f\to\infty} G(m;-\omega^2{}_L f^2) = -\frac{(m_1 + m_2 + m_3)!}{\gamma m_1!m_2!m_3!}\left(\frac{1}{12f^2}\right)^{m_1 + m_2 + m_3 + 1} \tag{169}$$

In Eq. (168), $A_0(m)$ and $A_1(m)$ are numerical constants

whose values depend upon the separation distance of the bulk atoms. It is due to the fact that the Green's function has a finite value at f = 0 and decays rapidly for large values of f that the first term in Eq. (136) vanishes.

(1) Low-temperature limit. To evaluate the lowest order temperature-dependent term in the interaction free energy at low temperatures [see Chapter 2, Eq. (385)] we must determine values up to the fifth derivative of the natural logarithm of Eq. (152). In the region where $f \ll f_0$, we may write this logarithm as

$$\ell n\ \Delta\left(-\omega_L^2 f^2\right) = -\left(\frac{\gamma' f^2}{\gamma f_0^2}\right)^2 \Big[B_0^2(m_1m_2)$$

$$+ 2|f|B_0(m_1m_2)B_1(m_1m_2) + \dots\Big] \qquad (170)$$

where we have written the surface Green's function as

$$U\left(001, m_1m_2 1; -\omega_L^2 f^2\right) = \gamma^{-1}\Big[B_0(m_1m_2) + B_1(m_1m_2)|f| + \dots\Big] \qquad (171)$$

and where

$$B_0(m_1m_2) = A_0(m_1m_2 0) + A_0(m_1m_2 1) \qquad (172)$$

$$B_1(m_1m_2) = A_1(m_1m_2 0) + A_1(m_1m_2 1) \qquad (173)$$

The derivatives of Eq. (170) are easily found to be

$$\Omega(0) = 0 \tag{174}$$

$$\Omega^{(II)}(0) = 0 \tag{175}$$

$$\Omega^{(IV)}(0) = -(240)\left[\gamma'/\gamma f^2{}_0\right]^2 B_0(m_1m_2)B_1(m_1m_2) \tag{176}$$

Hence, from Eq. (385) in Chapter 2, the free energy in the low-temperature region becomes

$$F(T) = \Delta E_0 + \frac{16\pi^5}{21}\left(\frac{\gamma'}{\gamma}\right)^2 \frac{\omega_0}{\omega_L} B_0(m_1m_2)B_1(m_1m_2)\frac{(k_BT)^6}{(\hbar\omega_0)^5} + \ldots \tag{177}$$

For the bulk Green's function (41), we have for all values of m

$$A_0(m) < 0 \quad A_1(m) > 0$$

Thus, we have that

$$B_0(m_1m_2) < 0 \quad B_1(m_1m_2) > 0$$

(2) High-temperature limit. For large values of the temperature, the argument of Δ in Eq. (388) of Chapter 2 becomes large, and the large f expansion of the surface Green's function is needed. By making the identification

$$\omega_L{}^2 f^2 = a_n^2 T^2 \tag{178a}$$

where

$$a_n = 2\pi k_B n/\hbar \tag{178b}$$

and by using Eqs. (77) and (169) we obtain for the leading term

$$U\left(001, m_1 m_2 1; -a_n^{\,2} T^2\right) = -\frac{(m_1 + m_2)!}{\gamma m_1! m_2!} \cdot \left(\frac{\hbar^2 \omega_L^{\,2}}{48\pi^2 k_B^{\,2} T^2 n^2}\right)^{m_1 + m_2 + 1} \tag{179}$$

Using this Green's function, the logarithm of $\Delta(-a_n^{\,2}T^2)$ can be easily found by using Eq. (155) and (178)

$$\ln \Delta\left(-a_n^{\,2}T^2\right) = -\left(\frac{\gamma'}{\gamma}\right)^2 \left(\frac{(m_1 + m_2)!}{m_1! m_2!}\right)^2 \left(\frac{\hbar^2 \omega_L^{\,2}}{48\pi^2 k_B^{\,2} T^2 n^2}\right)^p \tag{180}$$

where we have set

$$p = 2(m_1 + m_2 + 1) \tag{181}$$

Substituting Eq. (180) into Eq. (389) of Chapter 2 gives

$$\Delta F(T) = -3k_B T \left(\frac{\gamma'(m_1 + m_2)!}{\gamma m_1! m_2!}\right)^2 \left(\frac{\hbar^2 \omega_L^{\,2}}{4k_B^{\,2} T^2}\right)^p \cdot \frac{1}{3^p (2\pi)^{2p}} \sum_{n=1}^{\infty} \left(\frac{1}{n}\right)^{2p} \tag{182}$$

To evaluate this sum, we use the definition of the Bernoulli number

$$B_p = \frac{2(2p)!}{(2\pi)^{2p}} \sum_{n=1}^{\infty} \left(\frac{1}{n}\right)^{2p} \tag{183}$$

Thus, we finally obtain for the interaction free energy in the high-temperature limit

$$\Delta F(T) = -\frac{3B_p}{2(2p)!3^p} \left[\frac{\gamma'(m_1 + m_2)!}{\gamma m_1!m_2!}\right]^2 \left(\frac{\hbar\omega_L}{2k_BT}\right)^{2p} k_BT \tag{184}$$

Frequencies of a Localized Mode for a Defect Near a Surface

In the two cases studied above, i.e., one isotopic substitutional defect close to the free surface and one adatom, there may occur localized modes above the band of bulk frequencies. Let us now calculate the possible localized frequencies as a function of M'/M.

Isotopic Substitutional Defect. The localized frequencies of the isotopic defect of mass M' defined above [Eq. (139)] are the solutions of (see Eq. 343, p. 170)

$$\Delta(\omega^2) = 1 - \varepsilon M\omega^2 U(\ell,\ell;\omega^2) \tag{185}$$

where $U(\ell,\ell;\omega^2)$ is defined by Eq. (77).

Due to the fact that there is no coupling between x_1, x_2, and x_3 in the Montroll-Potts model, we will have a triply degenerate localized mode even near a crystal surface. In a more sophisticated model this degeneracy may be split when the impurity interacts with the surface, as discussed on pp. 170-74. Nevertheless, the Montroll-Potts model enables us to study the depression of the localized mode frequencies with increasing distance of the impurity from the free surface. For M' << M, the localized mode frequency can be obtained easily by an expansion of the Green's function. Using Eq. (48) for a semi-infinite Bravais lattice, we can write

$$M\omega^2 U(\ell,\ell;\omega^2) = \left[\left(\overleftrightarrow{I} - \omega^{-2}\,\overleftrightarrow{D}\right)^{-1}\right]_{\ell,\ell} \tag{186}$$

It is easy to transform this expression into

$$M\omega^2 U(\ell,\ell;\omega^2) = [1 - S(\ell)]^{-1} \tag{187}$$

where

$$S(\ell) = \sum_{p=1}^{\infty} a_p(\ell)\omega^{-2p} \tag{188}$$

The first coefficients of this expansion are

$$a_1(\ell) = D_{\ell\ell} \tag{189}$$

$$a_2(\ell) = \sum_{\ell' \neq \ell} D_{\ell\ell'} D_{\ell'\ell} \tag{190}$$

The next terms are defined in a similar way. By substituting Eq. (187) into Eq. (185), one easily obtains the frequency of the localized mode

$$\omega^2 = \frac{a_1(\ell)}{\lambda} + \frac{a_2(\ell)}{a_1(\ell)} + 0(\lambda) \tag{191}$$

where

$$\lambda = M'/M \tag{192}$$

This expansion is valid for $(\omega/\omega_L) >> 1$ and was first used to study the frequencies of a substitutional defect near a surface by Ashkin (65). He studied this problem for a (001) surface of a simple cubic crystal with nearest and next-nearest neighbor, central force interactions. We give here only the results for the Montroll-Potts model

$$\frac{\omega^2}{\omega_L^2} = \frac{5}{12\lambda} + \frac{1}{12} + O(\lambda) \qquad \text{for } \ell_3 = 1 \tag{193}$$

$$\frac{\omega^2}{\omega_L^2} = \frac{6}{12\lambda} + \frac{1}{12} + O(\lambda) \qquad \text{for } \ell_3 \geq 2 \tag{194}$$

These results are given in dotted lines in Fig. 4.13. They are valid only for $\omega^2 >> \omega_L^2$. It is then interesting to compare them to the exact values. The bulk Green's functions have been calculated (41) in the forms with $f = \omega/\omega_L$

$$G(\ell) = \frac{(-1)^{\ell_1 + \ell_2 + \ell_3}}{2\gamma} \, I(\ell;\beta) \tag{195}$$

$$\beta = 2f^2 - 1, \quad \omega > \omega_L$$

$$G(\ell) = -\frac{1}{2\gamma} \quad I(\ell;\beta) \tag{196}$$

$$\beta = 2f^2 + 1, \quad \omega \to -i\omega$$

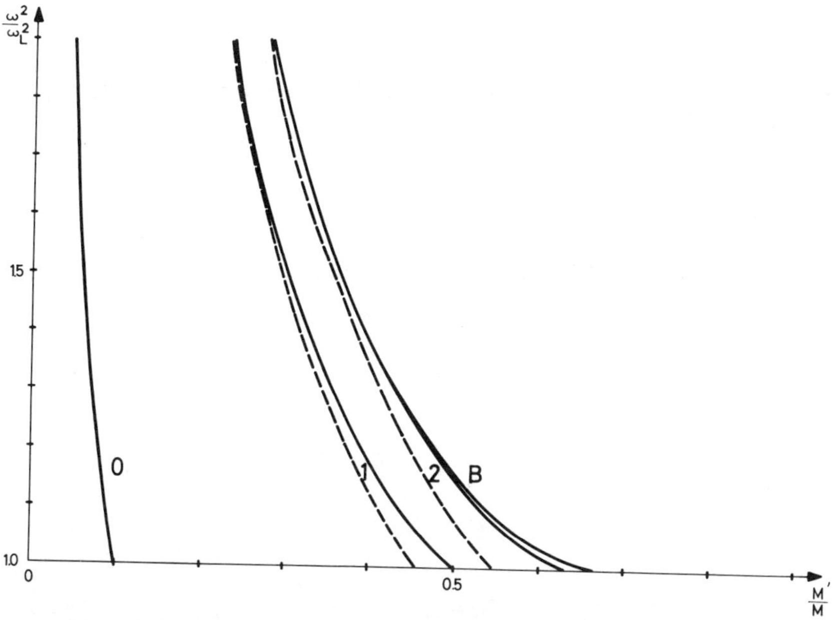

Fig. 4.13.--Localized frequencies ω plotted against the maximum bulk frequency ω_L for a substitutional isotopic defect of mass M' in a simple cubic monoatomic crystal with a (001) free surface. M is the mass of the bulk atoms. The full lines give the exact results for the defect, respectively, adsorbed (0), in the surface layer (1), in the plane ℓ_3 = (2), and in the bulk (B). The dotted lines give the high-frequency expansions for $\ell_3 = 1$ and $\ell_3 \geq 2$, respectively [see also Eqs. (193) and (194)].

and the functions $I(\ell;\beta)$ are tabulated (41). With the help of Eqs. (77) and (195), Eq. (185) takes the form

$$6f^2\varepsilon[I(0,0,0;f^2) - I(0,0,2\ell_3 - 1;f^2)] = 1 \quad (197)$$

Using the tabulated values (41) of $\vec{I}(f^2)$, it is then straightforward to plot the ratio of the localized frequency ω and the maximum bulk frequency ω_L as a function of $\lambda = M'/M$. In Fig. 4.13 the full lines represent the results for the impurity in the surface plane $\ell_3 = 1$, in the plane $\ell_3 = 2$, and in the bulk. When comparing these exact results with the approximate ones (dotted lines), one sees that the simple results [from Eq. (193)] hold with a precision for f^2 of better than 10%, for $f^2 \geq 1.1$.

Adatom. The localized frequencies due to one adatom give the solution of Eq. (152)

$$\Delta_1(\omega^2) = 1 - \frac{\gamma'\omega^2}{\omega^2 - \omega_0^2} U(001,001;\omega^2) = 0 \qquad (198)$$

With the help of Eqs. (77) and (195), this equation takes the form

$$M/M' = 6f^2\left[\frac{2\gamma}{\gamma'} - I(0,0,0;f^2) + I(0,0,1;f^2)\right] \qquad (199)$$

In the same manner as above, we can solve it numerically. The ratio f of the localized frequency to the bulk maximum frequency ω_L was plotted for $\gamma = \gamma'$ as a function of λ in Fig. 4.13.

The large difference for a given M'/M between the localized frequency of the same isotopic atom when it is adsorbed (curve 0) or substituted (curve 1) in the surface layer is due here to the fact that for simplicity we assumed the adatom was interacting only with one (rather than probably four) surface atoms. The substituted surface atom interacts with five neighbors.

In general, one expects an adsorbed atom to have a weaker interaction γ' than a substituted atom. Adatom interactions are in general weaker than those of a surface substituted atom. Then one may expect higher localized frequencies for an adatom than for the same atom substituted in the surface layer.

Let us also note that when the coverage of adsorbed atoms increases towards one monolayer, one expects the localized level to broaden into a band of localized modes centered about the localized level of one adatom (66). These localized modes are easily observable in the infrared spectrum, especially when they are well above the bulk bands. In this case, it is easy to calculate their frequency by the approximative methods described, once the adsorption site and the force constants are known. These force constants for an adsorbed atom have not yet been determined.

SCATTERING OF PHONONS BY A CRYSTAL SURFACE

In calculations of lattice thermal conductivities at low temperatures, one needs to know the inverse relaxation time $\tau_{\vec{k}j}^{-1}$ for the scattering of phonons, in the mode $(\vec{k}j)$, from crystal surfaces. The general theory was given on pp. 195-202; we will apply it here, as a simple example, to a (001) surface of the Montroll-Potts model. In this case, as we have a threefold degenerate phonon branch, Eq. (404) from Chapter 2 takes the following form

$$\tau_{\vec{k}}^{-1} = \frac{\pi}{2}\frac{1}{\omega^2(\vec{k})}\sum_{\vec{k}'(\neq\vec{k})}\left|t\left(\vec{k}';-\vec{k};\omega^2(\vec{k})+i\varepsilon\right)\right|^2\delta\left(\omega(\vec{k})-\omega(\vec{k}')\right) \qquad (200)$$

as a function of the coefficients $t(\vec{k}',\vec{k};\omega^2)$ of a double Fourier series expansion of the scattering matrix $\overleftrightarrow{T}$ [see Chapter 2, Eq. (263)]. These coefficients can be calculated once the perturbation matrix $\overleftrightarrow{\delta L}$ due to the creation of the two free surfaces is known [Eq. (7)]. The corresponding expressions are given in Chapter 2, Eq. (405). Let us calculate them here for our triply degenerate phonon branch

$$t(\vec{k};\vec{k}';\omega^2) = V(\vec{k};\vec{k}';\omega^2) + \sum_{\vec{k}''} \frac{V(\vec{k};-\vec{k}'';\omega^2)}{\omega^2 - \omega^2(\vec{k}'')} t(\vec{k}'';\vec{k}';\omega^2) \tag{201}$$

$$V(\vec{k};\vec{k}';\omega^2) = \frac{1}{N}\sum_{\ell}\sum_{\ell'} \frac{\delta L(\ell;\ell')}{M} e^{-i\vec{k} \cdot \vec{x}(\ell) - i\vec{k}' \quad \vec{x}(\ell')} \tag{202}$$

For the model used here, $\overleftrightarrow{\delta L}$ was given by Eq. (7). As all the removed interactions are along $\hat{x}_3$, the expression given by this equation is formally the same before or after the Fourier transformation [of Eq. (6)] parallel to the surface. Then one easily obtains

$$V(\vec{k};\vec{k}';\omega^2) = \frac{1}{L} \delta(k_1 + k_1')\delta(k_2 + k_2')u(k_3)u(k_3') \tag{203}$$

where

$$u(k_3) = 2\left(\frac{\gamma}{M}\right)^{\frac{1}{2}} e^{-ik_3/2} \sin(k_3/2) \tag{204}$$

and L is the number of planes parallel to the surface.

In order to solve Eq. (201), let us try a solution of the form

$$t(\vec{k};\vec{k}';\omega^2) = \frac{1}{L} u(k_3)A(k_1,k_2;\omega^2)u(k_3') \cdot \delta(k_1 + k_1')\delta(k_2 + k_2') \tag{205}$$

By simple substitution of this solution into Eq. (201),

one finds that

$$\overleftrightarrow{A} = (\overleftrightarrow{I} - \overleftrightarrow{M})^{-1} \qquad (206)$$

where

$$M(k_1, k_2; \omega^2) = \sum_{k_3''} \frac{u(k_3'')\,u(-k_3'')}{\omega^2 - \omega^2(k_1, k_2, k_3'')} \qquad (207)$$

Using Eq. (204), $\overleftrightarrow{M}$ becomes

$$M(k_1, k_2; \omega^2) = \frac{4\gamma}{M} \sum_{k_3''} \frac{\sin^2 \frac{k_3''}{2}}{\omega^2 - \omega^2(k_x, k_y, k_3'')} \qquad (208)$$

In order to obtain $\tau_{\vec{k}}^{-1}$, using Eq. (4) for $\omega^2(\vec{k})$, we can calculate

$$M(k_1, k_2; \omega^2(\vec{k}) + i\varepsilon) =$$

$$\frac{a_0}{\pi} \int_0^{\pi/a_0} dk_3'' \quad \frac{\sin^2\left(\frac{a_0}{2}\right) k_3''}{\sin^2\left(\frac{a_0}{2}\right) k_3 - \sin^2\left(\frac{a_0}{2}\right) k_3'' + i\varepsilon}$$

$$= -1 + \frac{2a_0}{\pi} \sin^2 \frac{a_0}{2} k_3$$

$$\cdot \int_0^{\pi/a_0} \frac{dk_3''}{\cos a_0k_3'' - \cos a_0k_3 + i\varepsilon}$$

$$= -1 - i\left|\tan(a_0/2)k_3\right| \qquad (209)$$

Consequently, from Eqs. (204)-(206) we obtain

$$t(\vec{k}';-\vec{k};\omega^2(\vec{k}) + i\varepsilon)^2 = \left(\frac{4\gamma}{M}\right)^2 \delta(k_1 - k_1')\delta(k_2 - k_2')$$

$$\cdot \cos^2(a_0/2)k_3 \sin^2(a_0/2)k_3' \qquad (210)$$

Introducing this expression into Eq. (200) gives

$$\frac{1}{\tau_{\vec{k}}} = \frac{a_0}{9L}\frac{\omega_L^4}{\omega(\vec{k})} \cos^2(a_0/2)k_3 \int_0^{\pi/a_0} dk_3'' \sin^2(a_0/2)k_3''$$

$$\cdot \delta\left[\omega^2(k_1k_2k_3) - \omega^2(k_1k_2k_3'')\right] \qquad (211)$$

or

$$\frac{1}{\tau_{\vec{k}}} = \frac{c}{La_0}\frac{\omega_L}{3^{\frac{1}{2}}}\frac{|\sin(a_0/2)k_3 \cos(a_0/2)k_3|}{\omega(\vec{k})} \qquad (212)$$

where $c = a_0\omega_L/(2\sqrt{3})$ is the speed of sound for this crystal model. If we note that

$$\frac{\partial\omega(\vec{k})}{\partial k_3} = a_0 \frac{\omega^2_L}{6} \frac{\sin(a_0/2)k_3 \cos(a_0/2)k_3}{\omega(\vec{k})} \tag{213}$$

we can express $\tau_{\vec{k}}^{-1}$ equivalently as

$$\frac{1}{\tau_{\vec{k}}} = \frac{1}{La_0} \left| \frac{\partial\omega(\vec{k})}{\partial k_3} \right| \tag{214}$$

Thus the inverse relaxation time for the scattering of phonons in the mode $(\vec{k})$ from crystal surfaces has the form

$$\tau_{\vec{k}}^{-1} = \frac{c(\vec{k})}{La_0}$$

where $c(\vec{k})$ is just the x_3-component of the phonon group velocity and La_0 the thickness of the slab.

REFERENCES

1. H.B. Rosenstock and G.F. Newell, J. Chem. Phys., 21, 1607 (1953).

2. E.W. Montroll and R.B. Potts, Phys. Rev., 102, 72 (1956).

3. W. Ludwig and B. Lengeler, Solid State Commun., 2 83 (1964).

4. L. Dobrzynski and G. Leman, J. Phys., A30, 119 (1969).

5. S.L. Cunningham, Surf. Sci., 33, 139 (1972).

6. S.L. Cunningham, L. Dobrzynski, and A.A. Maradudin, Phys. Rev., B7, 4643 (1973).

7. L. Dobrzynski and D.L. Mills, J. Phys. Chem. Solids, 30, 1043 (1969) and 30, 2797 (1969).

8. U.T. Höchli, J. Phys. Chem. Solids, 32, 2231 (1971).

9. L. Dobrzynski and D.L. Mills, Phys. Rev., B7, 1332 (1973).

10. G. Toulouse, Solid State Commun., 4, 593 (1966).

11. B.S. de Witt, Phys. Rev., 103, 1565 (1956).

12. See, Eq. 17, p. 24.

13. W. Ludwig and B. Lengeler, Solid State Commun., 2, 83 (1964).

14. A.A. Maradudin and R.F. Wallis, Phys. Rev., 148, 945 (1966).

15. R.E. Allen and F.W. de Wette, J. Chem. Phys., 51, 4820 (1969).

16. A.A. Maradudin, E.W. Montroll, G.H. Weiss, and I.P. Ipatova, Theory of Lattice Dynamics in the Harmonic Approximation (Academic Press, New York, (1971).

17. P. Lenglart, L. Dobrzynski, and G. Leman, Ann. Phys., 7, 407 (1972).

18. L. Dobrzynski and J. Friedel, Surface Sci., 12, 469 (1968).

19. L. Dobrzynski, Ann. Phys., 4, 637 (1969).

20. L. Salter, Proc. Roy. Soc. (London), A233, 418 (1956).

21. L. Dobrzynski and A.A. Maradudin, Phys. Rev., B7, 1207 (1973).

22. M.R. Schafroth, Helv. Phys. Acta, 24, 645 (1951).

23. P. Masri, G. Allan, and L. Dobrzynski, Journal of Physique, 33, 85 (1972).

24. F. Cyrot-Lackmann, J. Phys. Chem. Solids, 29, 1235 (1968) and Thesis, Orsay, 1968.

25. G. Thibaudier, Thesis, Orsay, 1972.

26. L. Dobrzynski, Le Vide, 139, 8 (1969).

27. G. Armand, P. Masri, and L. Dobrzynski, J. Vac. Sci. Tech., 9, 705 (1972).

28. L. Dobrzynski, J. Phys. Chem. Solids, 30, 2395 (1969).

29. J. Friedel, Journal de Physique, 35, L-59 (1974).

30. M. Born, Report Prog. Phys., 9, 294 (1942).

31. R.J. Reid, Phys. Stat. Sol. (a),4, K211 (1971).

32. C.B. Duke and G.E. Laramore, Phys. Rev., B2, 4765 (1970).

33. P. Masri and L. Dobrzynski, Journal de Physique, 32, 939 (1971).

34. J.B. Theeten and L. Dobrzynski, Phys. Rev., 5, 1529 (1972).

35. P. Masri and L. Dobrzynski, Surface Sci., 32, 623 (1972).

36. L. Dobrzynski and P. Masri, J. Phys. Chem. Solids, 33, 1603 (1972).

37. P. Masri and M. Bienfait, Solid State Comm., 11, 919 (1972).

38. P. Masri, Le Vide, 164, 92 (1973).

39. P. Masri, J. Phys. Chem. Solids, 34, 435 (1973).

40. G. Pepe, P. Masri, M. Bienfait, and L. Dobrzynski, Acta Cryst., 30A, 290 (1974).

41. A.A. Maradudin, E.W. Montroll, G.H. Weiss, R. Herman, and H.W. Milnes, Mem. Soc. R. Sci. Liege, 14, 1709 (1960).

42. See also p. 428 of Ref. 16; J. Oitmaa, Solid State Comm., 9, 745 (1971); and T. Horiguchi, J. Phys. Soc. Jap., 30, 1261 (1971).

43. L. Dobrzynski and J. Lajzerowicz, Phys. Rev., B12, 1358 (1975).

44. B. Djafari-Rouhani, L. Dobrzynski, and G. Allan, Surface Sci., 55, 663 (1976).

45. P. Masri and L. Dobrzynski, Journal de Physique, 32, 295 (1971).

46. A. Blandin, D. Castiel, and L. Dobrzynski, Solid State Comm., 13, 1175 (1973).

47. S. Trullinger and S. Cunningham, Phys. Rev. Lett., 30, 913 (1973).

48. S. Trullinger and S. Cunningham, Phys. Rev., B8, 2622 (1973).

49. M. Ashkin, Phys. Rev., 136, A821 (1964).

50. L. Dobrzynski, Surface Sci., 20, 99 (1970).

51. M. Lannoo and L. Dobrzynski, J. Phys. Chem. Solids, 33, 1447 (1972).

52. D. Lagersie, M. Lannoo, and L. Dobrzynski, Journal de Physique, 32, 964 (1971) and in "Phonons," Edited by M.A. Numisovici (Flammarion Sciences) 377 (1971).

53. P. Masri and L. Dobrzynski, J. Phys. Chem. Solids, 34, 847 (1973).

54. See, for example, G.A. Somorjai, Principles of Surface Chemistry, Prentice Hall, Englewood Cliffs, New Jersey (1971).

55. G. Armand and J.B. Theeten, Phys. Rev., B9, 3969 (1974).

56. See for references, the Proceedings of the Yülich International Conference on Vibrations in Adsorbed Monolayers, June 1978, Editors S. Lehwald and H. Ibach, Zentralbibliothek der K.F.A. Yülich.

57. L. Dobrzynski and D.L. Mills, Phys. Rev., B7, 2367 (1973), L. Dobrzynski, D.L. Mills, and S.L. Cunningham, Le Vide, 164, 102 (1973).

58. W. Ho, S.L. Cunningham, W.H. Weinberg, and L. Dobrzynski, Solid State Commun., 18, 429 (1976) and Applications of Surf. Sci., 1, 33 and 44 (1977).

59. L.D. Landau and E.M. Lifshitz, Statistical Physics, p. 469, Pergamon Press (1959).

60. R.E. Allen and F.W. de Wette, J. Chem. Phys., 51, 4820 (1969).

61. J.B. Theeten, L. Dobrzynski, and J.L. Domange, Surface Sci., 34, 145 (1973).

62. J.P. Coulomb, J. Suzanne, M. Bienfait, and P. Masri, Solid State Commun., 15, 1585 (1974).

63. S.L. Cunningham, L. Dobrzynski, and A.A. Maradudin, Phys. Rev., B7, 4643 (1973).

SUBJECT INDEX

AUTHOR INDEX

Numbers in parentheses are reference numbers and indicate that an author's work is referred to, although his name is not cited in the text. Numbers underlined show the page on which the complete reference is listed.

AUTHOR INDEX

AUTHOR INDEX